普通高等教育"十三五"规划教材
Web应用&移动应用开发系列规划教材

Android YINGYONG KAIFA ANLI JIAOCHENG

Android应用开发案例教程

编　著　吴志祥　柯　鹏　张　智　胡　威
参　编　周　兵　曾　辉　曾　诚　马　杰
肖　念　鲁屹华　董剑波　蔡晓鸿
赵小丽　张新华　王新颖　王向丽

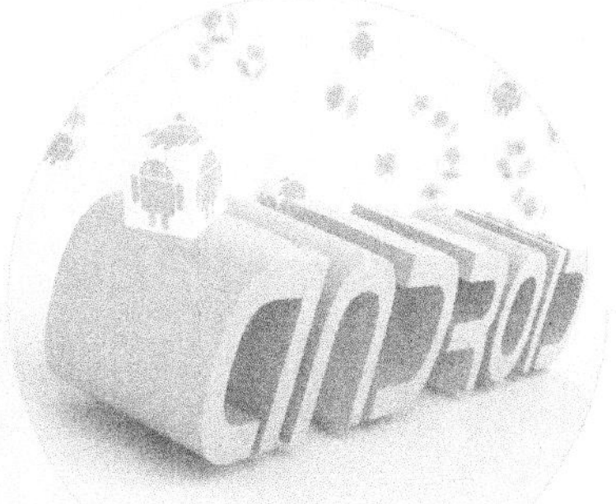

U0303231

华中科技大学出版社
http://www.hustp.com
中国·武汉

内 容 提 要

本书系统地介绍了Android应用开发的基础知识和实际应用。全书共分11章,包括Android开发的基础知识、开发环境搭建、Android工程的文件系统分析、Android程序的运行原理、用户界面设计、服务与广播接收组件、SQLite数据库的存储与访问、内容提供者组件、Android近距离通信、位置服务与地图应用、Android网络编程(访问Web服务器与手机客户端、消息推送)等,其内容从简单到复杂,循序渐进,结构合理,逻辑性强。

本书以实用为出发点,以介绍Android的四大组件为主线。对于章节中的很多知识点,本书都精心设计了典型例子以说明其用法,每章配有习题及实验。与本书配套的教学网站,包括了教学大纲、实验大纲、各种软件的下载链接、课件和案例源代码下载、在线测试等,极大地方便了教与学。

为了方便教学,本书还配有电子课件等教学资源包,任课教师和学生可以登录"我们爱读书"网(www.ibook4us.com)免费注册并浏览,或者发邮件至hustpeiit@163.com免费索取。

本书可以作为高等院校计算机专业和非计算机专业学生学习"Android应用开发"和"移动开发"等课程的教材,也可以作为Android初学者的入门参考书。

图书在版编目(CIP)数据

Android应用开发案例教程/吴志祥等编著.—武汉:华中科技大学出版社,2014.11(2023.1重印)
ISBN 978-7-5680-0531-9

Ⅰ.①A… Ⅱ.①吴… Ⅲ.①移动终端-应用程序-程序设计-教材 Ⅳ.①TN929.53

中国版本图书馆CIP数据核字(2014)第275432号

Android应用开发案例教程 吴志祥 柯 鹏 张 智 胡 威 编著

策划编辑:张凌云
责任编辑:史永霞
封面设计:龙文装帧
责任校对:马燕红
责任监印:朱 玢
出版发行:华中科技大学出版社(中国·武汉) 电话:(027)81321913
武汉市东湖新技术开发区华工科技园 邮编:430223
录 排:华中科技大学惠友文印中心
印 刷:武汉邮科印务有限公司
开 本:787mm×1092mm 1/16
印 张:27
字 数:671千字
版 次:2023年1月第1版第6次印刷
定 价:58.80元

本书若有印装质量问题,请向出版社营销中心调换
全国免费服务热线:400-6679-118 竭诚为您服务
版权所有 侵权必究

前　言

Android 一词的本义是“机器人”，是由 Google 公司于 2007 年 11 月对外发布的一种以 Linux 为基础的开源操作系统，主要用于移动设备。近年来，Android 平台得到了广大手机厂商和移动运营商的广泛支持。目前，3G 智能手机的强大功能和广泛普及，促使各高校纷纷开设 Android 移动平台的开发与设计课程。

目前，市场上关于 Android 移动开发相关的书籍比较多，几乎都是针对有一定基础的行内研发人员而编写的，而符合高校 Android 教学需要、真正从零基础开始教学的教材并不多见。为此，笔者组织一线相关教师编写了这本符合高校教学需要的教材。

本书系统地介绍了 Android 应用开发的基础知识和实际应用。全书共分 11 章，包括 Android 开发的基础知识、开发环境搭建、Android 工程的文件系统分析、Android 程序的运行原理、用户界面设计、服务与广播接收组件、SQLite 数据库的存储与访问、内容提供者组件、Android 近距离通信、位置服务与地图应用、Android 网络编程（访问 Web 服务器与手机客户端、消息推送）等，其内容从简单到复杂，循序渐进，结构合理，逻辑性强。

本书以实用为出发点，以介绍 Android 的四大组件为主线。对于章节中的很多知识点，本书都精心设计了典型例子以说明其用法，每章配有习题及实验。与本书配套的教学网站，包括了教学大纲、实验大纲、各种软件的下载链接、课件和案例源代码下载、在线测试等，极大地方便了教与学。

本书写作特色鲜明，一是教材结构合理，对教材目录的设置进行了深思熟虑，多次推敲，在正文中指出了相关章节知识点之间的联系；二是知识点介绍简明，作者精心设计的例子紧扣理论；三是采用大量的截图，清晰地反映 jar 包、软件包、类（或接口）三个软件层次；四是通过综合案例的设计与分析，让学生综合使用 Android 应用开发的各个知识点；五是有配套的上机实验网站，包括实验目的、实验内容、在线测试（含答案和评分）和素材的提供等。

本书可以作为高等院校计算机专业和非计算机专业学生学习“Android 移动平台应用开发”等课程的教材，也可以作为 Android 初学者的入门参考书。

为了方便教学，本书还配有电子课件等教学资源包，任课教师和学生可以登录“我们爱读书”网(www.ibook4us.com)免费注册并浏览，或者发邮件至hustpeiit@163.com免费索取。

获取本书配套的课件、案例源代码等教学资料，可访问作者教学网站http://www.wustwzx.com及本课程网站http://www.wustwzx.com/android。其中，所有项目导入eclipse android后可直接运行，也可导入android studio里运行。

由于编者水平有限，书中错漏之处在所难免，在此真诚欢迎读者多提宝贵意见，读者通过访问作者的教学网站与作者QQ联系，以便再版时更正。

编　者

2014年12月于武汉

目　　录

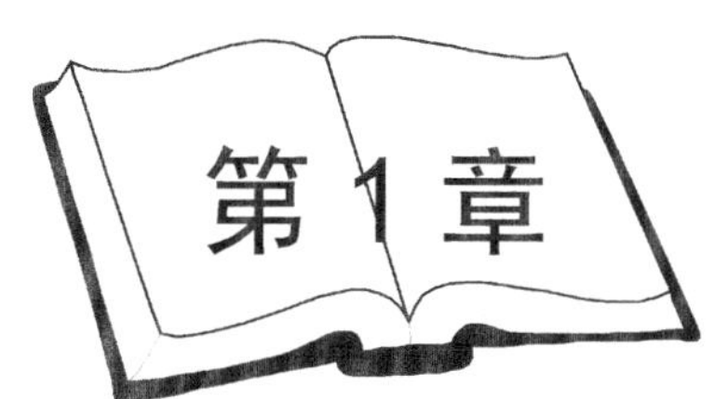

第1章 Android应用开发概述及技术基础

随着3G智能手机时代的到来，人们对Android应用开发的需求日趋增多。Android作为3G智能手机的操作系统，是新一代基于Linux的开源手机操作系统。手机应用软件的开发方式和环境与传统的Windows应用程序或者Web程序有很大的不同。本章主要介绍了Android应用开发的一些预备知识，其学习要点如下：

- 了解3G智能手机的系统结构、与传统手机在使用上的区别；
- 了解Android手机与普通计算机的区别与联系；
- 掌握Android开发中常用的Java编程技术；
- 掌握Android设备的软件系统架构；
- 掌握Android手机的配置，尤其是手机内存的划分；
- 理解Android移动开发是Web应用开发的延伸。

1.1 移动开发与3G智能手机

1.1.1 移动开发概述

3G，全称为3rd Generation，中文含义是第三代数字通信。所谓3G，是指将无线通信与国际互联网等多媒体通信相结合的新一代移动通信系统。3G只是一种通信技术标准，符合这个标准的技术有WCDMA、CDMA2000、TD-SCDMA三种制式。中国联通使用的是WCDMA(世界上大部分3G网络采用的是该标准)；中国电信使用的是CDMA2000 (日、韩和北美的一些国家使用)；中国移动使用的是具有自主知识产权的TD-SCDMA(只有中国才使用)。

相对第一代模拟制式手机(1G)和第二代GSM、CDMA等数字手机(2G)，3G网络能处理图像、音乐、视频等多种媒体形式，提供包括网页浏览、电话会议、电子商务等多种信息服务。第三代数字通信与前两代的主要区别是在传输声音和数据的速度上有很大的提升。

3G商用是一项相当浩大的工程，要从目前的2G迈向3G不可能一下就衔接得上，因此，前几年2.5G的手机就出现了。符合2.5G标准的技术有CDMA2000 1X和GPRS，中国联通使用的是CDMA2000 1X标准，中国移动使用的是GPRS(general packet radio service，通用分组无线服务技术)标准。目前，我们可以把2.5G移动通信技术看作是2G迈向3G的衔接性技术，在2.5G网络下出现了如WAP、蓝牙(Bluetooth)等技术。

Android 一词最早出现于法国作家利尔亚当于 1886 年发表的科幻小说《未来夏娃》中，将外表像人的机器命名为 Android，它是一个全身绿色的机器人。

作为手机操作系统的 Android，最早由 Google 公司工程副总裁安迪·罗宾(Andy Rubin)研发完成，Google 公司于 2005 年收购了原 Android 公司，并于 2007 年 11 月发布了基于 Linux 的开源手机平台。该平台包括操作系统、中间键和应用软件，是一个可以完全定制、免费、开放的手机平台。

注意：不同手机制造商所采用的操作系统可能不同，参见第 1.1.3 小节。

1.1.2 Android 智能手机的使用特点

Android 智能手机除了具备模拟手机的打电话、发短信、蓝牙、上网等基本功能外，还具有用户定制操作系统功能，可以像普通的计算机一样，安装或卸载应用程序。

智能手机本质上也是一台计算机，但与普通计算机有一定的差别。普通计算机的键盘、鼠标对应于较多的操作(如翻页、双击等)，而手机支持各种手势对应的事件(例如长按事件)。一般计算机(包括台式计算机和笔记本计算机)适合于软件编程，但手机不适合。

- 手机上只有 Home(用于返回桌面)、Return(用于退出主界面或返回到上一级界面)和位于 Home 左侧的 Task 键(常用于结束任务)或 Menu 键(快捷菜单)。
- 手机的用户操作可分为按键和触屏两种。其中，触屏事件(如滑屏、长按等)是 Android 设备常用的。
- 手机进入文本录入时，使用的是软键盘(不同于普通计算机)。
- 手机系统集成了众多的硬件，如摄像头、录音机、GPS 芯片、蓝牙芯片、WiFi 网卡等。

手机的存储系统分为运行内存、手机内存和扩展存储三部分。其中，手机内存主要指系统区(包括最底层的 Linux 系统、自带的应用程序和用户应用程序)。此外，手机厂商通常会从手机内存中划分一部分存储用户数据(如照片、音乐等)，即标准 SD 卡。扩展存储通常是指手机用户额外添加的 SD 卡。

Android 手机的软件系统包括操作系统、中间件和一些主要应用，是基于 Java 的系统，运行在 Linux 2.6 内核上。此外，Android 手机还具有如下特点：

- Android SDK 提供多种开发所必需的工具与 API，例如提供访问硬件的 API 函数，简化了如摄像头、GPS 等硬件的访问过程；
- 具有自己的运行时和虚拟机 Dalvik；
- 提供丰富的界面控件供使用者之间调用，加快用户界面的开发速度，保证 Android 平台上程序界面的一致性；
- 提供轻量级的进程间通信机制 Intent，使跨进程组件通信和发送系统级广播成为可能，提供了 Service 作为无用户界面、长时间后台运行的组件；
- 支持高效、快速的数据存储方式。

注意：很多人说，手机内存相当于计算机的硬盘。此话有一定的道理，但不够细腻。实际上，手机内存中除标准 SD 卡之外的部分，手机用户通常是无法直接使用的，它只能由 Android 系统管理和使用(例如安装新的 Android 应用程序)；而使用 Windows 系统的计算机的硬盘，对于计算机用户是可以任意读写的。

1.1.3　智能手机操作系统及其分类

早期的手机没有操作系统 OS，内部所有的软件都是由生产商在设计时定制的，手机在设计完成后基本没有扩展功能。

为了提高手机的可扩展性，很多手机都使用了专为移动设备开发的操作系统 OS，使用者可根据需要安装不同类型的软件。

智能手机制造商所使用的半导体芯片并不都是相同的，不同的手机所采用的操作系统也可能不同。目前，主流的手机操作系统有如下几种。

- iPhoneOS：由苹果公司开发的手机操作系统。
- Android：由谷歌发布的基于 Linux 的开源手机平台。
- Symbian：由 Symbian 公司开发和维护，后被诺基亚收购，不开源。

注意：Android 手机应用开发是移动开发的一种。

1.2　360 手机助手软件的使用

1.2.1　创建 Android 手机与计算机的连接

为了在 Android 手机上运行应用程序，通常需要将手机与计算机相连接。在物理上将手机通过手机数据线与计算机连接前，应打开手机的 USB 调试开关。

打开手机 USB 调试开关，是在计算机上安装手机驱动程序的前提。打开手机 USB 调试开关的方法是：运行手机的“设置”程序→开发人员选项→勾选“USB 调试”。

注意：Android 手机与计算机首次连接时，如果计算机安装了 360 软件，则 360 软件自动安装手机的驱动程序，且有一个等待过程。以后连接时，就不必安装手机的驱动程序了。

手机与计算机相连后，使用 360 手机助手软件可以方便地管理手机上的文件和软件,其界面如图 1.2.1 所示。

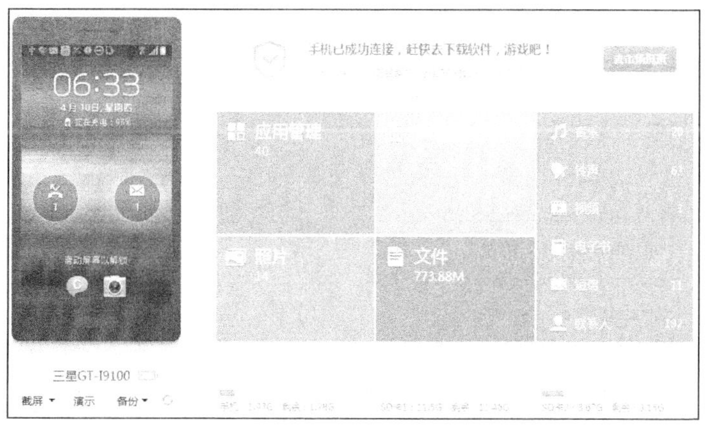

图 1.2.1　作者三星手机的系统配置

注意：有些手机助手软件，不仅能方便地管理手机软件，还能检测出手机的硬件信息。

1.2.2 Root Android 手机取得 Root 权限

为了查看手机里的系统文件夹，需要先 Root 手机取得 Root 权限，其操作方法是使用 360 软件提供的一键 Root 软件，操作界面如图 1.2.2 所示。

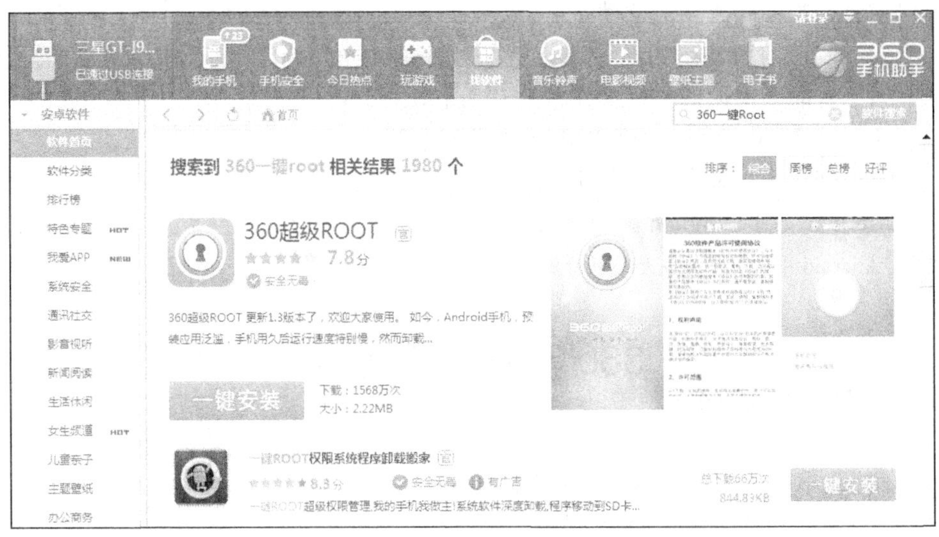

图 1.2.2　通过 360 手机助手软件查找并安装手机 Root 软件

注意：如果将 Android 应用程序部署在 Android 模拟器(参见第 2.4 节)上，查看模拟器内的系统文件夹则不需要先进行 Root 操作。

1.2.3 使用 Root Explorer 程序浏览手机上的系统文件夹

已经 Root 过的手机，使用 Root Explorer 软件可以查看手机内存里的所有文件夹及文件。例如，浏览手机根目录下的 data 文件夹，其效果如图 1.2.3 所示。

名称	大小	日期	时间	权限	
data					
app		2014-07-11	19:20	drwxrwx...	
ApiDemos.apk	4514195	2013-10-29	12:01	-rw-r--r--	com.example.android.apis
ApiDemos.odex	954584	2013-10-29	12:01	-rw-r--r--	
data		2014-07-11	19:20	drwxrwx--x	
cn.itcast.sqlite		2014-05-09	18:37	drwxr-x--x	
com.android.backupconfirm		2014-04-19	09:00	drwxr-x--x	
com.android.browser		2014-04-22	08:42	drwxr-x--x	
com.android.calculator2		2014-04-20	08:24	drwxr-x--x	
com.android.calendar		2014-04-19	09:11	drwxr-x--x	
introduction.android.fileDemo		2014-07-12	06:03	drwxr-x--x	
introduction.android.mydbDemo		2014-07-12	06:03	drwxr-x--x	

图 1.2.3　手机里两个重要的文件夹

注意：

(1) 使用手机自带的程序“文件管理”或“我的文件”是无法浏览内存(不是 SD 卡，也不是运行内存)中的文件夹(例如 data)的。

(2) 目录或文件权限共有 10 位，第 1 位表示文件、目录和超链接(分别用-、d 和小写字母 l 表示)；第 2～4 位表示文件所有者的权限；第 5～7 位表示文件所有者所属组成员的权限；第 8～10 位表示所有者所属组之外的用户的权限。

(3) 每个 Linux 文件具有四种访问权限：可读(r)、可写(w)、可执行(x)和无权限(-)。

1.3 Android 移动开发技术基础

1.3.1 Android 移动开发与 Java Web 开发

Java EE 的开发主要以 Web 开发为代表，Java Web 的主要内容是建立 JSP 网站和 JSP 页面设计。Web 开发中的用户界面 UI (user interface)与 Android 中的用户界面 Activity 是平行的。

网站开发和 Android 都有 UI 设计。例如，ASP.NET 网站的窗体设计主要是在 VS 中摆放控件，在 Android 中，将窗体中的控件及其布局放在一个 XML 文件中进行描述(参见第 4.1 节)。

Android 开发是 Java 开发的一个分支。Android 类库包含了一些 Java 包(类)，比如 String、Integer 等类，但没有包含 Java 的 Swing 类，即 Android 只用了 Java 中的常用的类库。

如果有 Java 开发的基础，那么 Android 上手是比较容易的，因为只是熟悉 Android API 的一个过程而已，而且 Android 开发比 Java Web 开发要简单容易很多。

总之，具有 Web 开发经验的人，特别是熟悉 JSP 编程的人，掌握 Android 应用开发相对容易些。

注意：Android 应用开发与 Java Web 开发尽管有相似之处，但其开发方式也有很多不同之处，特别是工程文件结构(参见第 3.1 节)。

1.3.2 使用面向对象的程序设计方法

面向对象就是将要处理的问题(对象)抽象为类，并将这类对象的属性和方法封装起来，通过对象的事件来访问该类对象的属性和方法来解决实际问题。

在 Java 语言里，一切皆对象，并以 Object 作为所有对象的超类。类是面向对象编程方式的核心和基础，通过类可以将零散的用于实现某项功能的代码进行有效管理。

Android 提供了一个用于开发的 android.jar 文件，其内的各个软件包内包含了 Android 各个方面应用的相关类(含抽象类和接口)。

面向对象软件开发的流程，可分为如下四个阶段。

1. 面向对象需求分析(OOA，object oriented analysis)

此阶段需要系统分析员对用户的需求做出分析和明确的描述，包括从客观的事物和它们之间的关系归纳有关的类及类之间的关系，并将具有相同属性和行为的对象用一个类来表示。

2. 面向对象设计(OOD，object oriented design)

此阶段在需求分析的基础上，对每一部分分别进行设计，首先是类的设计，可能包括多个层次，利用继承和组合等机制设计出类的层次关系，然后提出程序设计的具体思路和方法。

3. 面向对象编程(OOP，object oriented programming)

此阶段需要选择适当的编程语言，如 Java、C++和 C#等。选用工具开发，设置开发环境，进行代码的编写工作。

4. 面向对象测试(OOT，object oriented test)

此阶段会对程序进行严格的测试，这个过程包括单元测试、集成测试及系统测试等。最后还要对程序进行维护管理。

注意：学习面向对象编程,最重要的是其思想，而不是其语法。在 Android 系统提供的软件包里，包含了大量的类(或抽象，或接口)，每个类又封装了很多的属性与方法。对于这些属性与方法，特别是构造方法，死记硬背其参数是不行的，要利用面向对象的思想和联机支持功能，灵活地根据实际需要在相应的类中选用。

1.3.3 Android中常用的Java编程技术

Java 编程是 Android 移动开发的重要基础，下面介绍在 Android 开发中经常使用的 Java 编程知识。

1. 类继承、内部类

使用类继承，主要是实现方法继承，这样可以减少重复代码。当然，子类还可以重写父类的方法。

内部类是在一个类或方法的定义中定义的类。内部类用法的一个示例，参见例 7.2.1 中的类 MyDbOpenHelper。

2. 接口、内部接口和匿名类

定义一个实现某种接口的类，需要使用 implements 子句，并且要实现该接口中定义的所有方法。

在 Eclipse 中，使用 implements 子句修改一个类去实现某个接口时，在该类名前会自动提示 。单击该提示图标，先引入该接口所在的包名(含接口)，然后再次单击该提示图标，双击“Add unimplemented methods”，即可自动添加需要实现的方法。

当创建实现某个接口的类只使用一次时，使用匿名类，可使程序更加简洁。此时，不需要使用 implements 子句。

一个使用匿名类方式创建实现某个接口对象的示例代码如下：

```
import android.content.ServiceConnection;   //引包(接口)
// 创建服务连接对象，需要实现ServiceConnection接口
private ServiceConnection conn = new ServiceConnection() {
    @Override
    public void onServiceConnected(ComponentName name, IBinder service) {
        // TODO Auto-generated method stub
    }
    @Override
    public void onServiceDisconnected(ComponentName name) {
        // TODO Auto-generated method stub
    }
};
```

在 Android 应用开发中，经常需要创建一个实现某个类的内部接口的匿名类。例如，给一个视图（View）对象添加单击事件监听器（实现 View 的内部接口 OnClickListener 的匿名类的实例），参见例 4.3.2。

Android API 中视图类 View 的定义，如图 1.3.1 所示。

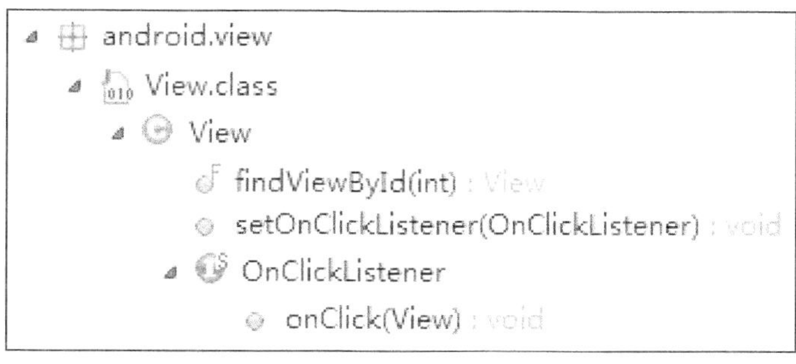

图 1.3.1　Android 视图类 View 的定义

3. 抽象类

使用抽象类时，由继承类改写其定义的抽象方法(只有方法声明，而没有方法体定义)。

抽象类不能直接使用 new 运算符创建其实例(对象)，它通常作为某个类的基类，参见第 8.1.1 小节。在 Android 应用开发中，很多抽象类的实例(对象)是通过使用另外某个类的某种方法而得到的，参见第 8.1.2 小节。

4. 流式文件的读写

由于 SD 卡中的文件采用 Windows 的 FAT 格式，因此 SD 卡里的文件读写仍然使用 Java 的文件读写方式，即使用文件输入/输出流 FileInputStream 和 FileOutputStream。

注意：

(1) Android 提供了内部文件读写的多种方法，参见第 4.6 节、4.7 节和第 7 章。

(2) 基于网络流的读写，分别参见第 11.1.3 小节和第 11.4 节。

5. 类的静态成员和静态方法

类的静态方法，直接使用“类名.静态方法名()”的方式，无须先创建类的实例。

6. Java 进程、多线程及其相关

进程(Process)属于操作系统的范畴，它是程序执行的一个实例。每一个进程都有自己的独立的一块内存空间、一组资源系统，其内部数据和状态都是完全独立的。多进程必须使用操作系统提供的复杂机制(如消息队列、管道和共享内存等)进行交互。

Android 系统提供了一个进程类，提供获取进程 id 和清除进程的两个主要方法，如图 1.3.2 所示。

图 1.3.2 Android 进程类 Process

线程是进程的一个顺序执行流，一个进程内可以同时启动多个独立工作的线程，同类的多个线程共享一块内存空间和一组系统资源。一个进程的任务可以细分为线程来处理。例如，上网下载歌曲时，可以使用多个线程同时下载，以提高下载速度。

一个不可回避的问题是：如何使用线程中的数据更新用户界面？例如，使用多个线程同时下载某个歌曲文件，在下载完成后要向主线程发送下载完毕的消息，主线程接收到这个消息后再在用户界面中显示出来。Android 提供了多种处理方法。例如，使用 Handler 类(详见第 8.2.4 小节例 8.2.2)和异步任务类 AsyncTask(参见第 8.3.4 小节例 8.3.2)。

建立和使用 Java 多线程的一个简单方法是使用 java.lang.Thread 类来实现，Thread 类中的相关函数可以启动线程、终止线程、线程挂起等。但由于 Java 只支持单继承，为了达到多重继承的效果，Java 还提供了 java.lang.Runnable 接口用来生成多线程。

Thread 类与 Runnable 接口的定义，如图 1.3.3 所示。

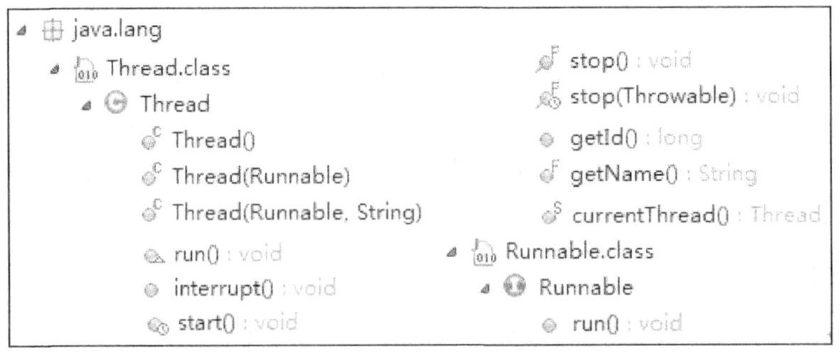

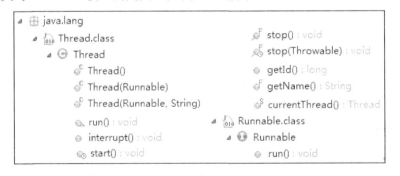

图 1.3.3 Java 处理多线程的类与接口

注意：子线程的代码含于方法 run()内，且通过 Thread 的 start()方法来启动该线程。

一个混合使用 Runnable 接口和 Thread 类的示例代码如下：

```
//创建线程并运行
new Thread(new Runnable(){
            @Override
            public void run() {
                // TODO Auto-generated method stub
                //实现接口（重写run()方法）的代码
            }
        }).start();
```

线程在 run()方法返回后，就自动终止了。如果使用 stop()方法在外部终止线程，则可能产生异常，因此，一般不推荐这种不安全的方法。最好的方法是调用 interrupt()方法通知线程准备终止，线程会释放它正在使用的资源，在完成所有的清理工作后自动终止。

使用子线程的一个示例，参见例 4.3.2 水平进度条对话框的设计。

7. 同步方法(代码块)

用 Java 关键字 synchronized 修饰某个方法(或代码块)后，相当于给这个方法(或代码块)加锁，即同一时间内只能有一个线程进入到这个方法(或代码块)里面，其他的线程需要等待，直到占用的那个线程离开。使用同步代码块的一个示例程序如下：

```
package multi_thread;
public class SaleTicket {
    /**
     * 民航售票2000张，分三个售票点
     * 保证线程安全问题：即同一张票不会被多次售出——使用对象锁解决并发问题
     * 使用同步锁保证块的原子性
     */
    public static void main(String[] args) {
        // TODO Auto-generated method stub
        TicketWindow tw=new TicketWindow();  //创建售票窗口对象
        Thread t1=new Thread(tw);  //创建线程对象
        Thread t2=new Thread(tw);  //创建线程对象
        Thread t3=new Thread(tw);  //创建线程对象
        t1.start();  //启动线程
        t2.start();  //启动线程
        t3.start();  //启动线程
    }
}
class TicketWindow implements Runnable{
    private int nums=2000;
    @Override
```

```
    public void run() {
        // TODO Auto-generated method stub
        try {      //一秒钟出一张票
            Thread.sleep(1000);   //设置线程休眠
        }
        catch (Exception e) {
            // TODO: handle exception
        }
        while(true){
            synchronized (this) {    //对代码块使用同步锁
                if(nums>0){
                    System.out.println("正在由窗口+Thread.
                            currentThread().getName()+"售出第"+nums+"张票...");
                    nums--;
                }
                else
                    break;
            }
        }
    }
}
```

程序的运行结果表明解决了并发问题，如图 1.3.4 所示。

```
<terminated> SaleTicket1 [Java Application]
正在由窗口Thread-2售出第2000张票...
正在由窗口Thread-2售出第1999张票...
正在由窗口Thread-2售出第1998张票...
正在由窗口Thread-2售出第1997张票...
正在由窗口Thread-2售出第1996张票...
正在由窗口Thread-2售出第1995张票...
正在由窗口Thread-0售出第2000张票...
正在由窗口Thread-0售出第1993张票...
```

图 1.3.4　多线程并发问题的示例

注意：上面的示例程序由于实现 Runnable 接口的对象使用了多次，因此不能写成匿名类（对象）的形式。

8. Java 对象集合框架与泛型

Java 对象集合框架由包 java.util 内的接口 Collection 和 Map 定义，其中，Collection 包含 List、Set 和 Queue 等子接口。Java 集合框架里的接口及其实现类都使用了泛型参数，如图 1.3.5 所示。

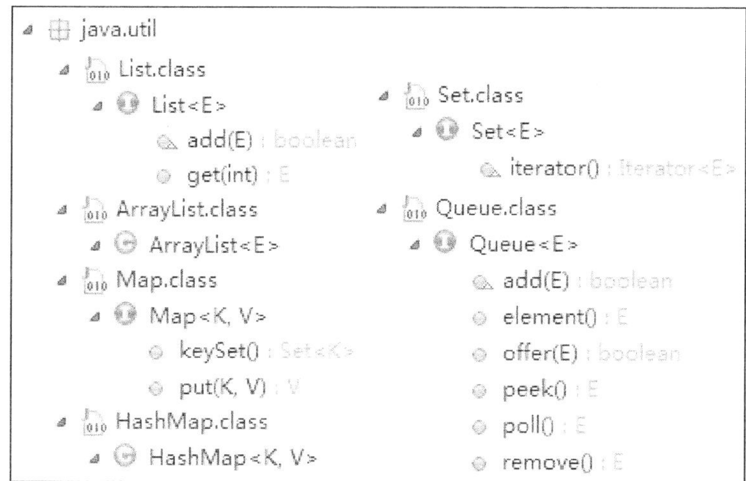

图 1.3.5　Java 对象集合框架的主要接口与类

注意：

(1) ArrayList 与 List、HashMap 与 Map 都是类与接口的关系，在 Android 开发中常用。

(2) 在使用 Java 集合类时，需要使用泛型，否则会出错警告性错误。例如，程序中使用

```
ArrayList al=new ArrayList();
```

时，会出现警告性错误；而使用

```
ArrayList<String> al=new ArrayList();
```

则不会出错警告。

(3) List 结构定义的是有次序关系且允许重复的对象集，而 Set 结构定义的是无次序且不能重复的对象集。

(4) Map 结构定义键值对形式的对象集， Queue 结构定义先进先出的对象集。

通过泛型机制可实现参数任意化，以提高程序通用性。

一个使用方法中使用泛型参数的示例代码如下：

```
package raw_type;
/*
 * 泛型方法的定义与使用示例，提高程序通用性
 * 泛型方法中的参数必须是对象类型
 * 泛型方法，不同于泛型类
 */
import java.util.Arrays;
public class Wu_fanxing1 {
    public static void main(String[] args) {
        Integer[]arr1=new Integer[]{1,2,3,4,5};
        //int[]arr1=new int[]{1,2,3,4,5}; //把Integer换成int时会报错
        changePosition(arr1,1,3);   //交换位置
        System.out.println(Arrays.toString(arr1));   //输出数组
```

```
        String[]arr2=new String[]{"aa","bb","cc","dd","ee"};
        changePosition(arr2,1,3);  //交换位置
        System.out.println(Arrays.toString(arr2));  //输出数组
    }
    //定义交换数组中两个元素的泛型方法
    public static <T>void changePosition(T[]arr, int index1, int index2) {
        T temp = arr[index1];
        arr[index1] = arr[index2];
        arr[index2] = temp;
    }
/*
    //不使用泛型，则需要根据数组元素值类型分别编写方法
    public static void changeposition(int[]arr,int index1,int index2){
        int tem=arr[index1];arr[index1]=arr[index2];arr[index2]=tem;
    }
    public static void changeposition(String[]arr,int index1,int index2){
        String tem=arr[index1];arr[index1]=arr[index2];arr[index2]=tem;
    }
*/
}
```

9. 反射机制

Java 程序运行时，对于任意一个类，都能够知道这个类的所有属性和方法；对于任意一个对象，都能够调用它的任意一个方法；这种动态获取以及动态调用对象的方法的功能称为 Java 语言的反射机制。

在跨进程的 Android 应用中，通常需要使用 Java 反射机制。Java 反射机制的一个应用，参见实验 1 内容 7 和例 6.4.1 中的服务程序 PhoneService.java 中的 getphoner()方法。

1.3.4 使用 XML 技术

XML 是 extensible markup language 的缩写，表示可扩展的标记语言。XML 文档以简单的文本格式存储具有层次结构的数据。目前，XML 技术在网站开发和 Android 中应用广泛，XML 格式实际上已成为 Internet 数据交换标准格式，XML 文件常用于解决跨平台交换数据的问题。

XML 文件允许自定义标记，并且标记必须成对出现。Android 开发中的布局文件、清单文件等都为 XML 格式。

一个 Android 应用工程的清单文件是 XML 格式，其代码如下：

```
<?xml version="1.0" encoding="utf-8"?>
<manifest xmlns:android="http://schemas.android.com/apk/res/android"
    package="com.example.test"
    android:versionCode="1"
    android:versionName="1.0" >
    <uses-sdk
        android:minSdkVersion="8"
        android:targetSdkVersion="19" />
    <application
        android:allowBackup="true"
        android:icon="@drawable/ic_launcher"
        android:label="@string/app_name"
        android:theme="@style/AppTheme" >
        <activity
            android:name="com.example.test.MainActivity"
            android:label="@string/app_name" >
            <intent-filter>
                <action android:name="android.intent.action.MAIN" />
                <category android:name="android.intent.category.LAUNCHER" />
            </intent-filter>
        </activity>
    </application>
</manifest>
```

一个.xml 文件的第一行是<?xml……?>，表示 XML 声明，version 属性指明遵循哪个版本的 XML 规范；encoding 属性指明使用的编码字符集；使用<!-- …… -->注释。

在一个.xml 文件中有且仅有一个根元素，并可以根据实际需要自定义语义标记。

XML 元素是可以嵌套的，嵌套在其他元素中的元素称为子元素。

属性用于给元素提供更详细的说明信息(但不是必需的)，它必须出现在起始标记中。属性以“名="值"”的形式出现。

注意：在 XML 文档中，标记嵌套体现层次关系，通常使用根节点、父节点和子节点等词语表达，但标记不能交叉。

1.3.5　文件型数据库 SQLite 及其操作软件 SQLiteSpy

SQLite 是 Android 系统手机自带(即内置)的轻量级数据库软件，类似于 Access 数据库。

SQLiteSpy 是一款查看和编辑 SQLite 数据库的软件，在作者的教学网站里提供了下载链接，解压后即可使用(无须安装)，其操作界面如图 1.3.6 所示。

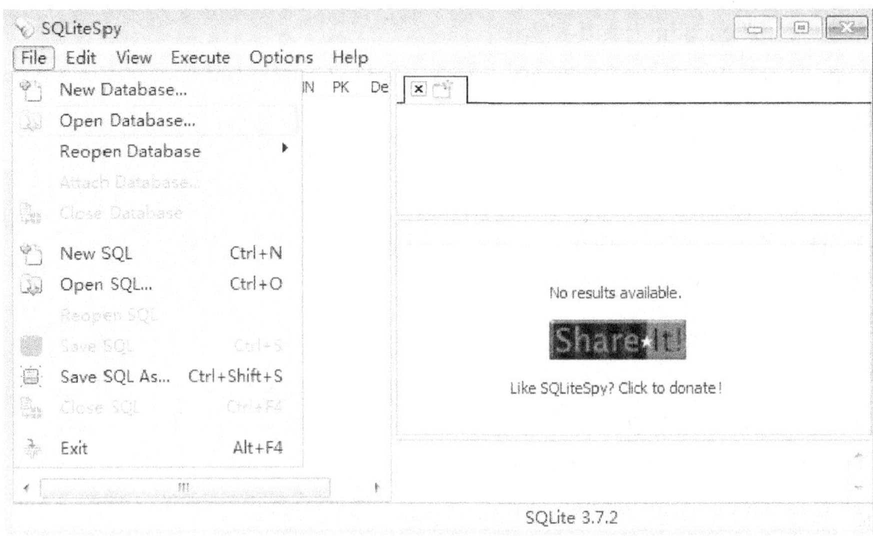

图 1.3.6 SQLiteSpy 的工作界面

注意：当涉及大量的数据存储时，需要使用专门的数据库服务器，参见第 11.3 节。

1.4 Android 系统架构

Android 是基于 Linux 内核的软件平台和操作系统，采用了软件堆栈架构，共分为四层，如图 1.4.1 所示。

图 1.4.1 Android 系统架构

1.4.1　Linux 内核及 Linux 文件系统

Linux 内核是硬件和其他软件堆层之间的一个抽象隔离层，提供由操作系统内核管理的底层基础功能，主要有安全机制、内存管理、进程管理、网络协议栈和驱动程序等。

Android 内核挂载/nfsroot/Androidfs 之后，根据 init.rc 和 init.goldfish.rc 来初始化并装载系统库、程序等直到开机完成。init.rc 脚本包括了文件系统初始化、装载的许多过程。init.rc 的主要工作是：

- 设置一些环境变量；
- 创建 system、sdcard、data、cache 等目录；
- 把一些文件系统 mount 到一些目录去，如，mount tmpfs tmpfs /sqlite_stmt_journals；
- 设置一些文件的用户群组、权限；
- 设置一些线程参数；
- 设置 TCP 缓存大小。

Android 源码编译后得到 system.img、ramdisk.img 和 userdata.img 映像文件。其中，ramdisk.img 是 emulator 的文件系统，system.img 包括了主要的包、库等文件，userdata.img 包括了一些用户数据。emulator 加载这 3 个映像文件后，会把 system 和 userdata 分别加载到 ramdisk 文件系统中的 system 和 userdata 目录下。

注意：Android 手机内存中的文件格式是 Linux，而 SD 卡里的文件格式是 FAT。

1.4.2　Android 函数库及运行时

Android 函数库和运行时是第二层，位于 Linux 内核之上，也称中间件层，由函数库和 Android 运行时构成。

由于 Linux 操作系统的内核使用及其组件使用 C 语言编写（少部分使用汇编语言），因此，开发人员可以通过应用程序框架调用一组基于 C/C++的函数库，主要包括以下几个。

- Surface Manager：支持显示子系统的访问，为多个应用程序提供 2D、3D 图像层的平滑连接。
- Media Framework：基于 OpenCORE 的多媒体框架，实现音频、视频的播放与录制功能。
- SQLite：关系型数据库引擎。
- OpenGL | ES：基于硬件的 3D 图像加速。
- FreeType：位图与适量字体渲染。
- WebKit：Web 浏览器引擎。
- SSL：数据加密与安全传输的函数库。
- libc：标准 C 运行库，是 Linux 系统中底层的应用程序开发接口。

Android 运行时由核心库和 Dalvik 虚拟机构成。核心库为开发人员提供了 Android 系统的特有函数功能和 Java 语言的基本函数功能，Dalvik 虚拟机采用适合内存和处理器受限的专用格式(参见第 3.3 节)。

注意：Java 是面向对象的编程语言，不能用来设计操作系统。

1.4.3 应用程序框架层

应用程序框架层提供了 Android 平台的管理功能和组件重用机制，包括 Activity 管理、资源管理、位置管理、通知消息管理、View 系统和内容提供者等。

Android 的三大核心功能如下。

(1) View：提供了绘制图形，处理触摸、按键事件等功能。

(2) ActivityManagerService：简称为 AMS，其主要功能是管理所有应用程序的 Activity、内存管理等。AMS 作为一种系统级服务管理所有 Activity，当操作(启动或停止)某个 Activity 时，必须报告给 AMS，而不能“擅自处理”。在内存不足时，AMS 可能主动杀死后台的 Activity。

(3) WindowManagerService：简称为 WMS，其主要功能是为所有应用程序分配窗口，并管理这些窗口。

1.4.4 应用程序层

应用程序层提供了一系列核心应用程序，如打电话、浏览器、联系人、相册、地图和电子市场等。

习　题　1

一、判断题

1. 苹果手机使用 Android 系统。
2. Android 手机通过数据线连接到计算机后，可作为 U 盘使用。
3. 中国移动与中国联通使用的 3G 技术是相同的。
4. 抽象类与接口是使用 new 运算符创建其实例(对象)的。
5. 在计算机上安装手机驱动程序之前，必须打开手机的 USB 调试开关。
6. 未安装 SIM 卡的手机将无法上网。
7. List 与 ArrayList 都是接口类型。
8. 泛型类中可以定义静态方法。
9. HashMap 与 Map、ArrayList 与 List，都是类与接口的关系。
10. 对 HashMap 对象集合应用 keySet()方法得到的是 Set 类型的对象集合。

二、选择题

1. 手机在户外上网，通常使用的方式是____。

   A. GPS　　B. WiFi　　C. GPRS　　D. Bluetooth
2. 与短信到来直接相关的管理器是____。

   A. Activity Manager　　B. Resource Manager

   C. Content Manager　　D. Notification Manager
3. 下列不属于 Android 应用程序框架的是____。

   A. Activity Manager　　B. SQLite

   C. Location Manager　　D. Notification Manager
4. 在使用 Java 泛型时，通常使用字母____作为类型通配符。

   A. N　　B. T　　C. K　　D. E
5. 要实现文本文件的读写，主要使用____软件包里的相关类。

   A. java.io　　B. java.text　　C. java.util　　D. java.sql
6. Android 移动设备底层(内核)使用____系统。

   A. DOS　　B. Linux　　C. Windows　　D. Unix
7. 关于类与接口，下列说法中不正确的是____。

   A. 接口没有构造函数

   B. 一个类可以实现多个接口

   C. 接口与类一样，可以被实例化

   D. 类实现接口时，需要实现该接口的所有方法
8. 下列软件包中，Java 系统自动引入(即不需要使用 import 语句)的是____。

   A. java.sql　　B. java.text　　C. java.lang　　D. java.util
9. 在集成环境中编辑 Java 源程序时，快速编辑 try…catch 块的方法：先选择要执行的

语句序列，然后在右击菜单里选择____菜单项。

A. Surround With　　B. Source　　C. Run As　　D. Refactor

三、填空题

1. 使用 Eclipse+ADT 开发时，查看某个类的继承关系所使用的快捷键是____。
2. 在 Android 中使用的 XML 文件里的标签必须成对出现或____。
3. 通常将 Android 手机操作系统的架构共分为____层。
4. 在 Android 系统中，Dalvik 虚拟机运行已经转换为____格式的 Java 应用程序。
5. 手机操作系统 Android 是建立在操作系统____之上。
6. Java 集合框架的相关类(接口)位于软件包 android.____里。
7. 取消程序里一段代码的注释，让它参与运行，应使用的快捷键是____。
8. 使用迭代器 Iterator 遍历 HashMap 所有对象，需要使用 Iterator 提供的____方法。
9. 在程序运行时动态创建某个类的实例，可使用类 java.lang.Class 提供的____方法。
10. 实现 Java 类方法的反射调用，需要使用类 java.lang.reflect.Method 提供的____方法。
11. 在 Eclipse 中快速输入代码“System.out.println( )”，其快捷方法是____。

实验 1　Android 应用开发技术基础

一、实验目的

1. 掌握手机及其管理软件的使用。
2. 掌握手机数据连接和个人热点的使用。
3. 掌握手机数据库管理软件 SQLite 的使用。
4. 掌握 Java 集合框架的使用，尤其是 ArrayList 和 HashMap。
5. 掌握 Java 泛型的使用。

二、实验内容及步骤

【预备】访问 http://www.wustwzx.com/android/index.html，单击“实验 1”超链接，下载本章实验内容的源代码并解压，得到文件夹 ch01，供研读和调试使用。

1. 手机及其管理软件的使用。

(1) 运行 Android 手机的设置程序，查看手机使用的 Android 系统版本。

(2) 运行 Android 手机的设置程序，在“开发人员选项”里勾选 USB 调试开关。

(3) 确保计算机已经安装了 360 安全卫士及其手机助手软件。

(4) 将手机通过数据线连接至计算机。首次连接时，有一个寻找驱动的过程。

(5) 在计算机上查看手机的配置信息及手机存储系统的使用情况。

(6) 安装 Root 手机的配置信息及手机存储系统的使用情况。

2. 利用手机个人热点，实现笔记本计算机上网。

(1) 断开笔记本计算机已有的上网连接(如果已经上网的话)。

(2) 关闭手机的 WiFi(如果已经打开的话)，打开手机的数据连接(GPRS)。

(3) 运行手机自带的“个人热点”程序，勾选“个人热点”复选框，选择“使用 WLAN 连接”，将出现连接手机热点的密码。

(4) 在笔记本计算机的网络服务列表中，选择本手机和“连接”按钮，输入刚才在手机上出现的连接密码。

(5) 打开笔记本计算机上的浏览器上网。

3. 掌握 SQLite 数据库软件的使用。

(1) 对于已经 Root 过的 Android 手机，使用 Root Explorer 软件查看手机自带的联系人数据库里的 data 表。

(2) 从作者教学网站里下载数据库操作软件 SQLiteSpy，得到压缩文件 SQLiteSpy.zip。

(3) 解压 SQLiteSpy.zip 至硬盘，建立程序 SQLiteSpy.exe 的桌面快捷方式(无须安装)。

(4) 从作者教学网站里下载一个数据库文件 contacts2.db，并使用程序 SQLiteSpy 打开，查看该数据库的表 data 里的记录。

4. 掌握 Java 集合架构的使用，尤其是 ArrayList 和 HashMap 的使用。

(1) 访问作者的教学网站 http://www.wustwzx.com，下载纯 Eclipse(即未安装任何插件的 Eclipse)，得到压缩文件 eclipse_3.7_cn.rar。

(2) 解压 eclipse_3.7_cn.rar 至硬盘，创建 eclipse.exe 的桌面快捷方式(无须安装)。

(3) 启动 Eclipse，导入文件夹 ch01 里的 Java 工程 Java_Foundation。

(4) 分别调试 collection 包里的两个源程序。

(5) 分析 Collection 接口、List 接口、Map 接口、HashMap 类和 ArrayList 类之间的关系。

5. 掌握 Java 泛型方法与泛型类的使用。

(1) 在 Eclipse 中，分别调试工程 Java_Foundation 的源程序包 raw_type 里的源程序。

(2) 掌握泛型方法的定义与使用，提高程序通用性。

(3) 掌握泛型类的定义与使用，提高程序通用性。

6. 掌握 Java 多线程型与同步锁 synchronized 的使用。

(1) 在 Eclipse 中，调试工程 Java_Foundation 的源程序包 multi_thread 里的源程序。

(2) 掌握多线程的创建与使用。

(3) 掌握使用 Java 关键字 synchronized 解决多线程并发冲突问题的处理方法。

7. 掌握 Java 反射机制的使用。

(1) 在 Eclipse 中，调试工程 Java_Foundation 的源程序包 reflection 里的源程序。

(2) 掌握三个类 java.lang.Object、java.lang.Class 与 java.lang.reflect.Method 之间的关系。

三、实验小结及思考

(由学生填写，重点写上机中遇到的问题。)

Android 开发环境及运行调试方法

Android 开发环境的搭建是 Android 应用开发的第一步，也是具体了解 Android 系统的一个途径。本章学习要点如下：

- 掌握搭建 Android 应用开发环境的方法；
- 掌握 Android SDK Manager 及手机模拟器的使用；
- 了解 Android ADT 插件的作用；
- 掌握 Android 的工程文件结构；
- 掌握 DDMS 的使用方法及 LogCat 在调试 Android 应用程序中的方法。

2.1 安装 JDK、下载 Android 集成开发环境

2.1.1 在使用 Eclipse 前确保已经安装 JDK

由于在 Eclipse 中是用 Java 语言编写应用程序的，因此，启动 Eclipse 时要求已经安装了 Java 运行环境 JRE。事实上，安装了 JDK，就一定安装了完整的 Java JRE。

如果计算机没有安装 JDK，则运行 Eclipse 时，会出现警告信息，如图 2.1.1 所示。

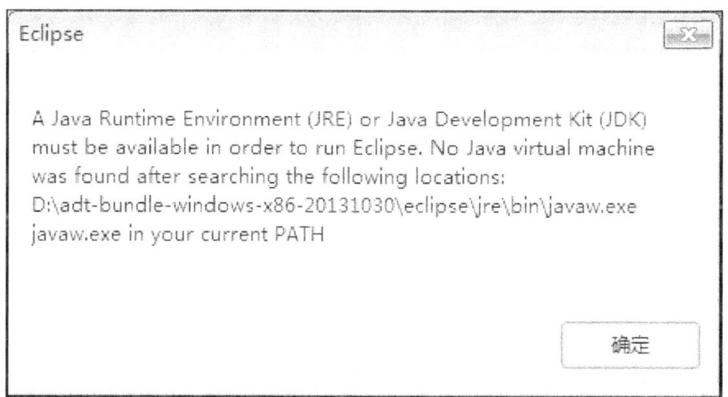

图 2.1.1　没有安装 JDK 时运行 Eclipse 的警告信息

注意：

(1) 由于 Sun 公司已经被 Oracle 公司收购，如果要下载 JDK，则可以访问 Oracle 公司网站

http://www.oracle.com。

(2) 计算机安装的 Windows 操作系统分为 32 位和 64 位两个版本，在作者教学网站 http://www.wustwzx.com 里下载的 JDK 是适合 32 位的操作系统。

(3) 搭建 Android 开发环境(参见第 2.1.2 小节)也分 32 位和 64 位两种。在 32 位操作系统的计算机上只能使用 32 位的 JDK 版本和 Android 开发环境。

(4) 在命令行方式下，输入命令 java -version，如果显示版本信息，才表明 JDK 已安装好。

2.1.2 下载集成包 Android SDK+ADT for Windows

早期搭建 Android 开发环境的方法是：在安装了 JDK 后，安装 Eclipse(可访问 http://www.eclipse.org/downloads 来下载)，然后安装 Eclipse 插件 ADT(Android Development Tools)，最后添加 Android SDK。

目前，搭建 Android 开发环境的最简便方法是使用捆绑了 ADT 的 Eclipse 集成环境。作者于 2013 年 10 月 30 日访问 http://developer.android.com/sdk/index.html，下载了较新的开发版本 Android 4.4(对应于 Android API 19)，然后上传至作者的百度云盘并做了分享的公开链接。

输入网址“http://pan.baidu.com/s/1gdnDup1”进入下载页面，解压下载后的压缩文件 adt-bundle-windows-x86-20131030.zip 至硬盘，会看到有两个子文件夹和一个 Android 下载管理器程序 SDK Manager.exe，如图 2.1.2 所示。

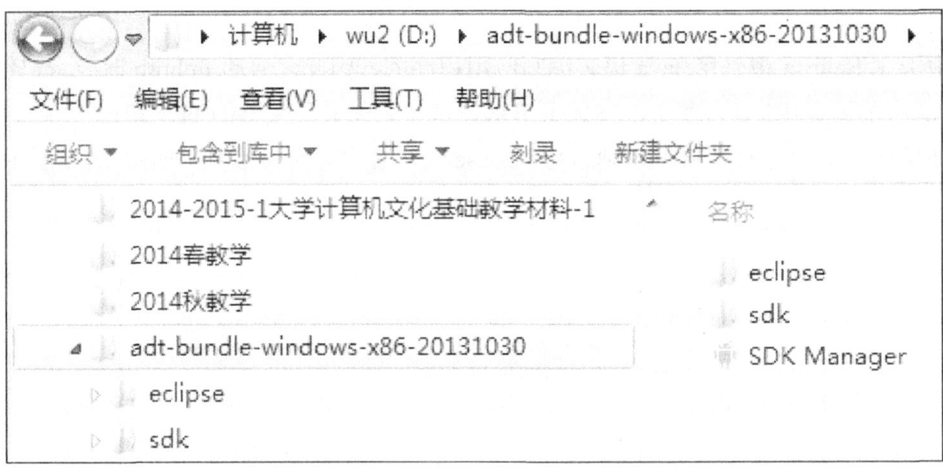

图 2.1.2 Android 集成包的目录结构

注意：为方便使用，通常在 eclipse 文件夹里建立启动 Eclipse 程序的桌面快捷方式。

Eclipse 启动后，单击“Windows→Preferences→Java→Compiler”，可以看到 Eclipse 所引用的 Java 运行环境 JRE，如图 2.1.3 所示。

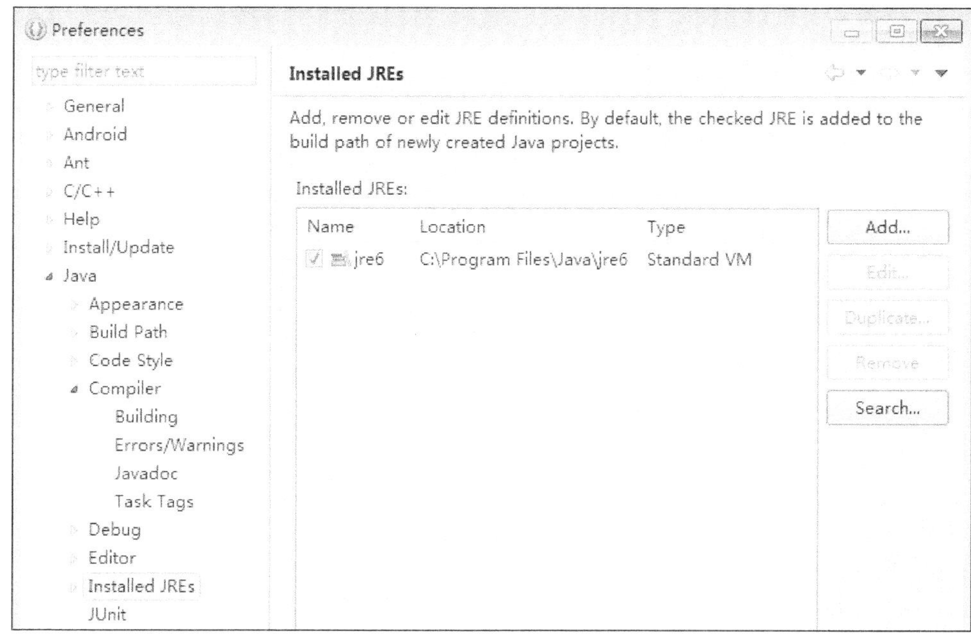

图 2.1.3 Eclipse 引用的 Java 运行环境

注意：若集成包是 32 位的，则必须安装和使用相应的 32 位 JDK 环境。

2.2 Android ADT 与 SDK 介绍

2.2.1 ADT 作为 Eclipse 的一个插件

ADT 是 Android Development ToolKit 的英文缩写，表示开发工具包，是 Eclipse 的一个插件。在 Eclipse 中使用 ADT 插件可以快速地创建 Android 工程、用户界面和基于 Android API 的组件，还可以在 Eclipse 中使用 Android SDK(Software Development Kit，SDK)提供的工具进行程序调试或对.apk 文件进行签名等。

ADT 也可以在 Eclipse 中输入其下载地址进行安装。

注意：

(1) 在 Eclipse 中单独安装 ADT 插件的方法是使用菜单 “Help→Install New Software” 。

(2) 在 Eclipse 的首选项中，只有安装了 ADT，才会出现做 Android 应用开发的选项 “Android” 。

2.2.2 SDK 与 SDK Manager

Android SDK 是 Android 软件开发包，是 Android 整体开发中所用到的工具包，提供了库文件以及其他开发所用到的工具。

启动 Eclipse 后，单击“Windows→Preferences→Android”，可以看到 Android SDK 的位置、已经安装(下载)的 Android API 版本等信息，如图 2.2.1 所示。

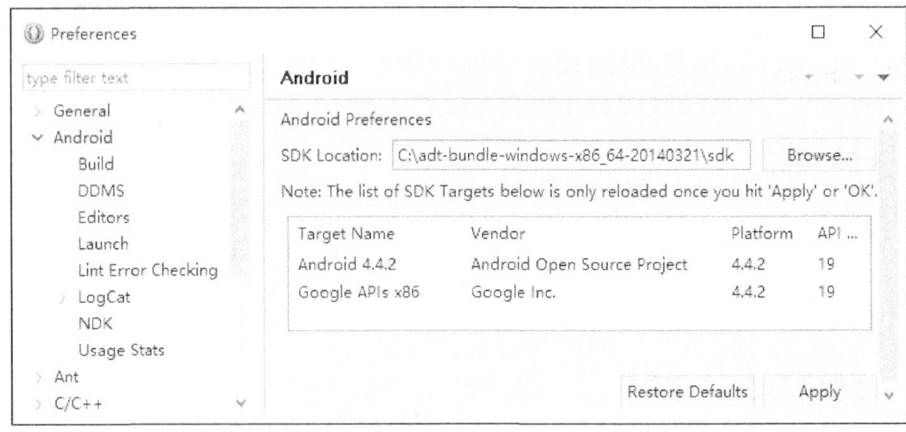

图 2.2.1　查看 Eclipse 中的 SDK 插件

注意：

(1) 如果在 Eclipse 中不关联 Android SDK，则无法开发 Android 应用程序。

(2) 如果不用 Eclipse 作为 Android 开发工具，就不需要下载 ADT，因为 SDK 在 Linux 环境下通过 make 命令进行编译，没有使用 “ADT+Android SDK” 的方式方便。

(3) Android API 是 Android SDK 的主体，其他内容参见第 2.2.3 小节。

单击 Eclipse 工具栏上的 ，会打开 Android SDK Manager 程序，可以完成不同的 Android 版本下载、Google APIs(用于地图开发)下载和 Android API 源码下载等功能，如图 2.2.2 所示。

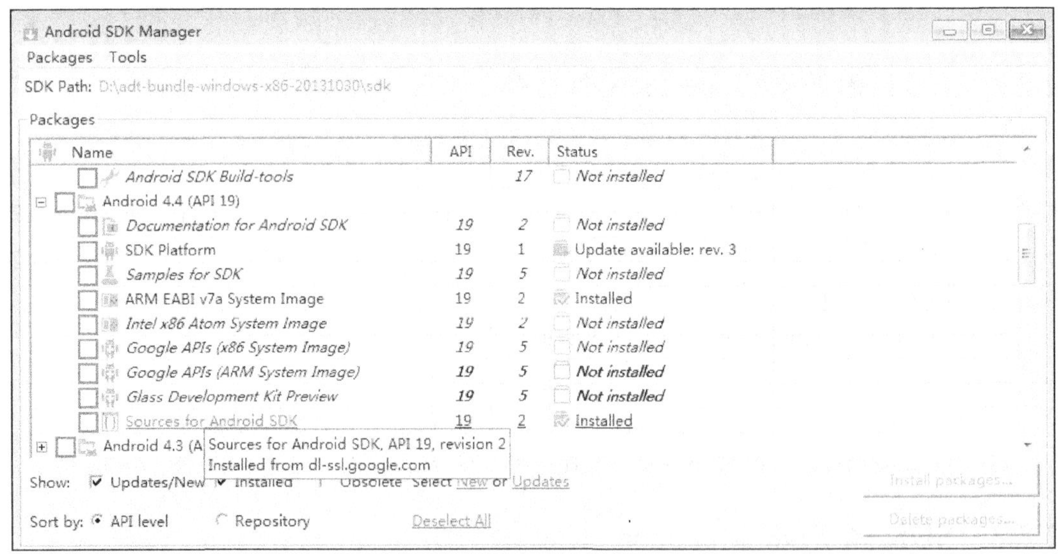

图 2.2.2　Android SDK Manager

2.2.3　Android SDK 目录结构

Android SDK 目录包含的子目录，如图 2.2.3 所示。

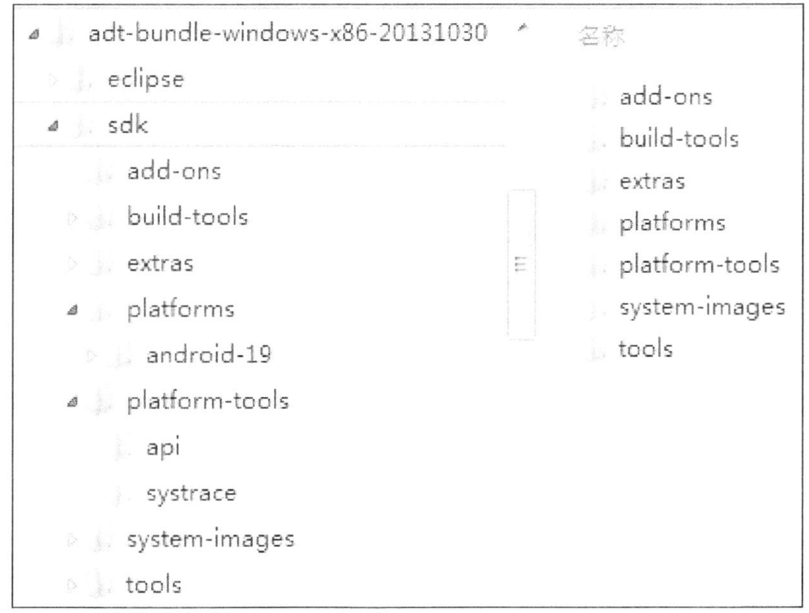

图 2.2.3 Android SDK 目录结构

下面介绍 sdk 文件夹内的子目录及文件。

- build-tools ：包含了 Android 应用的编译工具，如将.class 字节码文件转换成 Android 字节码.dex 文件的批处理程序 dx.bat、生成 Android 设备进程间通信代码的应用程序 aidl.exe(参见第 6.2.2 小节)等。
- platforms：存放 Android SDK Platforms 平台相关的文件，包括 android.jar、字体、res 资源、模板等。
- add-ons：存放 Android 的扩展库，如用于地图开发的 Google Maps。
- sources：存放 Android API 的源码。
- platform-tools：存放平台工具，例如用于将 Android 手机连接到 PC 端的 Android 调试桥(Android Debug Bridge，简称 ADB)程序 adb.exe。
- tools：包含了用于操作 SQLite 数据库的程序 sqlite3.exe、模拟器管理程序 AVD Manager.exe、对 Android 应用程序进行调试和模拟服务的批处理程序 ddms.bat 等工具。

注意：

(1) 当使用 Android SDK Manager 下载 Android API 的源码后，将会在文件夹 sdk 里自动新建一个名为 sources 的子文件夹，并将源码存放在 sources 文件夹里。

(2) 设置 Android 应用的图标时，可以使用系统自带的图标库，这些图标文件就位于文件夹 sdk\platforms\android-19\data\res\drawable-hdpi 里。

(3) ADB(Android Debug Bridge)是用于连接 Android 设备的工具，负责将应用程序安装到 Android 设备，它是一个客户端/服务器程序。

2.2.4 Android API 核心包

标准的 Android API 包含在许多软件包里，这些软件包又包含在文件 android.jar 里。下面介绍 Android 开发中常用的软件包。

- android.util：包含一些辅助类，如时间、日期的操作。
- android.text：包含文本处理类。
- android.text.method：提供为各种控件输入文本的类。
- android.os：提供基本的操作服务、消息传递和进程间通信。一些重要的类包括 Binder、Handler、FileObserver、Looper 和 PowerManager。Binder 类支持进程间通信。FileObserver 监视对文件的更改。Handler 类用于运行与消息线程有关的任务，Looper 用于运行消息线程。
- android.app：实现 Android 的应用程序模型，主要的类包括 Application(表示开始和结束语义)，以及众多与 Activity 相关的类、控件、对话框、提醒和通知。
- android.view：提供基础的用户界面接口框架，是 Android 的核心框架，包含类 Menu、View、ViewGroup 以及一系列监听器和回调函数。
- android.widget：包含在应用程序的屏幕中使用的各种 UI 元素，通常派生自 View 类，包括 Button、Checkbox、Chronometer、AnalogClock、DatePicker、DigitalClock、EditText、ListView、FrameLayout、GridView、ImageButton、MediaController、ProgressBar、RadioButton、RadioGroup、RatingButton、Scroller、ScrollView、Spinner、TabWidget、TextView、TimePicker、VideoView 和 ZoomButton。
- android.webkit：默认浏览器操作接口，包含表示 Web 浏览器的类，主要有 WebView、CacheManager 和 CookieManager。
- android.content：包含各种对设备上的数据进行访问和发布的类。
- android.content.pm：实现与包管理器相关的类。包管理器知道各种权限、安装的包、安装的提供程序、安装的服务、安装的组件(比如 Activity)和安装的应用程序。
- android.content.res：用于访问结构化和非结构化资源文件。主要的类包括 AssetManager(用于结构化资源)和 Resources。
- android.database：实现抽象数据库的理念，提供了 Cursor 接口。
- android.database.sqlite：将 SQLite 用作物理数据库。主要的类包括 SQLiteCursor、SQLiteDatabase、SQLiteQuery、SQLiteQueryBuilder 和 SQLiteStatement。
- android.provider：提供类，访问 Android 的 ContentProvider，如 Contacts、MediaStore、Browser 和 Settings。
- android.media：提供一些类，管理多种音频、视频的媒体接口，包含类 MediaPlayer、MediaRecorder、Ringtone、AudioManager 和 FaceDetector。
- android.hardware：实现与物理照相机相关的类。android.graphics.Camera 表示一种图形概念，与物理照相机完全无关。
- android.bluetooth：提供一些类来处理蓝牙功能。主要的类包括 BluetoothAdapter、

BluetoothDevice、BluetoothSocket、BluetoothServerSocket 和 BluetoothClass。

- android.net：提供帮助网络访问的类，实现基本的套接字级网络 API。主要的类包括 Uri、ConnectivityManager、LocalSocket 和 LocalServerSocket。
- android.net.wifi：管理 WiFi 连接。主要的类包括 WifiManager 和 WifiConfiguration。WifiManager 负责列出已配置的网络和目前处于活动状态的 WiFi 网络。
- android.telephony：提供手机设备的通话接口，包含类 CellLocation、PhoneNumberUtils 和 TelephonyManager。TelephonyManager 可用于确定手机位置、电话号码、网络运营商名称、网络类型、电话类型和 SIM(subscriber identity module，用户身份模块)序列号。
- android.telephony.gsm：可用于根据基站来收集手机位置，还包含负责处理 SMS 消息的类。这个包名为 GSM，是因为全球移动通信系统是最初定义 SMS 数据消息标准的技术。
- android.location：定位和相关服务的类，包含类 Address、GeoCoder、Location、LocationManager 和 LocationProvider。
- com.google.android.maps：包含类 MapView、MapController 和 MapActivity，它们在本质上是处理 Google 地图所需的类。
- android.gesture：包含处理用户定义的手势所需的所有类和接口。主要的类包括 Gesture、GestureLibrary、GestureOverlayView、GestureStore、GestureStroke 和 GesturePoint。Gesture 是 GestureStrokes 和 GesturePoints 的集合。手势都收集在 GestureLibrary 中。手势库存储在 GestureStore 中。手势都具有名称，这样可以将其标识为动作。
- android.graphics：底层的图形库，包含画布、颜色过滤、点、矩形，可以将它们直接绘制到屏幕上，包含类 Bitmap、Canvas、Camera、Color、Matrix、Movie、Paint、Path、Rasterizer、Shader、SweepGradient 和 TypeFace。
- android.graphics.drawable：实现绘制协议和背景图像，支持可绘制对象动画。
- android.graphics.drawable.shapes：实现各种形状，包括 ArcShape、OvalShape、PathShape、RectShape 和 RoundRectShape。
- android.view.animation：提供对补间动画的支持。主要的类包括 Animation、一系列动画插值器，以及一组特定的动画绘制类，包括 AlphaAnimation、ScaleAnimation、TranslationAnimation 和 RotationAnimation。
- android.opengl：提供 OpenGL 的工具，3D 加速。

2.2.5 关于 Google APIs

Google APIs 除了包括标准的 Android API 提供的 android.jar 外，还包括了用于地图开发的 maps.jar 等，它是 Android API 的扩展包，例如 Google APIs 创建地图应用工程，参见第 10.1 节图 10.1.3。

2.3 创建一个 Hello 工程

2.3.1 设置 Eclipse 工作空间

每个 Android 工程对应一个文件夹，其内又含有若干子文件夹，它们用来存放程序文件、资源文件、工程配置文件等。

首次启动 Eclipse 时，Eclipse 会出现一个用来指定存放 Android 工程的位置(也称工作空间 workspace)确认对话框，如图 2.3.1 所示。

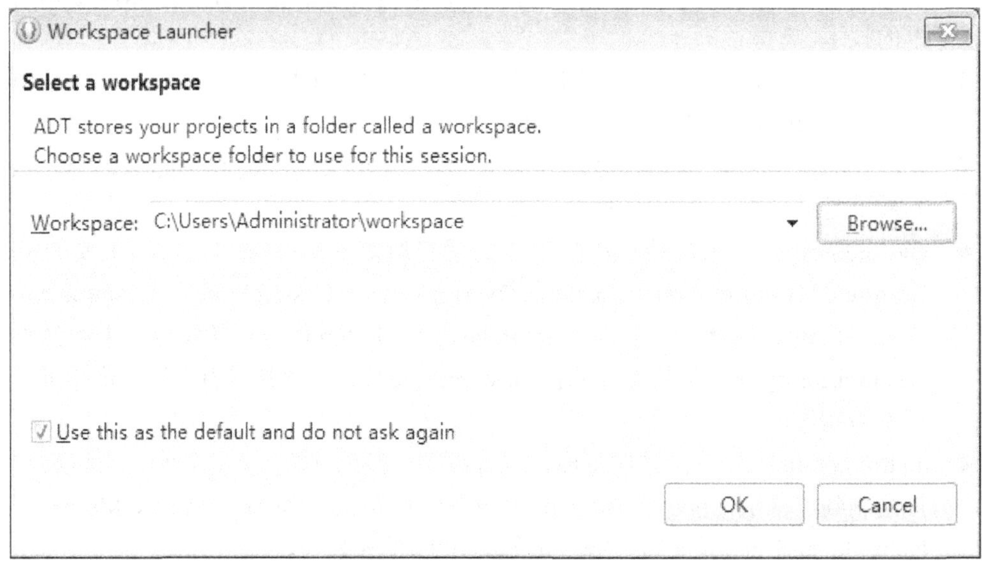

图 2.3.1　设置工作空间

所有工程文件夹会放置在这个 Eclipse 工作空间内。Eclipse 工作空间也可以通过使用 Eclipse 菜单“File→Switch Workspace”重新设置。

2.3.2 创建一个简单的 Android 工程

在 Eclipse 环境中，使用菜单“File→New→Android Application Project”后，出现创建 Android 应用工程的对话框。输入应用名称为“Hello”后的效果如图 2.3.2 所示。

其中，输入的应用名称的首字母一般大写，工程名称、程序包名和编译版本等一般采用默认值。

在接下来的对话框中，勾选创建 Activity、设置应用的图标、Activity 名称和对应的布局文件名等。

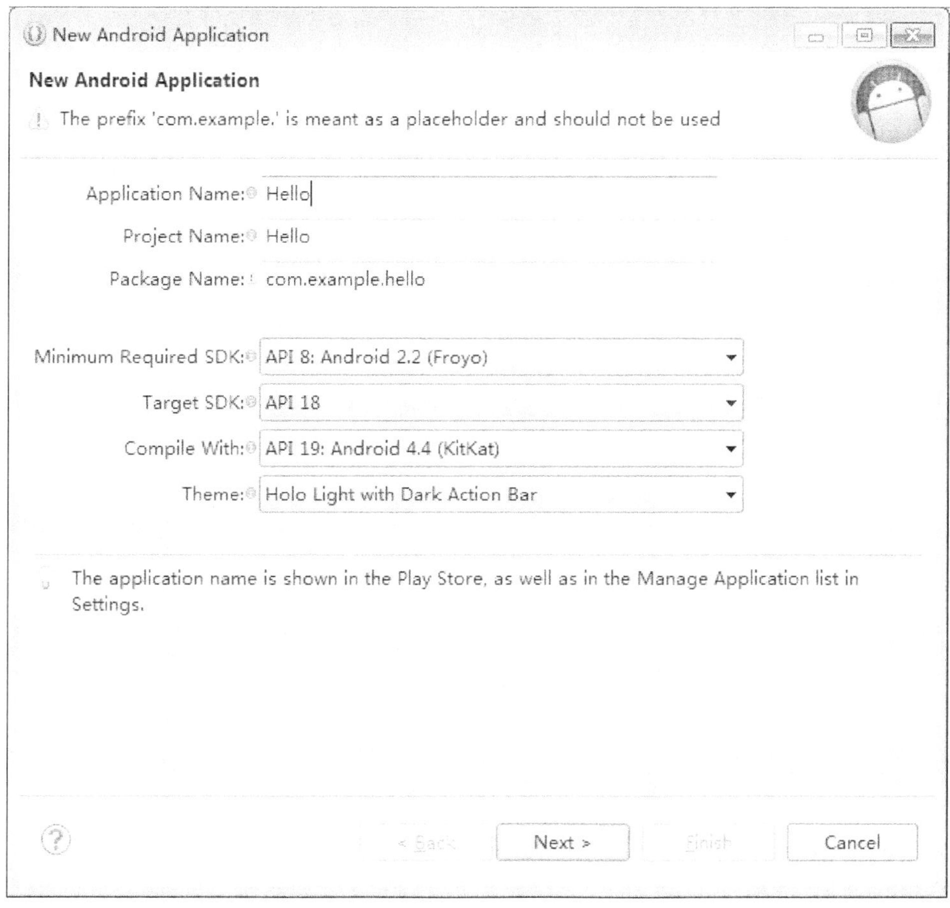

图 2.3.2 新建 Android 应用工程对话框

注意：

(1) 通常情况下，一个 Android 应用至少需要建立一个作为用户界面的 Activity，显示的控件在布局文件里定义。

(2) Android 应用工程的保存路径，一般指定为 Eclipse 默认的工作空间。

2.4 部署和运行 Android 应用程序

2.4.1 创建 Android 手机模拟器 AVD

AVD(android virtual device)是 Android SDK 提供的最重要的工具之一，它使得开发人员在没有物理设备的情况下，在计算机上对 Android 程序进行开发、调试和仿真。

单击 Eclipse 工具栏上的 ，在出现的 AVD 管理器窗口里，单击“New”按钮， 即可出现创建 AVD 的对话框，如图 2.4.1 所示。

Edit Android Virtual Device (AVD)

AVD Name:	AVD1
Device:	Nexus 4 (4.7", 768 × 1280: xhdpi)
Target:	Android 4.4 - API Level 19
CPU/ABI:	ARM (armeabi-v7a)
Keyboard:	☑ Hardware keyboard present
Skin:	☑ Display a skin with hardware controls
Front Camera:	None
Back Camera:	None
Memory Options:	RAM: 512 VM Heap: 64
Internal Storage:	200 MiB
SD Card:	◉ Size: 128 MiB
	○ File: Browse...

图 2.4.1 创建 Android 模拟器对话框

注意：CPU/ABI 选项里的 ARM，是英国的半导体芯片商。

AVD 创建后(名为 AVD1)，选中它，然后单击 AVD 管理器窗口里的“Start”按钮，即可启动该模拟器，效果如图 2.4.2。

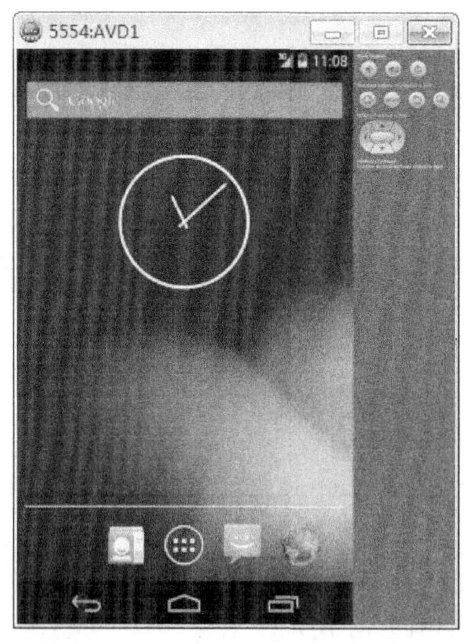

图 2.4.2 开启一个名为“AVD1”的模拟器

注意：启动模拟器相当于手机开机，通常需要较长的时间。

2.4.2 部署 Android 应用到模拟器

在 Android 应用工程名的快捷菜单中，选择“Run As→Android Application”，即可将工程部署到 AVD 并运行。

注意：

(1) 实际调试时，也可以先选中工程名，然后单击 Eclipse 工具栏上的 图标。

(2) 如果建立了多个模拟器(包括手机)，则会出现选择对话框，如图 2.4.3 所示。

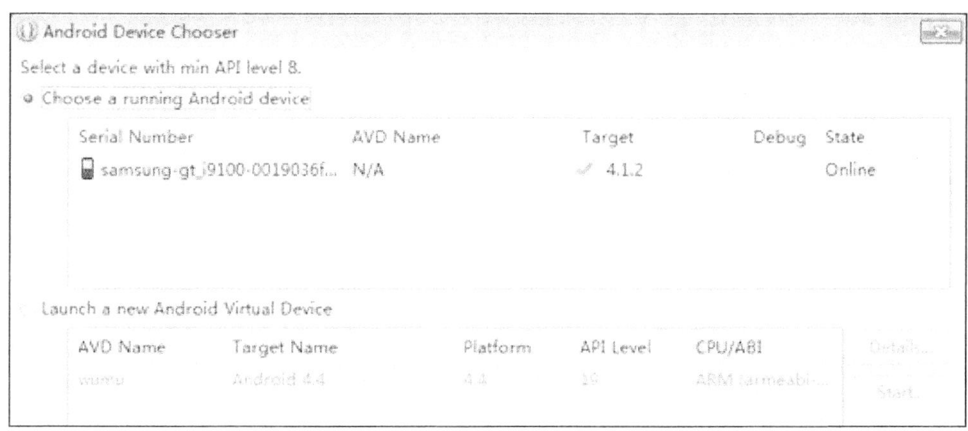

图 2.4.3 选择调试和运行的设备

2.4.3 部署 Android 应用到手机

部署 Android 应用到手机与到模拟器的方法相同，但要注意的是，手机版本不能低于建立 Android 应用工程时指定的最低版本。否则，在部署该应用时，找不到手机(尽管手机已经与计算机正常连接)。

注意：

(1) 手机较 AVD 而言，响应速度更快。

(2) AVD 并不能模拟手机的全部功能(如真实电话呼叫、摄像和蓝牙等)。

(3) 当手机内存不足时，可能造成部署不成功。

2.4.4 Android 工程的导入与管理

初学者导入别人开发的工程，在自己的开发平台中测试、验证和研究，是学习 Android 的一条非常好的捷径。导入项目时，若文档出现中文乱码，则表明项目文档的编码与 Eclipse 的编码不一致，此时修改项目文档的编码或 Eclipse 编码即可。

导入 Android 工程与导入 Java 工程的方法是类似的，如图 2.4.4 所示。

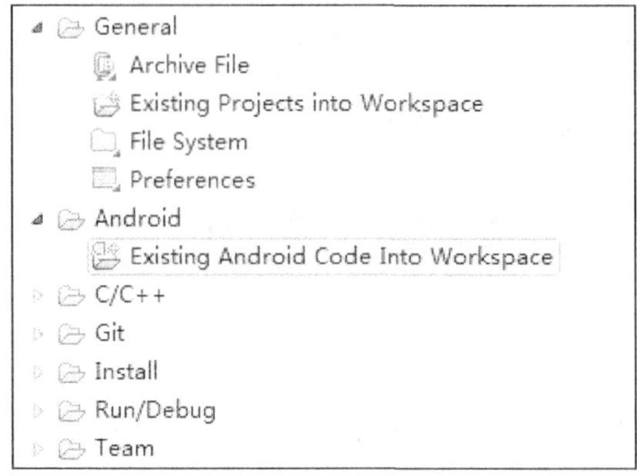

图 2.4.4　导入一个已经存在的 Android 工程

注意：通常在导入工程时，勾选“Copy projects into workspace”，以避免对原工程文件夹的依赖，即将所有工程文件夹统一放置到工作空间里。

如果工程的源文件只有一个包名，默认情况下，应用的包名与之相同。实质上，应用的包名可以任意修改，推荐的方法是使用工程名的快捷菜单，如图 2.4.5 所示。

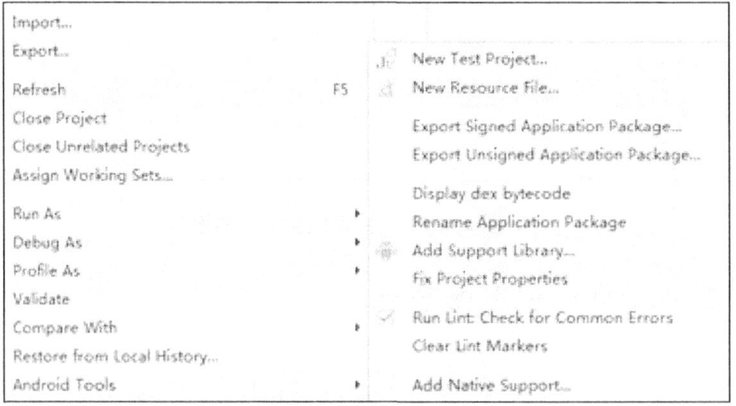

图 2.4.5　推荐的工程包名更改方法

注意：由于 Android 应用工程的包名有多处使用，因此不能只在清单工程中修改；否则，会造成工程异常。上面推荐的方法可以进行彻底的修改，不会造成工程异常。

在 Android 学习和实际开发过程中，会接触大量的工程。当 Android 工程数量增加时，Eclipse 的启动时间会相应地延长，因为 Eclipse 启动时会加载每一个工程。因此，当 Android 工程调试完毕后，通过工程的右键菜单或系统的“Project”菜单关闭该工程可以节约下次启动 Eclipse 的时间。

此外，对工程的管理还有删除、复制、jar 包的配置等。工程的快捷菜单提供了管理工程的常用功能。

注意：

(1) 导入别人的工程时，可能存在与自己的 Android 版本不一致的情况，此时尽管程序代码没有错误，但工程前面出现 。解决的办法是打开工程属性文件 Project.properties，修改编译版本为自己已经安装的

Android 版本。

(2) 导入别人的工程时，如果文件中出现中文乱码，表明别人工程使用的文件编码与本机的编码不一致。解决办法是将别人的文件以本机 Eclipse 使用的编码格式保存后再导入。

(3) 如果导入的工程含有对该工程以外的 jar 包的引用，则导入后也可能在导入的工程前面出现 ，此时可以使用工程名的快捷菜单，选择“Build Path→Configure Build Path”。

2.5　Android 平台的调试方法

Eclipse 启动后，除了编辑源代码外，还经常要进行 Eclipse 视图切换和若干信息查看及调试工作。

2.5.1　Eclipse 常用的两种视图切换

在 Eclipse 工具栏右边，有 Java 和 DDMS 用于 Eclipse 视图模式的切换。默认情况下，Eclipse 处于前一模式，即 Java 模式。在 Java 模式下，可以看到所有的项目(包括 Android 项目和 Java 项目)，还可以对项目里的源文件进行编辑。

2.5.2　查看所有工程的 Problem 报告

单击“Problem”选择卡时，会显示所有项目中错误和警告的条目数，如图 2.5.1 所示。

图 2.5.1　Eclipse 中所有项目的 Problem 报告

2.5.3　查看工程部署和运行的控制台输出

在 Java 视图模式下，当一个 Android 工程成功部署后，会在 Eclipse 控制台输出一些相关信息，如图 2.5.2 所示。

```
Console
Android
[2014-07-06 09:09:01 - Dialing] Automatic Target Mode: Unable to detect device compatibility.
[2014-07-06 09:09:04 - Dialing] Uploading Dialing.apk onto device '0019036f40c66e'
[2014-07-06 09:09:05 - Dialing] Installing Dialing.apk...
[2014-07-06 09:09:11 - Dialing] Success!
[2014-07-06 09:09:11 - Dialing] Starting activity com.example.dialing.MainActivity on device
[2014-07-06 09:09:11 - Dialing] ActivityManager: Starting: Intent { act=android.intent.action
```

图 2.5.2　成功部署一个 Android 工程后的控制台输出

2.5.4　DDMS 视图及其 LogCat

DDMS(Dalvik debug monitor service)是 Android 系统内置的调试工具，用于对 Android 应用程序的调试和模拟服务，包括设备管理器、查看特定的进程、文件浏览、日志、广播状态、模拟电话呼叫、接收 SMS、虚拟地理位置、为测试设备截屏等，如图 2.5.3 所示。

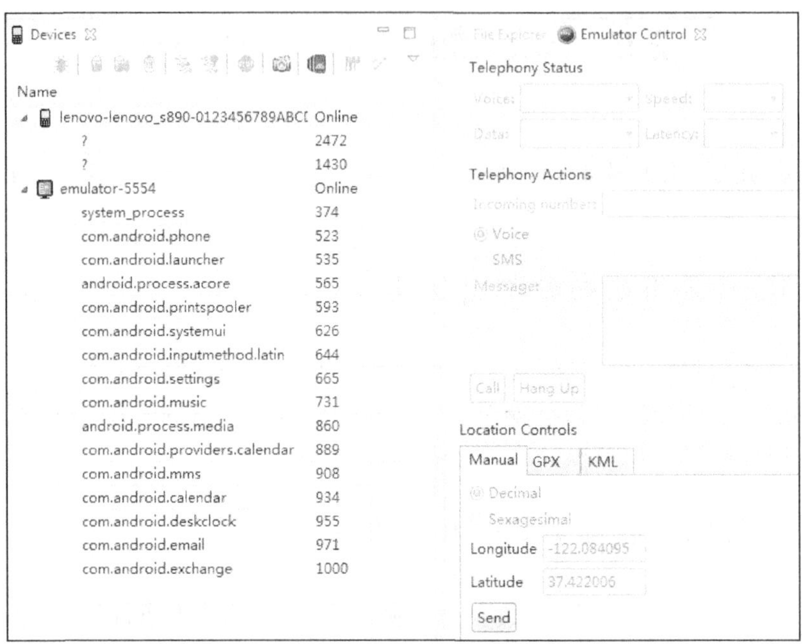

图 2.5.3　DDMS 中的文件浏览器和模拟器控制器

作为 DDMS 的日志浏览器 LogCat，可以浏览 Android 系统、Dalvik 虚拟机(参见第 3.3 节)和应用程序产生的日志信息，有助于快速定位应用程序产生的错误。

为了查看某个应用程序的日志信息，通常要设置 LogCat 过滤器，并按应用的包名过滤，如图 2.5.4 所示。

Logcat Message Filter Settings

Filter logcat messages by the source's tag, pid or minimum log level.
Empty fields will match all messages.

Filter Name: cn.itcast.sqlite

by Log Tag:

by Log Message:

by PID:

by Application Name: cn.itcast.sqlite

by Log Level: verbose

OK　Cancel

图 2.5.4　建立某个应用程序的 LogCat 过滤器

2.5.5 动态调试方法 Debug

如同 VC++和 Visual Studio 等开发环境，在开发 Android 应用程序的 Eclipse 环境中，也可以设置断点检查每个变量的运行输出，较适合一些大型项目的排除错误操作。

在某行代码的左边区域双击鼠标，即可设置或取消断点。选中工程名后，单击工具栏上的 ，即以调试模式运行程序。此时，屏幕出现多个窗口，其中右上角的窗口用于显示变量的值。

当程序运行到设置的断点时就会停下，这时可以选择如下操作。

- 按 F8：直接执行程序，直到下一个断点处停止。
- 按 F5：单步执行程序，遇到方法时进入。
- 按 F6：单步执行程序，遇到方法时跳过。
- 按 F7：单步执行程序，从当前方法跳出。

注意：

(1) 以调试模式运行程序的另一种方法是：右击工程名，选择“Debug As→Android Application”。

(2) 开发一个 Android 应用工程时，可能需要编写很多的业务逻辑类，其正确性可以单独测试而不必在整个应用中调试，它涉及 Android 的工作原理，因此在第 3.4 节里介绍。

2.5.6 软件设计的国际化与“I18n”警告性错误

国际化是指在软件设计过程中将特定语言及区域脱钩的过程。当软件移植到不同的语言及区域时，软件本身不需要做任何的修改。Android SDK 并没有提供专门的 API 来实现国际化，而是通过对不同的资源(resource)文件进行不同的命名来达到国际化的目的。

在 Hello 工程中，布局文件 res/layout/activity_main.xml 中文本框控件放在值文件 res/values/strings.xml 文件的成对标签<string>和</string>之间，并通过引用字符串变量的方式引用该值，即

```
<TextView ...android:text="@string/hello_world" />
```

这是符合国际化的做法。

如果不通过引用字符串变量的方式，而是直接把字符串常量写在 TextView 控件的 android:text 属性后，即

```
<TextView ...android:text="字符串常量" />
```

这种写法对程序运行没有任何影响，但它不符合国际化的做法，并且会导致“I18n”警告性错误——黄色的感叹号。

“I18n”是国际化的英文单词 Internationalization 的简称，表示字母 I 与 n 之间还有 18 个字符。

2.6　Android 签名策略

2.6.1　导出未经签名的应用程序

为了导出未经签名的 Android 应用，需要右击 Android 工程名，选择“Android Tools”，再选择“Export Unsigned ApplicationPackage”，操作如图 2.6.1 所示。

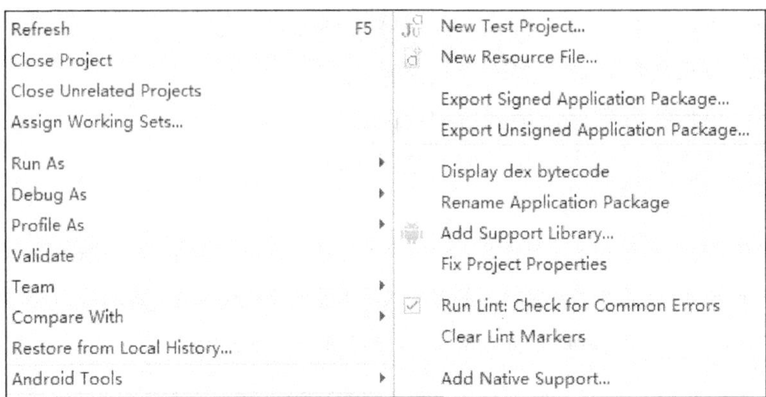

图 2.6.1　导出未经签名的应用程序

在出现的导出工程对话框中，需要指定 Android 应用工程的安装包文件的存放路径，如图 2.6.2 所示。

图 2.6.2　导出未签名的应用程序对话框

导出安装包文件(扩展名为.apk)后，出现警告信息，提示在发布本应用程序之前需要进行签名。警告信息如图 2.6.3 所示。

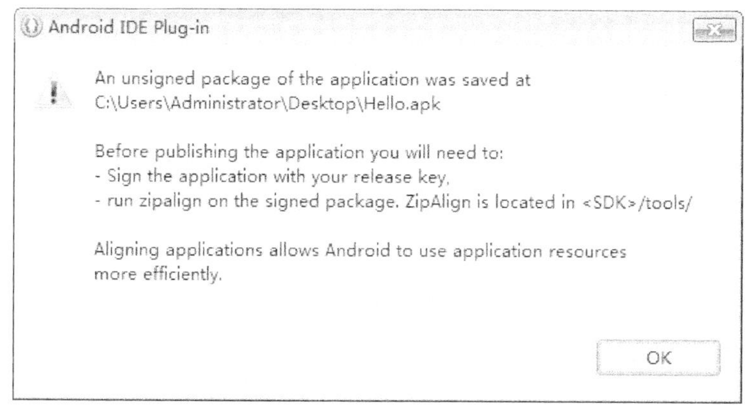

图 2.6.3　警告信息(对未签名的 Android 应用程序)

这种未经签名的.apk 文件不能被安装到手机上。在 Eclipse 集成开发环境中，如果试图安装，则出现图 2.6.4 所示的错误信息。

图 2.6.4　在集成开发环境中安装未经签名的安装包时的错误提示

注意：把该安装包文件复制到手机上，也是不能成功安装的。

2.6.2　导出经过数字签名的 Android 应用程序

为了将测试通过的 Android 应用程序发布到 Google Play Store，为了能让 Android 应用程序支持在线更新和版本升级，通常需要对 Android 应用程序进行签名。

由于开发 Android 应用的人很多，因此，很有可能出现相同的包名和类名，通过数字签名可以解决这个问题。Android 应用开发者都应使用一个证书来进行签名，证书的私钥归应用的开发者所有。数字签名可以保证包名相同但签名不同的包名不被替换。

创建经过数字签名的 Android 应用程序的步骤如下。

(1) 右击 Android 工程名，使用菜单“Android Tools→Export Signed Application Package”，单击“Next”按钮。

(2) 在出现的对话框中，勾选“Create new keystore”，指定证书的存放位置、文件名和密码，如图 2.6.5 所示。

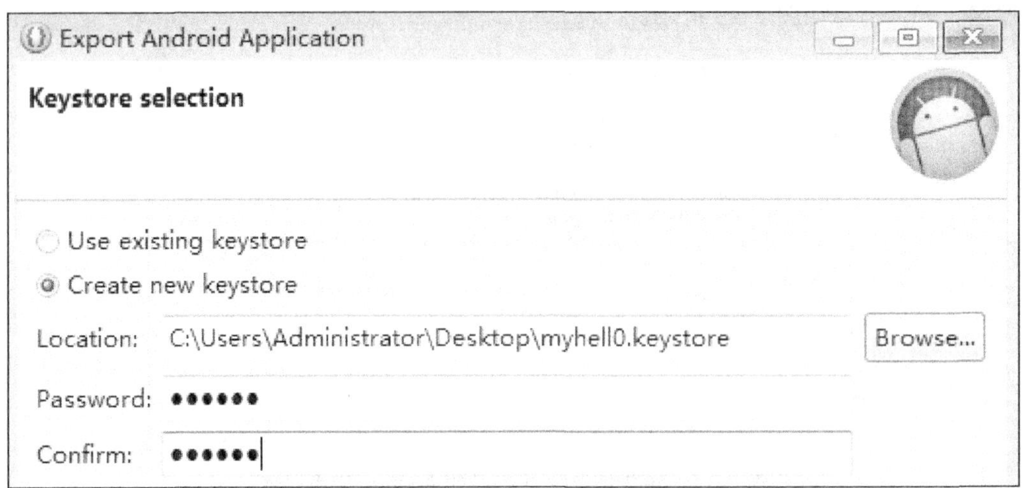

图 2.6.5　设置 Android 应用证书的存放位置、名称和密码

注意：证书文件已经加密，无法使用记事本等文本编辑软件打开，也不能成功安装。

(3) 输入开发者的相关信息(需要使用刚才设置的证书密码)，效果如图 2.6.6 所示。

Export Android Application

Key Creation

Alias: MyHello

Password: ●●●●●●

Confirm: ●●●●●●

Validity (years): 25

First and Last Name: KePeng

Organizational Unit: Computer Science

Organization: WUST

City or Locality: WuHai

State or Province: HuBei

Country Code (XX): 86

图 2.6.6　输入开发者信息

(4) 指定安装包文件的名称(可随意修改)及存放位置，如图 2.6.7 所示。

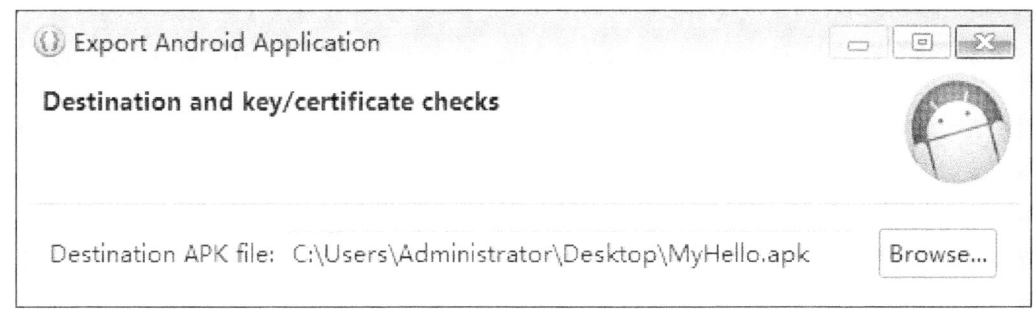

图 2.6.7 指定安装包文件的路径及名称

注意：

(1) 使用数字签字是为了辨别软件的开发者。

(2) 安装使用数字签字的 Android 应用程序后，该应用程序不会自动运行(不同于 “Run As” 方式)。

(3) 以前通过 “Run As” 方式发布应用到 Android 设备时，虽然没有签名，但实际上是使用系统提供的默认证书文件 debug.keystore。使用 Debug 证书的 Android 应用，不能在 Android 超市里销售，也不能进行产品的有效升级。Debug 证书文件的存放位置及密码(与开发环境相关)，如图 2.6.8 所示。

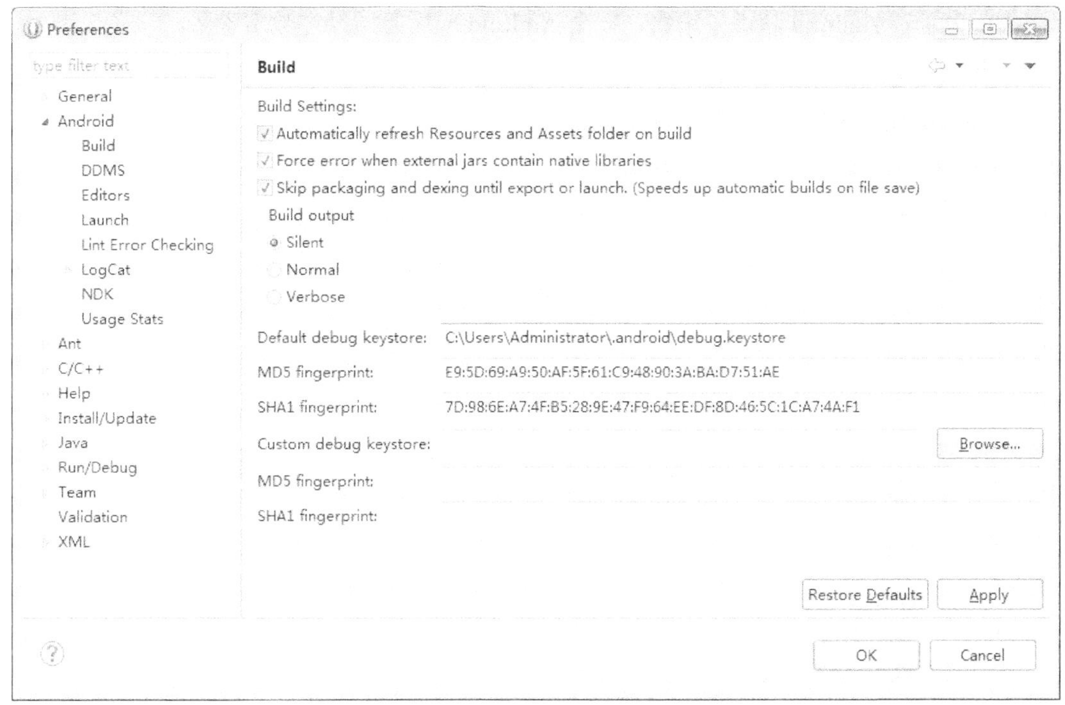

图 2.6.8 Debug 证书文件的存放位置及密码

习 题 2

一、判断题

1. 可以正常启动 Eclipse 的前提是已经安装了 JRE(或 JDK)。
2. 在 Eclipse 中，Android ADT 是一个插件。
3. 在 Android 中，SDK Manager 实质上是下载管理器。
4. android.jar 中不包含任何 Java 软件包。
5. 单击 Eclipse 工具栏上的工具图标启动模拟器后，再关闭 Eclipse 时将会关闭模拟器。
6. AVD 是 Android 虚拟设备的缩写。
7. 模拟器里的日期、时间与选择的时区相关。
8. 部署到手机或 AVD 的每个 Android 应用，都会在系统文件夹 dada/data 下建立一个与该应用包名相同的文件夹。
9. 新建一个 Android 应用项目时，其 Application Name 与 Project Name 必须相同。
10. 为了在 Android 超市中销售，一个 Android 应用必须经过数字签名。

二、选择题

1. Android 自带的图标文件存放在 Android SDK 的____文件夹中。
   A. platforms B. platform-tools C. system-images D. tools
2. 在 Eclipse 中创建 AVD 时，通常不可更改的选项是____。
   A. AVD Name B. Device C. CPU/ABI D. Target
3. 切换 Eclipse 的视图模式，应使用 Windows 菜单中的____菜单项。
   A. Open Perspective B. Show C. Navigation D. Preferences
4. 用于 Android 应用开发的 android.jar 文件，位于 sdk 文件夹的____子文件夹里。
   A. add-ons B. platforms C. system-images D. tools
5. 修改 Android 工程应用的包名，应使用工程名的快捷菜单中的____选项。
   A. Properties B. Run As C. Debug As D. Android Tools

三、填空题

1. ____压缩包是一个标准的压缩包，其内容包含的是编译后的 class，包含了 Android 的全部 API。
2. 使用 Android 平台的____方法，可以准确地排除错误。
3. Android 目前常用的开发环境是 Eclipse+Android ____+Android SDK。
4. 在 DDMS 视图方式下，为了加载所需要的面板，应使用 Windows 菜单中的____菜单项。
5. 在 Eclipse 中最先启动的 Android 模拟器的号码是____。

实验 2　Android 开发环境搭建及运行调试方法

一、实验目的

1. 掌握 Android 集成开发环境的搭建方法。
2. 掌握 Android 模拟器的创建与使用。
3. 使用向导创建一个简单的 Hello 示例工程。
4. 掌握 Android 工程的部署及运行方法。
5. 了解 Android SDK 的作用，掌握 Android SDK Manager 的使用方法。
6. 掌握 Android 平台的调试方法。
7. 了解 Android 的签名机制，掌握对 Android 应用签名的方法。

二、实验内容及步骤

1. 快速搭建 Android 集成开发环境。

(1) 确保计算机已经安装 JDK，若未安装，访问作者教学网站 http://www.wustwzx.com，在第三门课程(Android 应用开发)的相关资源里下载并安装。

(2) 输入网址“http://pan.baidu.com/s/1gdnDup1”进入下载页面，解压下载的压缩文件 adt-bundle-windows-x86-20131030.zip 至硬盘，在其内的 eclipse 文件夹内建立启动 Eclipse 程序的桌面快捷方式。

(3) 在硬盘的某个逻辑盘里新建文件夹(如 d:\android_kf)用以存放 Android 应用工程，双击桌面上 Eclipse 的快捷方式启动 Eclipse 程序，通过使用菜单“File→Switch Workspace”设置该文件夹为 Eclipse 的默认工作空间。

(4) 使用菜单“Windows→Preferences→Java→Compiler”，查看 Eclipse 所引用的 Java 运行环境 JRE。

2. 手机模拟器的使用。

(1) 单击 Eclipse 工具栏的模拟器工具，出现 Android Virtual Device Manager 窗口。

(2) 单击“New”按钮，出现创建模拟器的对话框。

(3) 输入 AVD 参数后，单击“OK”按钮。

(4) 选择已经建立的 AVD，单击“Start”按钮启动 AVD。

(5) 在 AVD 上运行系统自带的计算器程序。

(6) 运行 AVD 的 Settings(设置)程序，调整 AVD 的时区为 Beijing(北京)，以实现日期、时间与计算机的系统日期、时间一致。

3. 创建 Android 示例工程 Hello。

(1) 使用 Eclipse 菜单“File→New→Android Application Project”。

(2) 在出现的对话框的 Application Name 文本框里，输入“Hello”，Project Name 和 PackageName 使用默认值，单击“Next”按钮。

(3) 在出现的对话框里，再次单击“Next”按钮。

(4) 依次单击命令按钮“Clipart”和“Choose”，为应用程序选择一个图标后单击“Next”按钮。

(5) 单击“Next”按钮。

(6) 单击“Finish”按钮。

4. 部署 Android 示例工程并做运行测试、应用程序卸载。

(1) 使用 360 手机管家，确保手机与计算机已经连接。

(2) 右击 Android 示例工程名，选择“Run As→Android Application”，或者单击 Android 示例工程名后单击工具栏上的 工具图标，即可部署工程到手机并运行。

(3) 三星手机卸载某个应用程序的方法是使用“设定”程序中的“应用程序管理器”选项，而联想手机卸载某个应用程序的方法是长按该应用程序图标(不松手)。

(4) 部署工程到 AVD 并做运行测试，然后卸载该应用程序。

5. 使用 Android SDK Manager 下载 Android API 的源码。

(1) 单击 Eclipse 工具栏上的 SDK Manager 工具 。

(2) 在出现的窗口里，勾选“Android 4.4(API 19)”里的“ Sources for Android SDK”，单击“Install”按钮、选择“Accept License”后再次单击“Install”按钮，开始下载 Android SDK 的源码(存放在系统文件夹 sdk/sources 内，参见第 2.1.2 小节)。

(3) 与使用 Java 类的源码一样，在双击 Android 类(或接口)时通过单击“Attach Source”的方式可添加 Android API 的源码(考虑到 Android 源码下载的时间较长，读者可以从作者的教学网站里下载源码的压缩包，解压后将文件夹 sources 复制到 sdk 文件夹内即可)。

6. 掌握 Android 平台的调试方法。

(1) 部署 Hello 工程后，查看控制台(Console)的输出信息。

(2) 有意在工程的清单文件 AndroidManifest 中将 MainActivity 前的包名写错，再部署工程时程序自动挂断，观察控制台也无异常。

(3) 切换至“LogCat”。

(4) 以工程的包名(即使用 by Application Name)设置过滤器，观察输出的错误信息(路径错误)。

(5) 在源程序中使用 Log.i()指令输出 Log 信息。

7.了解 Android 的签名机制，掌握 Android 应用的签名方法。

(1) 使用 Eclipse 菜单“Windows→Preferences→android→Build”，查看 Debug 调试模式下使用的签名证书文件 debug.keystore。

(2) 右击“Hello”工程名→Android Tools→Export Unsigned Application Package，指定 APK 安装包名称。

(3) 验证该未经签名的.apk 文件不能安装到手机上。

(4) 右击“Hello”工程名→Android Tools→Export Signed Application Package，单击“Next”按钮后选择“Create new keystore”，然后按照向导操作。

(5) 双击生成的经过签名的.apk 文件，安装至手机，然后在手机上做运行测试。

三、实验小结及思考

(由学生填写，重点写上机中遇到的问题。)

第3章 Android 应用程序结构及运行原理

在第 1.4 节，我们宏观地介绍了 Android 系统应用的架构。Android 系统提供了 Activity、Service、BroadcastReceiver 和 ContentProvider 等四大组件和用于组件通信的相关类，同时还提供了用于通知、音频、视频、WiFi、Bluetooth、位置服务和通信等应用的管理器类。因此，Android 应用是基于组件的编程。

在通常情况下，一个 Android 应用有一个主界面，并使用 Android 的 Activity 组件来实现。本章将具体阐述 Android 应用工程的结构、组成部分和 Android 虚拟机的工作原理。本章学习要点如下：

- 掌握 Android 工程的文件系统结构；
- 初步了解 Android 四大组件，特别是 Activity 组件的作用；
- 初步了解 Activity 组件的相关类(如 View、Intent 等)的作用；
- 掌握 Android 工程的清单文件 AndroidManifest.xml 的作用；
- 了解 Java 虚拟机的工作原理；
- 掌握 Android 应用工程中的单元测试方法。

3.1 Android 工程的文件系统结构

由于 Android 将用户界面和资源从业务逻辑中分享出来，并使用 XML 文件进行描述形成独立的资源文件，因此 Android 应用工程的文件系统结构比 Java/Java Web 工程更为复杂，特别是资源文件的调用关系。

3.1.1 源程序文件夹 src

Android 以 Java 作为编程语言，因此其程序文件以.java 作为扩展名。Java 程序文件位于项目文件夹的 src 文件夹的某个包内。

注意：

(1) 包名习惯上小写，而类名首字母习惯上大写。

(2) src 文件夹里可以建立若干个包，用以分类存放 Java 源程序文件。

3.1.2 资源文件夹 res、assets 与 gen

在 Android 工程中，有字符串、位图、布局等资源，可以将它们划分为三种类型的文件：XML 文件、位图(图像)文件和 raw(声音)文件。

在 Android 工程中，有两个用于存放资源文件的文件夹，分别为 res 和 assets。其中，res 文件夹内的资源文件最终被打包到编译后的.java 文件中，res 文件夹内不支持深度的子目录；assets 中的资源文件不会被编译，而是直接打包到应用中，assets 文件夹支持任意深度的子目录。

Android 工程中使用的资源文件都会在 gen/包名/R.java 中生成对应项，由系统自动为每个资源分配一个十六进制的整型数，用以标明每个资源。

例如，Hello 工程的 R.java 文件代码如下：

```
package com.example.hello;
public final class R {
    public static final class attr {
    }
    public static final class dimen {
        public static final int activity_horizontal_margin=0x7f040000;
        public static final int activity_vertical_margin=0x7f040001;
    }
    public static final class drawable {
        public static final int ic_launcher=0x7f020000;
        public static final int wzx=0x7f020001;
        public static final int wzx2=0x7f020002;        }
    public static final class id {
        public static final int action_settings=0x7f080002;
        public static final int button1=0x7f080001;
        public static final int textView1=0x7f080000;        }
    public static final class layout {
        public static final int activity_main=0x7f030000;        }
    public static final class menu {
        public static final int main=0x7f070000;        }
    public static final class string {
        public static final int action_settings=0x7f050001;
        public static final int app_name=0x7f050000;
        public static final int hello_world=0x7f050002;        }
    public static final class style {
        public static final int AppBaseTheme=0x7f060000;
```

```
        public static final int AppTheme=0x7f060001;        }
}
```

注意：

(1) res 内的资源文件可以通过 R 资源类访问，而 assets 内的资源文件不能。

(2) 资源类文件 res/gen/R.java 是系统自动生成的，不需要手工修改。

3.1.3 布局文件夹 res/layout

工程的布局文件夹 res/layout 存放扩展名为.xml 的布局文件，每个布局文件对应一个 Activity。

3.1.4 值文件夹 res/values

值文件夹 res/values 里的 strings.xml 是最重要的文件，通常存放着布局文件中控件对象的属性值。

3.1.5 图片文件夹 res/drawable 与音乐文件夹 res/raw

与 Windows 应用程序一样，每个 Android 应用工程都有一个图标，相应的图标文件 ic_launcher.png 存放在 drawable 文件夹里。

如果工程使用音频资源文件，则需要存放在 res/raw 文件夹里。

3.1.6 编译文件夹 bin

使用 Debug 调试模式 (即使用 Run As 方式)开发 Android 应用程序时，包含了生成应用程序的安装包文件、安装到 Android 设备、在 Android 设备上运行等过程。在工程的 bin 文件夹里自动生成扩展名为.apk 的安装包文件。

3.1.7 使用扩展.jar 包文件夹 libs

当使用第三方库文件时，需要将其.jar 包复制到工程的 libs 文件夹里。做百度地图开发(参见第 10.3 节)和消息推送(参见第 11.4 节)需要使用第三方提供的软件包。

3.1.8 工程配置清单文件 AndroidManifest.xml

每个 Android 应用工程都包含有一个名为 AndroidManifest.xml 的文件，它是 XML 格式的文件，包含了 Android 系统运行前必须掌握的相关信息，如应用程序名称、图标、应用程序的包名、组件注册信息、授权(参见第 4.9 节)和运行设备的最低 Android 版本(严格地说，为 Android API 的版本)等。

3.2 Android 应用程序的基本组成

Android 应用程序是由组件组成的，组件可以调用相互独立的功能模块。根据完成的功能，组件可划分为四类核心组件，即 Activity、Service、BroadcastReceiver 和 ContentProvider。

注意：

(1) 在结构上，Android 应用程序与传统的 C 语言程序不同，它是基于组件的编程的。

(2) 四大组件中除 ContentProvider 组件外，都是通过 Intent 对象激活的(详见第 4.8 节)。

(3) 四大组件均需要在工程的清单文件中使用标签注册。

3.2.1 Activity 组件与视图 View

Activity 是 Android 最重要的组件，负责用户界面的设计。Activity 用户界面框架采用 MVC 模式(model view controller)。控制器负责接受并响应程序的外部动作；通过视图反馈应用程序给用户的信息(通常是屏幕信息反馈)；模型是应用程序的核心，用于保存数据和代码。

注意：Android 组件在清单文件中使用标签<activity>注册。

3.2.2 Service 组件

Service 是 Android 提供的无用户界面、长时间在后台运行的组件。

注意：Android 组件在清单文件中使用标签<service>注册。

3.2.3 BroadcastReceiver 组件

在 Android 系统中，当有特定事件发生时就会产生相应的广播。例如，开机启动完成、短信到来、电池电量改变、网络状态改变等。

为了通知手机用户有事件发生，在通常情况下，通知管理器 NotificationManager(详见第 4.5 节)会在手机的状态栏里产生一个具有提示音的通知，用户通过下滑手势可以查看其相关信息。

BroadcastReceiver，即广播接收者，用来接收来自系统或其他应用程序的广播，并做出回应。

注意：

(1) Android 组件在清单文件中使用标签<receiver>注册；

(2) 广播接收者组件 BroadcastReceiver(详见第 6.3 节)与 Service 组件一样，也没有 UI 界面。

3.2.4 ContentProvider 组件

为了跨进程共享数据，Android 提供 ContentProvider 接口，可以在无须了解数据源、路径的情况下，对共享数据进行查询、添加、删除和更新等操作。

注意：

(1) ContentProvider 组件在清单文件中使用标签<provider>注册。

(2) ContentProvider 组件的使用，详见第 8 章。

3.2.5　意图对象 Intent

Android 提供轻量级的进程间通信机制 Intent，使跨进程组件通信和发送系统广播成为可能，组件 Activity、Service 和 BroadcastReceiver 都是通过消息机制被启动(激活)的，其使用的消息就封装在对象 Intent 里。

Android 组件及通信机制如图 3.2.1 所示。

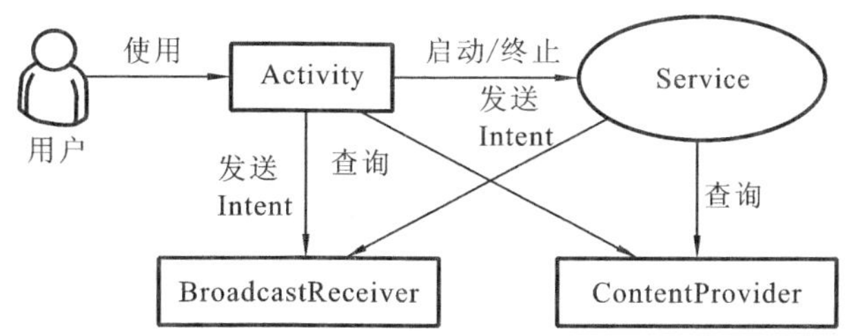

图 3.2.1　Android 组件及通信机制

注意：

(1) 启动或调用组件的消息不是通常的数据。

(2) Intent 类的使用，详见第 4.8 节。

3.2.6　Android 应用程序的运行入口

一个 Android 应用程序通常由多个 Activity 组成，但只有一个主 Activity。在工程清单文件中使用<activity> 标签注册主 Activity 时，还需要内嵌<intent-filter>、<action>和<category>标签，以此说明该 Activity 为 Android 应用程序的入口。定义一个 MainActivity.java 为应用的主 Activity 的代码如下：

```
<activity
    android:name="prg_packname.MainActivity"
    android:label="@string/app_name">
    <intent-filter>
        <action android:name="android.intent.action.MAIN"/>
        <category android:name="android.intent.category.LAUNCHER" />
    </intent-filter>
</activity>
```

其中，prg_packname 为程序 MainActivity.java 的包名。

3.3 Android 虚拟机 Dalvik

虚拟机(virtual machine)是通过软件模拟的具有完整硬件系统功能、运行在一个隔离环境中的完整计算机系统。

尽管 Android 的编程语言是 Java，但 Android 使用的虚拟机 Dalvik 与 Java 使用的虚拟机 JVM 并不兼容。因为 Dalvik 是基于寄存器的架构，而 JVM 是基于栈的架构。此外，Dalvik 能根据硬件实现更大的优化，更适合于移动设备。

3.3.1 Java 虚拟机执行的是字节码文件

Java 虚拟机(简称 JVM)包括一套字节码指令集、一组寄存器、一个栈、一个垃圾回收堆和一个存储方法域。 JVM 屏蔽了与具体操作系统平台相关的信息，使 Java 程序只需生成在 Java 虚拟机上运行的目标代码(字节码)，就可以在多种平台上不加修改地运行。

Java 语言是跨平台的，其跨平台的基石是字节码。字节码按照 Java 虚拟机规范的格式组成了.class 文件，Java SE 程序中的 Java 类会被编译成一个或者多个字节码文件，然后打包到 JAR 文件中。Java 虚拟机从相应的.class 文件和.jar 文件中获取相应的字节码来运行。JVM 在执行字节码时，实际上最终还是把字节码解释成具体平台上的机器指令执行。

3.3.2 Android 虚拟机的特点

Android 应用虽然也是使用 Java 语言进行编程，但是在编译成.class 文件后，还会通过一个批处理程序(dx.bat)将应用所有的.class 文件转换成一个名为 classes.dex 的文件。

Android 安装包文件的扩展名为.apk，一个工程只能放进一个.apk 文件内。Android 安装包文件包含了与某个 Android 应用工程相关的所有文件，包括工程的清单文件 AndroidManifest.xml、应用程序代码(classes.dex)、资源文件和其他文件打成的一个压缩包。解压.apk 文件后的文件(目录)如图 3.3.1 所示。

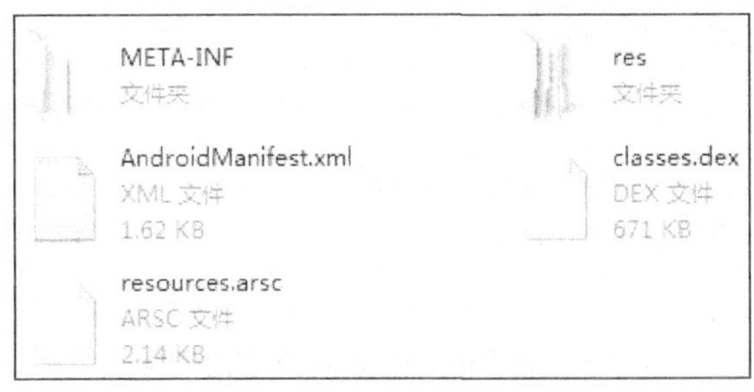

图 3.3.1 解压 APK 安装包后的文件(目录)

当用户部署一个自己开发的 Android 应用至手机内存(不是 SD 卡)后，在系统目录\data\app 里可以找到该应用的安装包文件(*.apk)。在系统目录\data\dalvik-cache 里可以找到对应的.odex 文件，它是从 classes.dex 文件优化而来的，或者说是从.apk 文件中提取出来的可运行文件。

运行用户开发的 Android 应用程序时，如果在系统目录\data\dalvik 里存放相应的.odex 文件，Android 的 Dalvik 虚拟机会直接从.odex 文件中加载指令和数据后执行；若没有 odex 的话，需要先从.apk 包中提取 classes.dex 并生成.odex 文件，然后再加载并执行。因为真正在 Android 虚拟机上运行的是.odex 文件，如果系统发现已经有了.odex 文件，那么就不会再从.apk 包里面去解压、提取。显然，这种预先提取方式可以加快软件的启动速度，减少对 RAM 的占用。

注意：

(1) Android 系统文件夹\system\app 存放的是系统的默认组件、相应的.apk 安装包文件和直接加载执行的.odex 文件，如图 3.3.2 所示。

(2) Android 系统文件夹\system\bin 用于存放 Linux 系统自带的组件，如包管理器\system\bin\pm。

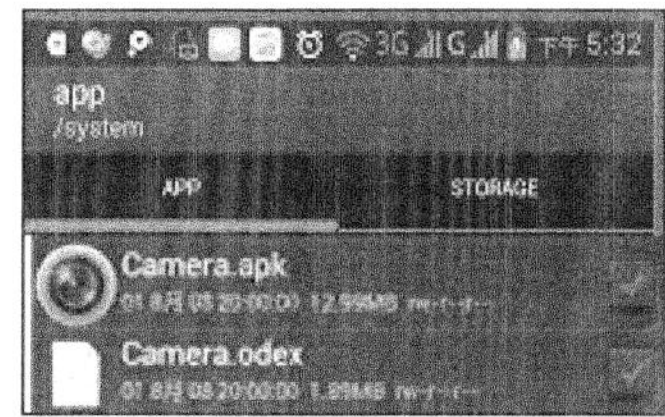

图 3.3.2　系统自带的相机程序

3.4　使用 AndroidTestCase 做 Android 单元测试

测试也是开发过程中的一个重要组成部分，Android 开发环境集成了一个测试框架，可以用来测试 Android 应用工程，包括单元测试和 UI 测试。

【例 3.4.1】使用类 android.test.AndroidTestCase 来测试一个 Android 应用工程中某个辅助类中方法的正确性。

【设计步骤】

(1) 新建名为 AndroidTest 的 Android 应用工程，在其对话框中均采用默认设置。

(2) 在 src 文件夹里新建名为 android.com.example.util 的包名，并在该包内建立一个名为 CalService.java 的类文件。

(3) 在 src 下新建名为 android.com.junit.test 的包名，并在该包内建立一个名为 TestCalService 且继承 android.test.AndroidTestCase 的类文件 TestCalService.java。

此时，Android 应用工程文件结构如图 3.4.1 所示。

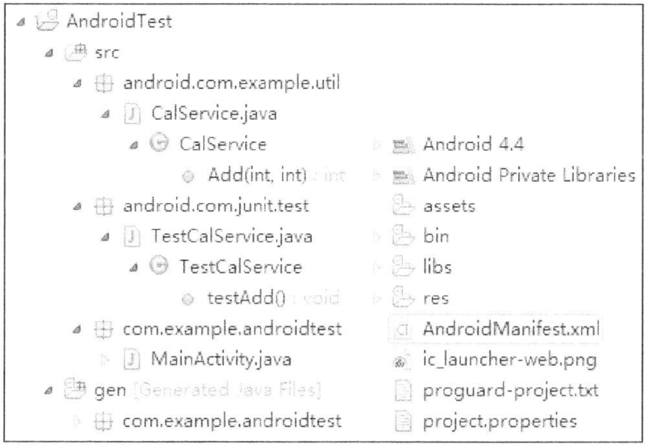

图 3.4.1　AndroidTest 工程文件结构

(4) 在清单文件的<manifest>节点内增加一条标签，用于指定 Android 提供的测试机和设置被测试的 Android 应用工程的包名，其代码如下：

```
<instrumentation
    android:name="android.test.InstrumentationTestRunner"
    android:targetPackage="com.example.androidtest" />
```

在清单文件的<application>节点内增加一条标签，用于指定与 Android 测试机配套的库，其代码如下：

```
<uses-library android:name="android.test.runner" />
```

(5) 编写被测试类 CalService.java，其代码如下：

```
package android.com.example.util;
//定义被测试类，它作为 Android 应用工程的一个辅助类
//测试本类的方法 Add()的正确性
public class CalService {
    public int Add(int a,int b){
        return a+b;
        //return a-b;  //测试时使用
    }
}
```

(6) 编写测试类 CalService.java，其代码如下：

```
package android.com.junit.test;
// AndroidTestCase是Android提供的测试框架
//是对Java单元测试( JUnit)的再封装
//测试类的testAdd()方法用于测试被测试类CalService的Add()方法
import android.com.example.util.CalService;  //
import android.test.AndroidTestCase;  //主类
public class TestCalService extends AndroidTestCase {
    public void testAdd() throws Exception{
        CalService service=new CalService();  //
        int result=service.Add(3, 5);
        AndroidTestCase.assertEquals(8,result);  //断言
    }
}
```

(7) 右击工程包浏览器里测试类 TestCalService 中的 testAdd()方法，选择“Run As→Android JUnit Test”，稍后出现测试结果，如图 3.4.2 所示。

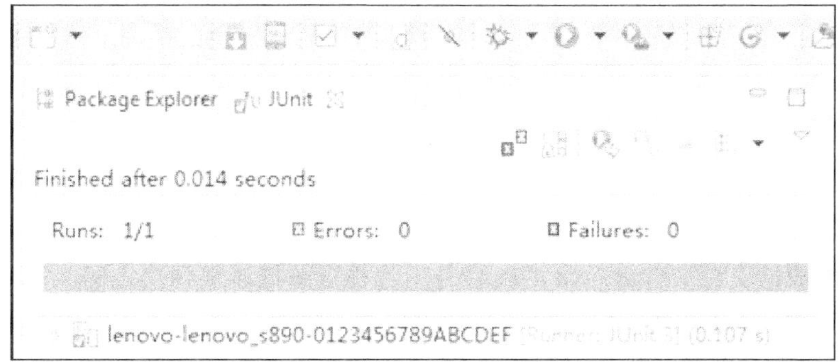

图 3.4.2　测试结果

其中，绿条表示 TestCalService 类的 testAdd()方法里的断言正确(因为 3+5=8)。如果将被测试类 CalService.java 里的 Add()方法中的“+”换成“-”，则测试结果为红条，表示断言错误(因为 3-5=-2≠8)。

注意：

(1) 在清单文件中添加的代码，可以从新建的 Android TestProject 工程(不是通常建立的 Android Application Project)的清单文件里复制。创建一个 Android TestProject 测试工程的方法是：使用菜单“File→New→Project→Android Test Project”，如图 3.4.3 所示。

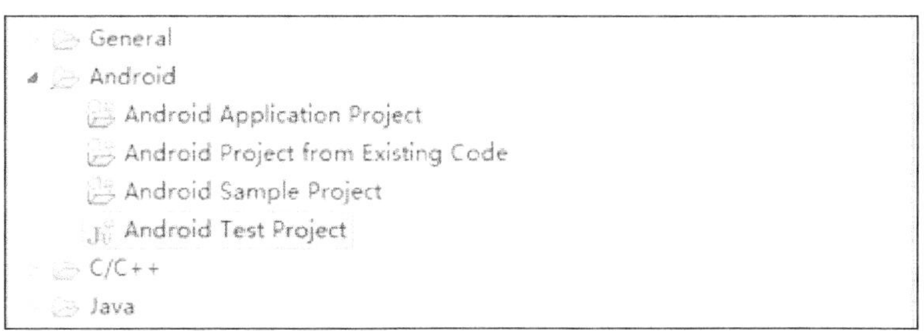

图 3.4.3　创建 Android Test Project

(2) Android 测试类是对 Java 测试类的再封装，用于测试 Android 应用工程中的类方法。

(3) 按照软件工程的理论，本测试为黑盒测试方法(即功能测试)。

(4) 在计算机的 Eclipse 环境中显示在 Android 设备上运行的结果，本质上是使用了 Android 的一个重要工具——调试桥 ADB(参见第 2.2.3 小节)。

习 题 3

一、判断题

1. res 和 assets 文件夹下的文件都能通过资源类 R 读取。
2. 自动生成的资源文件 R.java 的内容是由各种资源描述文件决定的。
3. 关闭 Android 应用程序的主界面，通常使用返回键。
4. Android 应用程序安装包文件(.apk)可以使用解压软件进行解压。
5. Intent 是 Android 的四大组件之一。
6. 工程清单文件里标签 application 的 label 属性值与标签 activity 的 label 属性值必须相同。
7. 一个 Android 应用工程中的每个 Activity 都有一个相应的布局文件。
8. 一个 Android 应用工程中的所有源程序必须放在一个包里。

二、选择题

1. 在 Android 应用项目的 res 文件夹里，存放音乐文件的子文件夹是____。

   A. drawable-hdpi　B. layout　C. raw　D. values

2. 下面关于 Android 应用工程的说法中，不正确的是____。

   A. Android 应用的 gen 目录下的 R.java 被删除后还能自动生成
   B. 文件夹 res 内包含了应用工程的全部资源
   C. 应用的包名就是工程清单文件里的标签< manifest>内的 xmlns 属性值
   D. 目录 assets 和 res 都能存放资源文件，但只有 assets 支持任意深度的子目录，并且它存放的文件不会在 R.java 里生成任何资源 ID

3. 右击 Android 工程名得到的属性信息中，不包含的信息是____。

   A. 工程创建时间　B. 源文件编码类型
   C. 最后修改时间　D. 工程的存放位置

4. 在 Android 工程的清单文件 AndroidManifest.xml 的根标签<manifest>里，与标签<use-sdk>层次相同的标签是____。

   A. <activity>　B. <application>　C. <intent-filter>　D. <action>

5. 在 Eclipse 中编辑工程清单文件时，要想对标签属性使用联机帮助，应按组合键____。

   A. Alt+/　B. Ctrl+/　C. Alt+\　D. Ctrl+\

6. 当 Android 应用发布后，其项目文件夹 bin 内不会出现的文件类型是____。

   A. .xml　B. .src　C. .apk　D. .dex

7. 在 Android 系统架构中，直接被 Android 组件调用的是____。

   A. Applications　B. Application Framework
   C. Libraries　D. Linux Kernel

8. 下列选项中，不属于 Android 四大组件的是____。

   A. Content　B. Server
   C. ContentProvider　D. BroadcastReceiver

9. 真正在 Android 虚拟机上运行的文件类型是____。

A. .exe　　B. .dex　　C. .class　　D. .odex

10. 将手机里的运行信息显示于计算机，是使用 Android 提供的____工具。

A. AVD　　B. SDK　　C. ADB　　D. ADT

三、填空题

1. Dalvik 是____公司开发的用于 Android 平台的 Java 虚拟机。

2. 位于 Android 手机操作系统的最底层的是____。

3. Android 程序运行时，出现在窗口上方的标题是由工程清单文件中的<application>标签使用____属性而指定的。

4. 在工程清单文件中注册了多个 Activity 时，只有____Activity 含有<intent-filter>标签。

5. Android 工程的 R.java 文件由若干用于描述各种资源的内部____组成。

6. 虚拟机 Dalvik 用于运行已转换为____格式化的 Java 应用程序。

7. Java 虚拟机 JVM 基于栈的架构，而 Dalvik 基于____的架构。

8. 使用 AndroidTestCase 测试 Android 应用里的某个类方法的正确性时，需要编写一个测试类并使用 AndroidTestCase 提供的静态方法____。

9. 在 Android 工程里，____文件夹的文件被原封不动打包到安装包文件里。

实验 3　Android 应用程序结构与运行原理

一、实验目的

1. 掌握 Android 应用工程的文件系统结构。

2. 初步掌握 Android 应用程序的基本组成部分。

3. 掌握 Android 虚拟机 Dalvik 的特点和应用程序的运行过程，特别是用户开发的 Android 应用程序的执行过程。

4. 掌握 Android 应用工程中的单元测试方法。

二、实验内容及步骤

【预备】访问 http://www.wustwzx.com/android/index.html，单击“实验 3”超链接，下载本次实验内容的源代码及视频资料并解压(得到文件夹 ch03)，供研读、调试和观看。

1. 掌握 Android 应用工程的文件系统结构。

(1) 在 Eclipse 中，打开上次实验创建的示例工程 Hello。

(2) 查看应用的图标文件 ic_launcher.png，它位于文件夹 res/drawable 内。

(3) 查看应用的标签存放在文件 res/values/strings.xml 里。

(4) 打开存放在文件夹 res/layout 里的布局文件(.xml 文件)，查看与应用程序运行窗口对应的布局及其代码(设计与代码两种视图)。

(5) 去掉主 Activity 里的方法 onCreateOptionsMenu()后，部署工程并做运行测试可知，此时手机的菜单键失效。

2. 初步掌握 Android 应用程序的基本组成部分。

(1) 在源程序 MainActivity.java 中使用 setTitle()方法动态地重设当前 Activity 的标题，然后部署工程并做运行测试。

(2) 在创建 Activity 的 onCreate()方法内的语句“setContentView(R.layout.activity_main);”前增加一条语句“requestWindowFeature(Window.FEATURE_NO_TITLE); ”，此时该语句前出现一个小红叉的错误提示，通过导入“android.view.Window”即可。

(3) 部署工程并运行，观察运行窗口中没有标题栏。

(4) 打开工程清单文件，查看定义 Android 应用包名的方法：<manifest package="…">。

(5) 查看清单文件中使用<activity>标签注册主 Activity 时对应用包名的引用：<activity android:name="…">。

(6) 右击工程名 Hello→Android Tools→Rename Application Package，输入一个新的包名后，再查看工程清单文件中本应用包名的定义和定义主 Activity 中使用包名的同步变化。

(7) 选中主 Activity 所在的包名，使用菜单“File→New→Other”,选择 Android 后再选择 Android Activity，输入名称(如 TestActivity)后单击“OK”按钮。

(8) 在清单文件中观察新定义的 Activity 的注册信息，并与主 Activity 的注册信息比较(主 Activity 使用了<intent-filter>标签定义为工程的主窗口，而新定义的 Activity 没有)。

(9) 对 src 文件夹使用右键菜单“New→Package”，创建一个新的程序包名，体会 Android

应用包名与程序包名的区别。

3. 了解 Android 虚拟机 Dalvik 和 Android 应用程序的运行过程。

(1) 解压工程 Hello 里的文件夹 bin 里的 Android 安装包文件 Hello.apk，可观察到只有一个文件 classes.dex，它是用户开发的 Android 应用工程中所有类文件的合并。

(2) 使用 360 手机助手软件，确保 Android 应用 Hello 安装在手机内存(不是 SD 卡)内。

(3) 运行手机上的 Root Explorer 程序，分别查看系统目录\data\app 下 Hello 工程的安装包文件(.apk)和系统目录\data\dalvik-cache 里的.odex 文件。

(4) 使用 360 手机助手软件，再次使用 Root Explorer 程序查看上述两个系统目录，与 Hello 工程相关的.apk 文件和.odex 文件已经被删除。

(5) 使用手机上的 Root Explorer 程序，查看系统目录\system\app 里系统自带的 Android 应用程序的安装包文件和对应的.odex 文件。

(6) 使用手机上的 Root Explorer 程序，查看系统目录\system\bin 里 Linux 系统自带的组件，如包管理器\system\bin\pm。

4. 从作者的教学网站里下载 Android 虚拟机的教学视频并观看、学习。

(1) 访问作者的教学网站 http://www.wustwzx.com，在 Android 课程里下载关于 Android 虚拟机的教学视频。

(2) 使用视频播放软件，重点掌握 Android 虚拟机的特点，并与 JVM 比较。

5. 掌握 Android 应用工程中单元测试的方法(参见例 3.4.1)。

(1) 导入 ch03 文件夹里的 Android 应用工程 AndroidTest。

(2) 分别查看 src 下的 android.com.example.util 包和 android.com.junit.test 包里的源文件。

(3) 查看清单文件对 Android 测试机的相关配置。

(4) 部署工程并做运行测试。

(5) 观看 ch03 文件夹里关于 Android 单元测试的视频文件。

三、实验小结及思考

(由学生填写，重点写上机中遇到的问题。)

第4章 Android 应用开发基础

通过使用 Activity 组件和 Widget 控件设计用户界面，是 Android 程序设计前的必需步骤。当程序调用手机自带的应用程序时，需要使用意图对象；当涉及手机的某些敏感功能时，需要注册用户权限。此外，三种形式的消息通知(Toast、AlertDialog 和 Notification)也是经常使用的，还有通知、偏好设定和文件读写。本章的学习要点如下：

- 掌握 Activity 的作用(作为装载可显示组件的容器和控制逻辑)；
- 掌握 Android 使用 XML 文件描述用户界面的特点；
- 掌握 Android 的视图模型；
- 掌握利用 Intent 对象实现应用程序间共享数据的实现方法；
- 掌握在工程清单文件 AndroidManifest.xml 中注册权限的方法；
- 掌握 android.content.Intent 类提供的常用动作名称及数据；
- 掌握单击事件监听器的多种使用方法；
- 掌握使用类 android.webkit.WebView 开发网络浏览器的方法。

4.1 用户界面 UI 设计

4.1.1 Android 界面视图类

Android 的图形化的用户界面(graphical user interface，GUI)采用 MVC(Model-View-Controller)模型，其含义包括：

- 提供了处理用户输入的控制器(Controller)；
- 显示用户界面的视图(View)；
- 保存数据和代码的模型(Model)。

在 Android MVC 中，控制器是由 Activity 组件(详见第 4.2 节)完成的，它能够接受并响应程序的外部动作，如按键动作或触摸屏动作等，每个外部动作作为一个对立的事件被加入队列中，按照“先进先出”的规则从队列中获取事件，并将这个事件分配给所对应的事件处理函数。

控制器负责接受并响应程序的外部动作；通过视图反馈应用程序给用户的信息(通常是屏幕信息反馈)；模型是应用程序的核心，用于保存数据和代码。

Android 界面元素的组织形式采用视图树(view tree)模型，如图 4.1.1 所示。

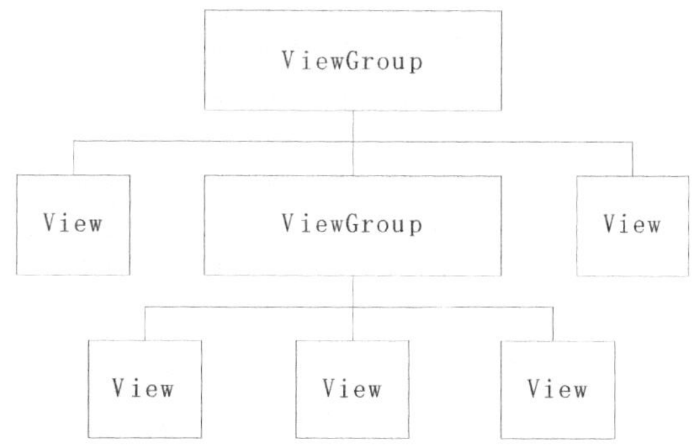

图 4.1.1　Android 界面元素的树状组织形式

Android 的视图类 android.view.View 提供了用于处理屏幕事件的多个内部接口 (如 OnClickListener 等)及常用方法(如 findViewById()等)，如图 4.1.2 所示。

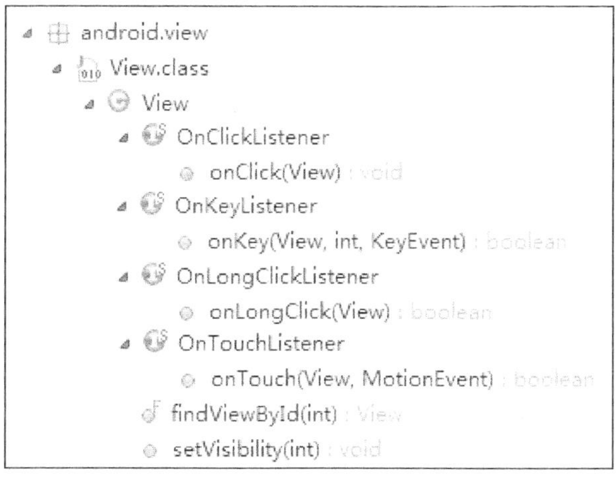

图 4.1.2　视图类 View 的内部接口及常用方法

注意：View 是一个矩形区域，它负责区域里元素的绘制，View Group 是 View 的子类，是一个容器，专门负责布局，本身没有可绘制的元素。

4.1.2　Android 用户界面事件

在 Android 系统中，各种屏幕手势的相关信息(如操作类别、发生时间等)被自动封装成一个 KeyEvent 对象，供应用程序使用。因此，在 Activity 的事件处理方法(函数)中，需要使用表示手势事件对象的事件参数 event。

Activity 提供了响应各种屏幕手势的方法，如按键方法 onKeyDown()、松开按键方法 onKeyUp()、长按方法 onKeyLongPress()、选择某个菜单项方法 onMenuItemSelected()等，如图 4.1.3 所示。

图 4.1.3 中的事件函数的返回值都是 boolean 型。当返回 true 时，表示已经完整地处理了这个事件，并不希望其他的回调方法再次进行处理；而当返回 false 时，表示没有处理完该事件，希望其他回调方法继续对其处理。

除了使用屏幕事件外，还有键盘事件。类 KeyEvent 定义了分别对应于 Back、Home 和 Menu 键的键盘扫描码(实质上是静态常量)，如图 4.1.4 所示。

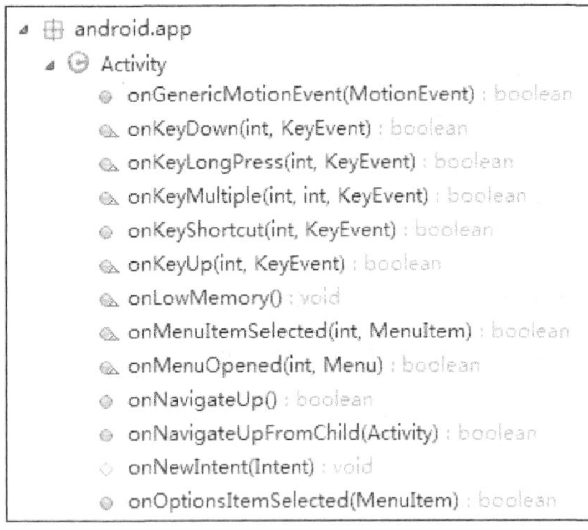

图 4.1.3　Activity 的屏幕手势处理方法

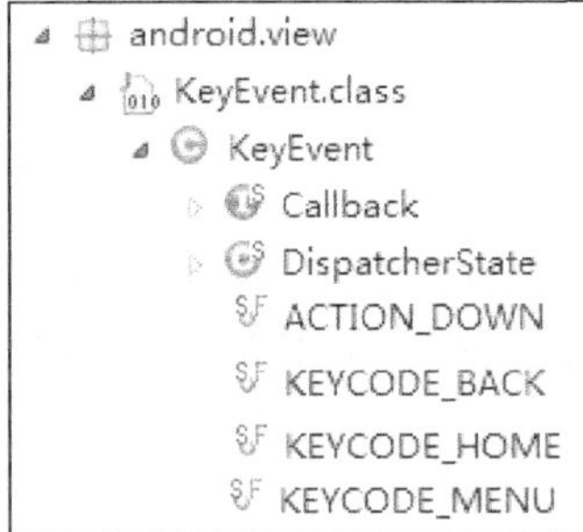

图 4.1.4　类 KeyEvent 的定义

Activity 提供的方法 onKeyDown(keyCode, event)用来捕捉手机键盘被按下(短按，不是长按)的事件。其中，参数 keyCode 表示键盘扫描码，通常使用类 KeyEvent 的静态常量表示。

默认情况下，用户按“返回”键将关闭一个 Activity 或返回上一级 Activity。有时，在一个 Activity 中，需要禁用返回键，这时应添加如下代码(参见例 6.1.1)：

```
@Override
public boolean onKeyDown(int keyCode, KeyEvent event) {
    // TODO Auto-generated method stub
    if(KeyEvent.KEYCODE_BACK == keyCode)    //返回键
        return false;    //屏蔽
    return super.onKeyDown(keyCode, event);    //其他按键默认处理
}
```

Android 程序通常需要侦听用户和应用程序之间交互的事件。对于用户界面中的事件，侦听方法就是从与用户交互的特定视图对象截获这些事件。

事件侦听器 Event Listeners 是视图 View 类的内部接口，包含一个单独的回调方法。这些方法将在视图中注册的侦听器被用户界面操作触发时由 Android 框架调用。下面这些回调方法被包含在事件侦听器接口中：

- onClick()包含于 View.OnClickListener，单击时调用；
- onLongClick()包含于 View.OnLongClickListener，长按时调用；
- onTouch()包含于 View.OnTouchListener，当用户执行的动作被当作一个触摸事件时被调用，包括按下、释放和在屏幕上进行的任何移动手势；
- onCreateContextMenu()包含于 View.OnCreateContextMenuListener，当正在创建一个上下文菜单时被调用。

注意：onClick()回调没有返回值，但是其他事件侦听器必须返回一个布尔值。返回 true 表示已经处理了这个事件而且到此为止；返回 false 表示还没有处理它并继续交给其他事件侦听器处理。

对于一个具备触摸功能的设备，一旦用户触摸屏幕，设备将进入触摸模式。自此以后，只有 isFocusableInTouchMode()为真的视图才可以被聚焦，比如文本编辑部件。其他可触摸视图，如按钮，在被触摸时将不会接受焦点；它们将只有被按下时才简单地触发 on-click 侦听器。任何时候用户按下方向键或滚动跟踪球，这个设备都将退出触摸模式，然后找一个视图来接受焦点，用户也许不会通过触摸屏幕的方式来恢复界面交互。

触摸模式的维护贯穿整个系统(所有窗口和活动)，用户可以调用 isInTouchMode()来查看设备当前是否处于触摸模式中。

框架将根据用户输入处理常规的焦点移动。这包含当视图删除或隐藏，或者新视图出现时改变焦点。视图通过 isFocusable()方法表明它们想获取焦点的意愿。要改变视图是否可以接受焦点，可以调用 setFocusable()。在触摸模式中，用户可以通过 isFocusableInTouchMode()查询一个视图是否允许接受焦点。用户可以通过 setFocusableInTouchMode()方法来改变它。焦点移动基于一个在给定方向查找最近邻居的算法。

想请求一个接受焦点的特定视图，需要调用方法 requestFocus()。

要侦听焦点事件(当一个视图获得或者失去焦点时被通知到)，使用 onFocusChange()。

4.1.3 界面与布局

res/layout 目录下存放定义 UI 设计的 XML 文件。UI 设计有两种方式：一种是对控件使用直观的拖拽方式，对应于 Graphical Layout 视图方式；另一种视图方式是在 XML 文件中直接写控件代码，如图 4.1.5 所示。

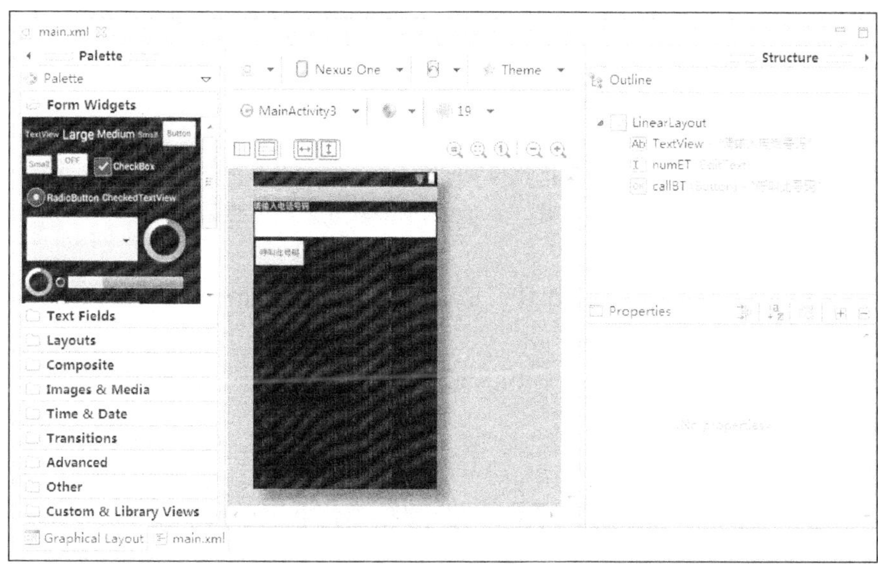

图 4.1.5 UI 设计

注意：在实际开发中，经常切换两种方式。纯代码方式要求用户对各种 UI 控件名称及属性非常清楚，而拖拽方式比较快捷和直观。

布局(相当于一个容器)属性 android:padding 等用于控制该容器内第一个元素与父布局(容器)之间的间隔，而控件属性 android:layout_margin 等用于设置同一布局内各元素之间的间隔。

布局内控件的常用布局属性还有以下几个。

- android:layout_width：控件的宽度。
- android:layout_height: 控件的高度。
- android:id：控件的标识。
- android:layout_below：将该控件的底部置于给定 id 控件之下。

其中，控件的宽度和高度以 dp(常用于图像)或 sp(常用于文本)等作为单位。

1. 线性布局

线性布局在布局文件中使用<LinearLayout>标签，其主要属性是 android:orientation，当取值为 vertical 时称为垂直线性布局，当取值为 horizontal 时称为水平线性布局。不管是水平线性布局还是垂直线性布局，单独使用时一行(列)里只能放置一个控件。

注意：调整线性布局内的各个控件之间的间距，使用 android:layout_weight 属性较方便。各个控件的 android:layout_weight 属性的默认值为 0，表示按照实际大小布局。

线性布局可以嵌套使用。例如，在垂直线性布局里嵌套一个水平线性布局时，就可以在一行内水平放置多个控件，参见例 4. 3. 1。

2. 相对布局

相对布局在布局文件中使用<RelativeLayout>标签，除了第一个元素外，其他元素需要参考另一个元素进行相对定位(含方向、偏移和对齐方式)，常用属性如下：

- android:layout_below　　//方向
- android:layout_marginTop　　//偏移
- android:layout_marginLeft
- android:layout_margin
- android:layout_alignRight　　//对齐

注意：

(1) 如果相对布局内的多个元素没有进行相对定位，则会重叠(只显示一个);

(2) 当布局内元素较多或布局不规则时，使用相对布局较方便且不需要嵌套。

3. 帧布局

帧布局像一层层画布，添加的控件一层层地放上去。帧布局添加的各个控件默认都是对齐到屏幕的左上角。

帧布局的一个实际应用，参见例 10.3.2(该帧布局共有两层，浮于上面的第二层是通过 background 属性设置透明度完成的)。

4.2 窗口组件 Activity

4.2.1 使用 Android 的 Activity 组件设计程序的运行窗口

Activity 作为 Android 最重要的组件，用于设计应用程序的用户界面，其内容来源于布局文件(在一个 Activity 的 onCreate()方法里，使用父类的方法 setContentView()时，以布局文件作为参数)。Activity 类位于软件包 android.app 里(参见图 4.2.2)，程序员开发的用户界面(类)都必须继承 Activity 类。

注意：

(1) 在使用 Android 的四大组件前，必须在工程清单文件中注册相应的权限。

(2) Activity 中包含了响应界面事件的代码，即具有控制器的功能。

4.2.2 Activity 作为上下文类 Context 的子类

在 Eclipse 中，展开 android.jar 的各相关软件包里的相关类(或接口)，按住“Ctrl”键，双击类(或接口)名可以查看该类(或接口)的定义。例如，通过此方法，可以得到 Activity 类的(直接)父类：

- public class Activity extends ContextThemeWrapper
- public class ContextThemeWrapper extends ContextWrapper
- public class ContextWrapper extends Context

抽象类 android.content.Context 表示上下文对象。显然，Context 是 Activity 的父类(但不是直接父类，Java 只支持单继承)。因此，Activity 继承(具有)Context 类的所有方法。

Context 类具有的主要方法，如图 4.2.1 所示。

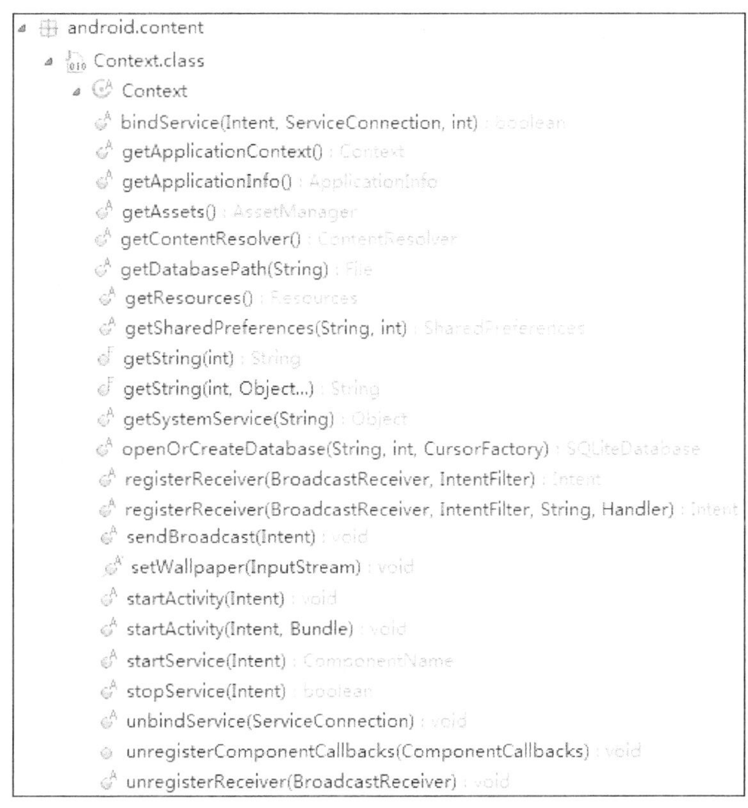

图 4.2.1 上下文类 Context 具有的主要方法

注意：

(1) 在上面的方法中，除了涉及 Activity 组件外，还涉及了 Android 的另外三个组件：服务组件 Service(将在第 6 章详细介绍)、广播接收者组件 BroadcasetReceiver(将在第 6 章详细介绍)和内容提供者组件 ContentProvider(将在第 8 章详细介绍)。

(2) 在第 4.3.3 小节里，创建 Toast 对象时需要使用上下文对象。上下文对象的使用实例，参见例 4.3.1。

4.2.3 Activity 类具有的基本方法

Activity 作为用户界面，提供了一些基本的方法，如设置内容视图方法 setContentView()、找控件方法 findViewById()、关闭当前窗口方法 finish()等。Activity 提供的几个基本方法如图 4.2.2 所示。

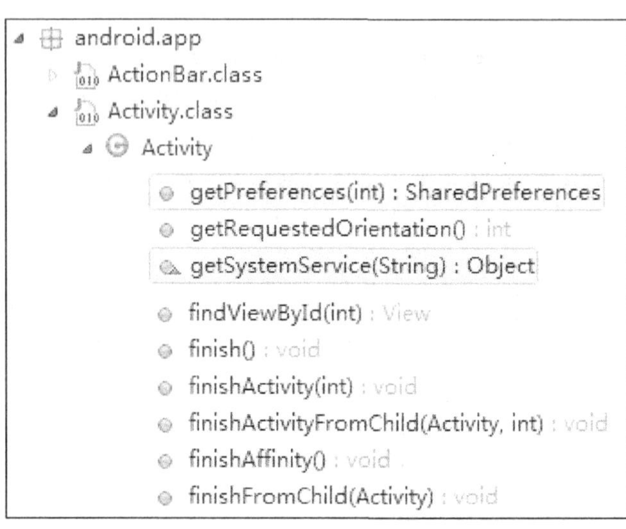

图 4.2.2　Activity 提供的几个基本方法

4.2.4 Activity 类具有的扩展方法

在 Android 应用开发中，除了必须掌握 Activity 具有的基本方法外，还需要掌握一些扩展的方法，如用于数据共享存储的 getPreferences()方法、用于动态加载页面布局的 getLayoutInflater()方法、开始后台服务的 startService()方法、发送广播的 sendBroadcast()方法、获得程序间数据共享的内容解析器的 getContentResolver()方法等，如表 4.2.1 所示。

表 4.2.1　Activity 类具有的扩展方法

方 法 名	功 能 描 述
getPreferences() getSharedPreferences()	得到 SharedPreferences 对象，使用 xml 文件存放数据，文件存放在/data/data/<package name>/shared_prefs 目录下，参见第 4.8 节
getLayoutInflater()	Activity 类的方法，将 layout 的 xml 布局文件实例化为 View 类对象，实现动态加载布局，参见例 4.3.3

续表

方 法 名	功 能 描 述
getMenuInflater()	Activity 类的方法，实现动态菜单，参见例 4.4.3
getSystemService()	根据传入的系统服务名称得到相应的服务对象，参见例 4.5.1
startActivity()	Activity 调用，参见第 4.8 节
startActivityForResult()	有返回值的 Activity 调用，参见第 5.5.1 小节、例 5.5.1 和例 9.2.1
startService()	继承得到的方法，启动某个服务，参见第 6 章
sendBroadcast()	继承的方法，发送广播，参见第 6 章
getContentResolver()	继承得到的方法，得到一个 ContentResolver 对象，参见第 8.1.2 小节
getPackageManager()	继承得到的方法，得到一个管理和查询系统所有应用程序的 PackageManager 对象，参见例 4.8.2

1. 动态布局膨胀器 LayoutInflater 及其相关类

抽象类 LayoutInflater 位于 android.view 包内，提供了加载布局的方法 inflate()，如图 4.2.3 所示。

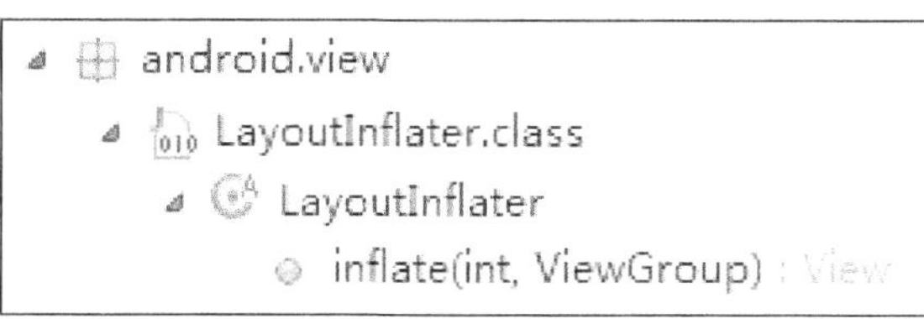

图 4.2.3 LayoutInflater 类的定义

LayoutInflater 通常在对话框里使用，并在得到 LayoutInflater 实例之后调用它的 inflate() 方法来加载布局，其代码如下：

```
LayoutInflater layoutInflater= MainActivity.this.getLayoutInflater();
View layout=layoutInflater.inflate(resourceId, root);    //
Builder.setView(layout);          //
```

inflate()方法一般接收两个参数，第一个参数就是要加载的布局 id，第二个参数是给该布局的外部再嵌套一层父布局，如果不需要就直接传 null。

注意：

(1) LayoutInflater 是用来加载 layout 里的.xml 布局文件的，而 findViewById()是查找.xml 文件里具体的 Widget 控件(如 Button、TextView 等)的。

(2) 获得 LayoutInflater 对象有多种方法。例如：

```
LayoutInflater layoutInflater = LayoutInflater.from(context);
```

(3) 通过调用系统服务也能得到 LayoutInflater 对象，代码如下：

```
LayoutInflater inflater = (LayoutInflater)
```

Context.getSystemService(LAYOUT_INFLATER_SERVICE);

2. 应用程序包管理器 PackageManager 及其相关类

每个 Android 应用的包名是唯一的，android.content.pm.PackageManager 用于应用程序的包的管理，它是一个抽象类。

类 PackageManager 及相关的组件名类 ComponentName 的定义，如图 4.2.4 所示。

```
android.content.pm
    PackageManager.class
        PackageManager
            queryIntentActivityOptions(ComponentName, Intent[], Intent, int) : List<ResolveInfo>
    ResolveInfo.class
        ResolveInfo
android.content
    ComponentName.class
        ComponentName
            ComponentName(Context, Class<?>)
            ComponentName(Context, String)
            ComponentName(Parcel)
            ComponentName(String, String)
```

图 4.2.4 类 PackageManager 及相关的组件名类 ComponentName 的定义

4.2.5 Activity 的生命周期

复杂的 Android 应用可能包含若干个 Activity。当打开一个新的 Activity 时，先前的那个 Activity 会被置于暂停状态，并压入历史堆栈中。用户可以通过返回键回退到先前的 Activity。

Activity 是由 Android 系统维护的，每个 Activity 除了有创建 onCreate()、销毁 onDestroy()两个基本方法外，还有停止方法 onStop()、激活方法 onStart()、暂停方法 onPause()和恢复方法 onResume()。

Activity 在其生命周期中存在三种状态：运行态、暂停态和停止态。运行态是指 Activity 调用 onStart()方法后出现在屏幕的最上层的状态，此时用户通常可以获取焦点；暂停态是指 Activity 调用 onPause()方法之后出现的状态，其上还有处于运行态的 Activity 存在，并且 Activity 没有被完全遮挡住，即处于暂停态的 Activity 有一部分视图被用户所见；停止态是指当前 Activity 调用 onStop()之后的状态，此时它完全被处于运行态的 Activity 遮挡住，即用户界面完全不被用户所见。

注意：

(1) 处于暂停态或停止态的 Activity 在系统资源缺乏时，可能被杀死，以释放其占用的资源。

(2) 在 Eclipse 中编辑 Activity 时，使用空白处的快捷菜单"Source→Override/Implement"，选择需要重写的父类方法或要实现的接口方法。

对于处于运行态的 Activity，当用户按返回键退出时，将调用方法 onStop()。

Activity 的生命周期如图 4.2.5 所示。

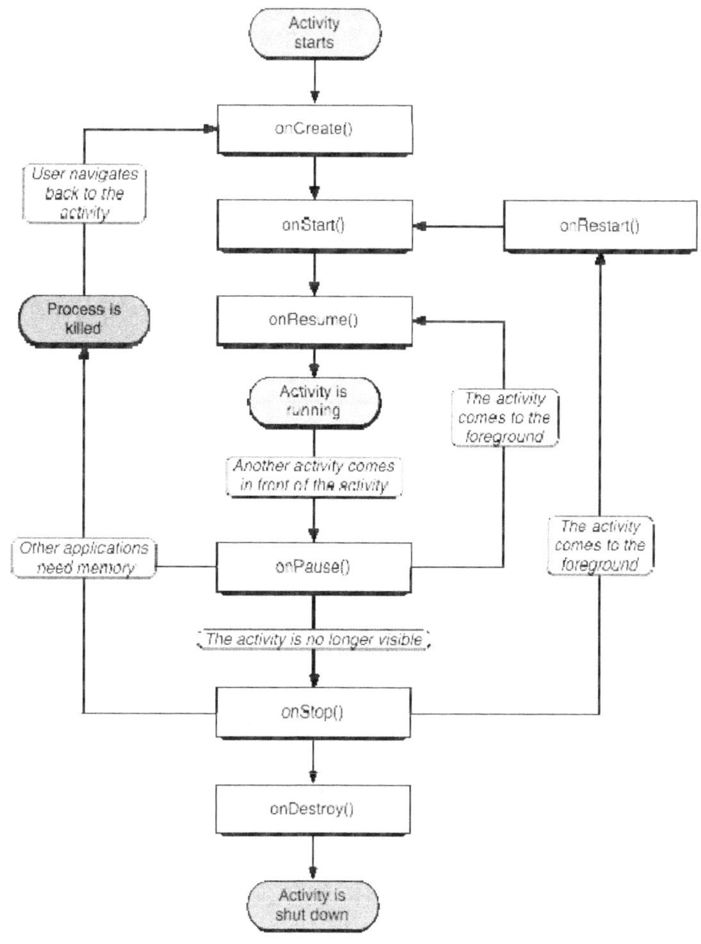

图 4.2.5　Activity 的生命周期

4.3　常用 Widget 控件的使用

android.widget 包内提供了大量的用于 UI 设计的控件。实际上，在创建 Android 实例工程时，在布局文件中就含有一个用于显示文本的文本框控件 TextView。下面分别介绍这些 Widget 控件的使用方法。

4.3.1　文本框控件 TextView 和 EditText

文本框控件包括 TextView 和 EditText。TextView 是用来显示字符的控件，而 EditText 是用来输入和编辑字符的控件。

类 TextView 的定义(部分)，如图 4.3.1 所示。

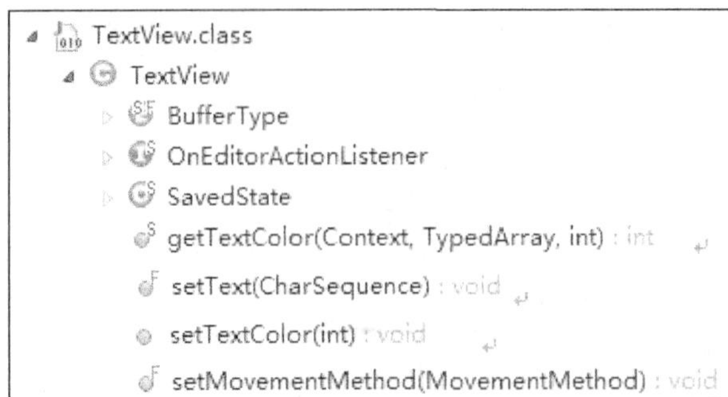

图 4.3.1　TextView 的定义(部分)

在布局文件中，可以使用的文本框控件的常用属性与方法，如表 4.3.1 所示。

表 4.3.1　文本框控件的常用属性与方法

方 法 名	功 能 描 述
text 属性	文本框内的文本
layout_width 属性	文本框的宽度
layout_height 属性	文本框的高度
password 属性	属性为“true”时，表示使用密码输入形式
singleLine 属性	属性为“false”时，表示多行文本

使用 EditText 控件的 hint 属性，以实现输入前的提示文本在输入任意字符后消失。

当 TextView 的内容特别多时，可使用它的一个重要方法 setMovementMethod(new ScrollingMovementMethod())实现滑动，即可以通过手指的上下滑动来查看内容。当在布局文件中使用属性 android:scrollbars="vertical"时，就会在文本框内的边上产生起提示作用的垂直滑块。

与 setMovementMethod()方法参数相关的类 ScrollingMovementMethod 的定义，如图 4.3.2 所示。

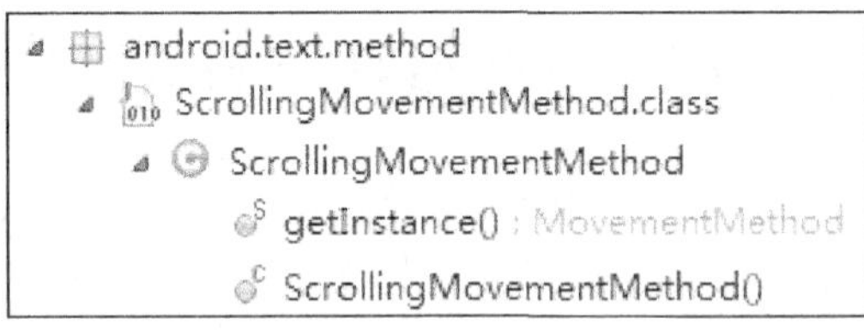

图 4.3.2　滚动方法类 ScrollingMovementMethod 的定义

注意：

(1) EditView 是 TextView 的子类，EditText 是一个具有编辑功能的 TextView 控件。

(2) 文本框内文字的大小，一般使用 sp 作为单位。

(3) 第 4.3.7 小节介绍的 ListView 控件与这里介绍的 TextView 控件，当其内容超过一屏的容量时，自动产生滚动条。

4.3.2　显示图像控件 ImageView

图像控件 ImageView 用来显示图像，在布局文件中，它的一个必填属性是 android:src，用来指定图像的来源。

4.3.3　快显信息与类 Toast

android.widget.Toast 类用于实现消息提醒，其信息在显示几秒后自动消失。Toast 只能是以程序代码的方式设计，这不同于其他可视化的 UI 控件(如文本框控件)。

Toast 可以看作是一个会自动消失的信息框，通常使用它自己提供的 makeText()方法创建一个 Toast 对象，再使用 show()方法显示。Toast 类的定义及主要方法如图 4.3.3 所示。

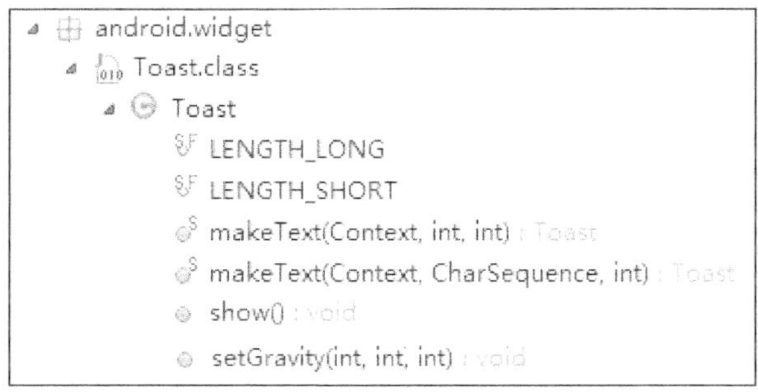

图 4.3.3　Toast 类的定义及主要方法

通常使用 Toast 提供的静态方法 makeText(Context,CharSequence,int)方法创建一个 Toast 对象。其中：第一参数是上下文对象；第二参数是要显示的信息；第三参数是显示的时长，可使用类常量 Toast.LENGTH_LONG 或 Toast.LENGTH_SHORT，还可以使用以毫秒为单位的正整数。

注意：可以在使用 show()方法前，通过 setGravity(int,int,int)设置 Toast 信息显示的位置(参见例 4.3.4)，否则，默认出现在手机屏幕偏下的位置。

4.3.4　命令按钮控件 Button、ImageButton 及其单击事件监听器设计

Button 是 UI 设计中使用相当频繁的一个控件，用来定义命令按钮。当用户单击按钮时，会有相应的动作，其动作代码放在按钮的单击事件监听器的 onClick()方法内。

Button 控件通过 setOnClickListener()方法设置单击事件监听器，该方法以一个实现了 android.view.View.OnClickListener 接口的对象为参数(见图 1.3.1)。

当一个 Activity 中需要定义单击事件的对象较多时，通常在定义时 Activity 通过子句 implements 实现接口 View.OnClickListener。在接口方法 public void onClick(View v) {…}中，每个 Button 都可以使用 onClick()方法定义自己的处理方法。

注意：

(1) OnClickListener 是类 android.view.View 的内部接口。

(2) 对 TextView 控件对象(显示文本)或 ImageView 控件对象(显示图像)也能设置单击事件监听器，尤其是文本的单击事件监听器在手机 App 中大量使用。

(3) 单击事件监听器有多种使用方式，分别参见例 4.3.1、例 4.3.2 和例 4.8.2。

【例 4.3.1】使用文本框和按钮控件制作用户登录界面。

【程序运行】程序运行时，在输入用户名和密码并单击“登录”按钮后，将在下方的文本框内显示用户的登录信息，程序运行界面如图 4.3.4 所示。

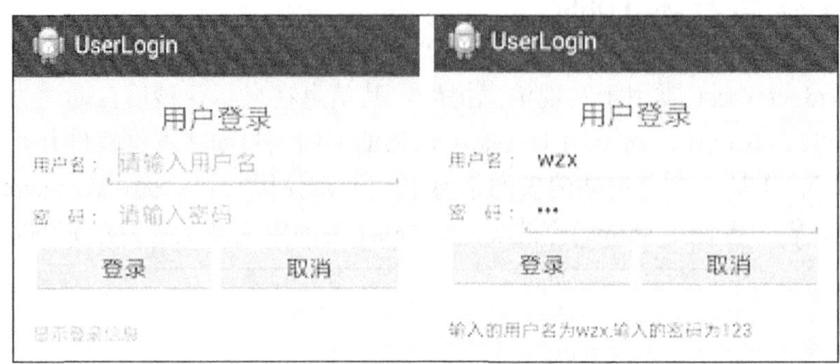

图 4.3.4 工程 UserLogin 的运行界面

【设计步骤】

(1) 新建名为 UserLogin 的 Android 应用工程，勾选创建 Activity，工程文件结构(主要部分)如图 4.3.5 所示。

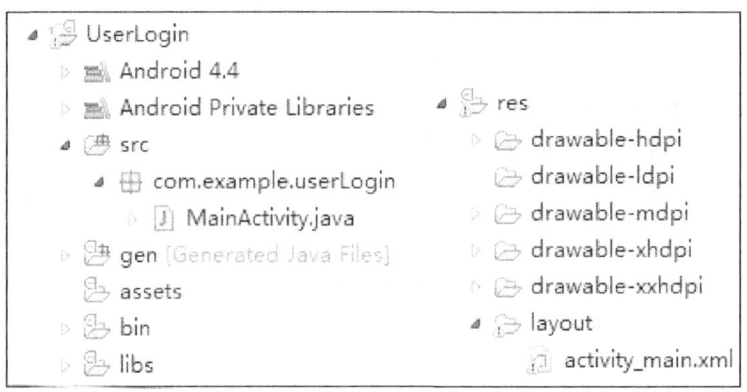

图 4.3.5 UserLogin 工程文件结构(主要部分)

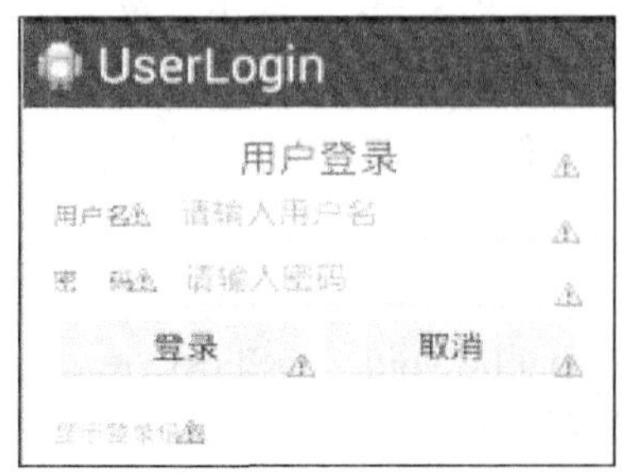

图 4.3.6 UserLogin 工程的视图效果(部分)

(2) 创建 MainActivity 对应的布局文件，采用垂直线性布局内嵌水平线性布局，对 EditText 控件采用了 android:hint 属性，以设定初始值且在用户输入后原来的初始值自动清除。对于按钮控件，使用 android:onClick 属性，用以指定响应按钮单击事件的方法名(方法在程序 MainActivity.java 中定义)。布局设计综合使用“代码”和“设计”两种模式，最终的视图效果(部分)如图 4.3.6 所示。

布局文件 activity_main.xml 的代码如下：

```
<LinearLayout xmlns:android="http://schemas.android.com/apk/res/android"
    xmlns:tools="http://schemas.android.com/tools"
    android:layout_width="match_parent"
    android:layout_height="match_parent"
    android:paddingBottom="@dimen/activity_vertical_margin"
    android:paddingLeft="@dimen/activity_horizontal_margin"
    android:paddingRight="@dimen/activity_horizontal_margin"
    android:paddingTop="@dimen/activity_vertical_margin"
    android:orientation="vertical"
    tools:context=".MainActivity" >
    <TextView
        android:layout_width="fill_parent"
        android:layout_height="wrap_content"
        android:gravity="center_horizontal"
        android:textSize="22sp"
        android:text="用户登录" />
    <LinearLayout
    android:layout_width="match_parent"
    android:layout_height="wrap_content"
    android:orientation="horizontal">
        <TextView
            android:layout_width="wrap_content"
            android:layout_height="wrap_content"
            android:text="用户名：" />
        <EditText
            android:id="@+id/edit1"
            android:layout_width="fill_parent"
             android:layout_height="wrap_content"
             android:hint="请输入用户名" />
      </LinearLayout>
      <LinearLayout
        android:layout_width="match_parent"
        android:layout_height="wrap_content"
        android:orientation="horizontal">
        <TextView
            android:layout_width="wrap_content"
            android:layout_height="wrap_content"
```

```
            android:text="密      码: " />
        <EditText
            android:id="@+id/edit2"
            android:layout_width="fill_parent"
            android:layout_height="wrap_content"
            android:hint="请输入密码"
            android:password="true"/>
    </LinearLayout>
    <LinearLayout
     android:layout_width="match_parent"
     android:layout_height="wrap_content"
     android:orientation="horizontal">
     <Button
            android:onClick="click1"
            android:layout_width="wrap_content"
            android:layout_height="40dip"
            android:layout_gravity="center_horizontal"
            android:layout_weight="1"
            android:text="登录" />
     <Button
            android:onClick="click2"
            android:layout_width="wrap_content"
            android:layout_height="40dip"
            android:layout_gravity="center_horizontal"
            android:layout_weight="1"
            android:text="取消" />
    </LinearLayout>
  <TextView
                android:id="@+id/text1"
                android:layout_marginTop="20dp"
            android:layout_width="wrap_content"
            android:layout_height="wrap_content"
            android:hint="显示登录信息" />
</LinearLayout>
```

(3) 编写程序 MainActivity.java，文件代码如下：

```
package com.example.userLogin;
import android.os.Bundle;
```

```
import android.app.Activity;
import android.widget.EditText;   //主类
import android.widget.TextView;   //主类
import android.view.View;
import android.widget.Toast;
public class MainActivity extends Activity {
    EditText et1;
    EditText et2;
    TextView tv;
    @Override
    protected void onCreate(Bundle savedInstanceState) {
        super.onCreate(savedInstanceState);
        setContentView(R.layout.activity_main);   //
        et1 = (EditText)this.findViewById(R.id.edit1);   //this可省略
        et2 = (EditText)this.findViewById(R.id.edit2);
        tv = (TextView)this.findViewById(R.id.text1);
    }
    public void click1(View view) {    //“登录”按钮的单击事件处理方法
        String s1=et1.getText().toString().trim();
        String s2=et2.getText().toString().trim();
        tv.setText("输入的用户名为"+s1+",输入的密码为"+s2);
    }
    public void click2(View view) {   //“取消”按钮的单击事件处理方法
         et1.setText("");
         et2.setText("");
         tv.setText("");
        //MainActivity.this表示MainActivity这个上下文，指示Toast的显示位置
         Toast.makeText(MainActivity.this, "取消登录",
                                        Toast.LENGTH_SHORT).show();
    }
}
```

4.3.5　单选控件 RadioGroup 及 RadioButton 与复选控件 CheckBox

RadioGroup 为单选按钮组控件，它包含若干单选按钮控件 RadioButton。通过这两个控件，可以为用户提供一种“多选一”的选择方式。

CheckBox 为复选框控件，通过使用父类 CompoundButton 提供的方法 isChecked()来判断用户是否勾选该复选框。

RadioGroup、RadioButton 和 CheckBox 的定义，如图 4.3.7 所示。

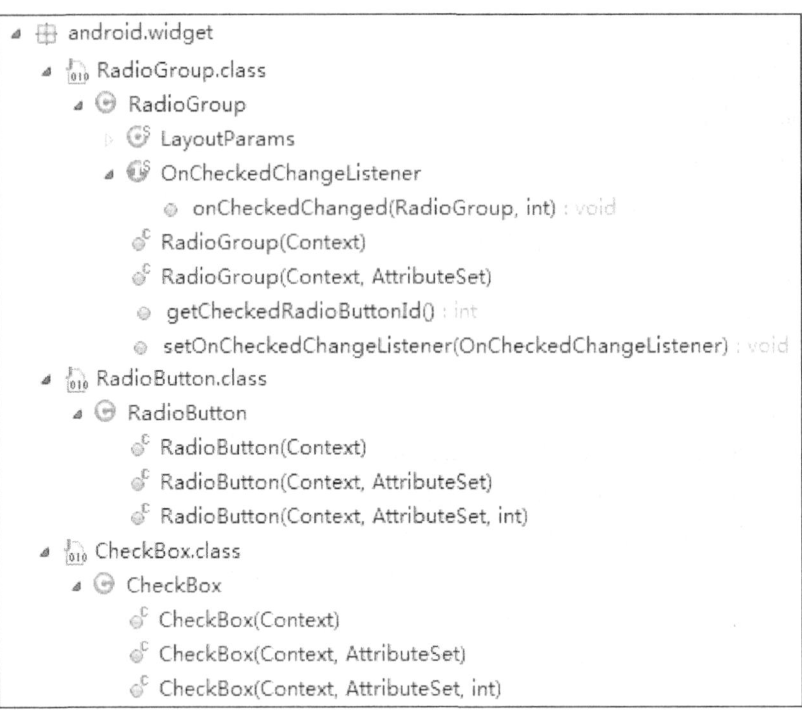

图 4.3.7　单选控件与复选控件的定义

注意：

(1) 使用单选按钮的布局文件中，一个 RadioGroup 控件内嵌若干 RadioButton 控件。

(2) RadioButton 和 CheckBox 控件，都有说明选项含义的 android:text 属性。

(3) RadioButton 和 CheckBox 都是继承抽象类 CompoundButton，CompoundButton 类的定义如图 4.3.8 所示。

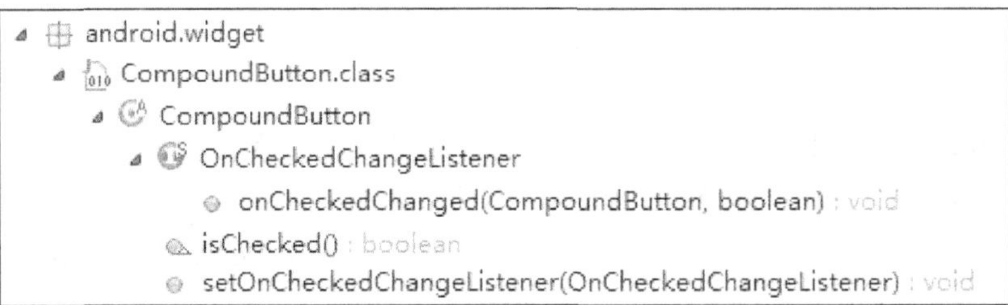

图 4.3.8　CompoundButton 类的定义

4.3.6　消息提醒对话框控件 AlertDialog 与进度控件 ProgressDialog

对话框与 Toast 一样，也属于消息提醒，但它是以对话框的形式出现在屏幕上的，用于需及时处理的通知。

对话框控件是特殊的控件，且有多种表现形式，它们不同于 TextView 等控件，在对话完成后才消失(即需要用户干预)。因此，对话框控件 AlertDialog 并不位于 android.widget 包内，而是位于 android.app 包内。

注意：

(1) AlertDialog 是 Dialog 的子类，ProgressDialog 是 AlertDialog 的子类。

(2) android.widget.ProgressBar 也是进度处理类。

【例 4.3.2】Android 对话框控件与进度控件的使用。

【程序运行】程序运行时，出现演示 6 种对话框的命令按钮，如图 4.3.9 所示。

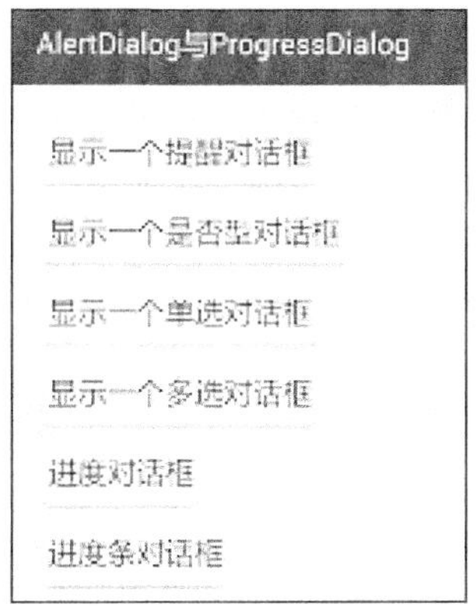

图 4.3.9　例 4.3.2 工程运行主界面

两种消息提醒对话框(提醒对话框和是否型对话框)的运行界面，如图 4.3.10 所示。

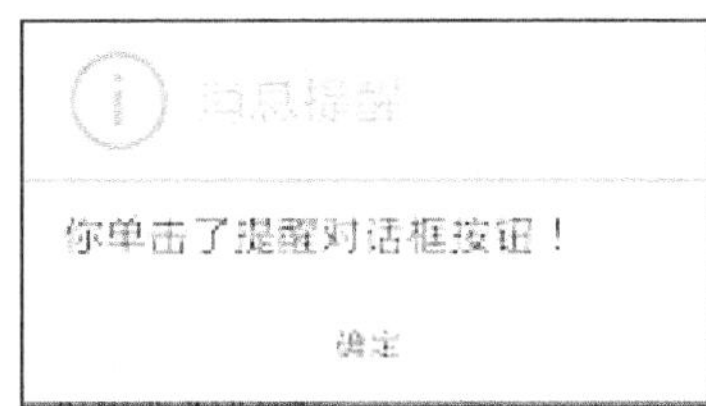

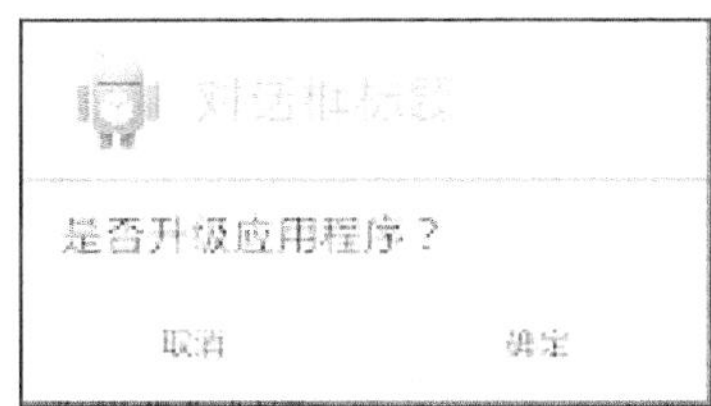

图 4.3.10　两种消息提醒对话框的运行界面

单选对话框和多选对话框的运行界面，如图 4.3.11 所示。

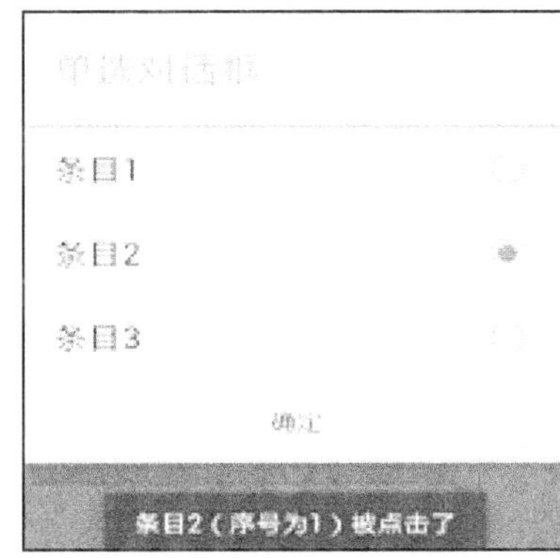

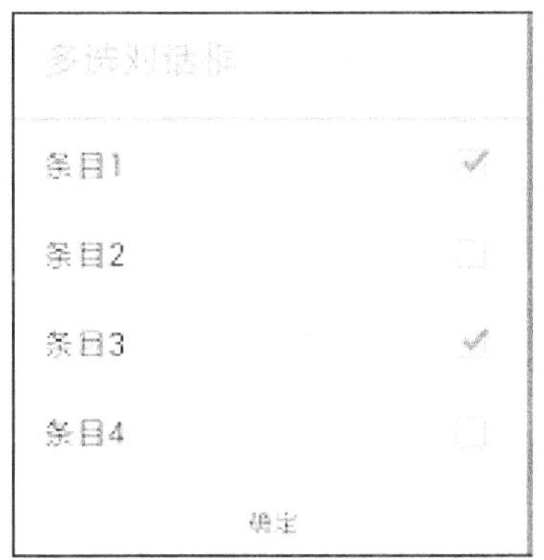

图 4.3.11　单选对话框与多选对话框的运行界面

两种常见的进度对话框(进度对话框和进度条对话框)的运行界面，如图 4.3.12 所示。

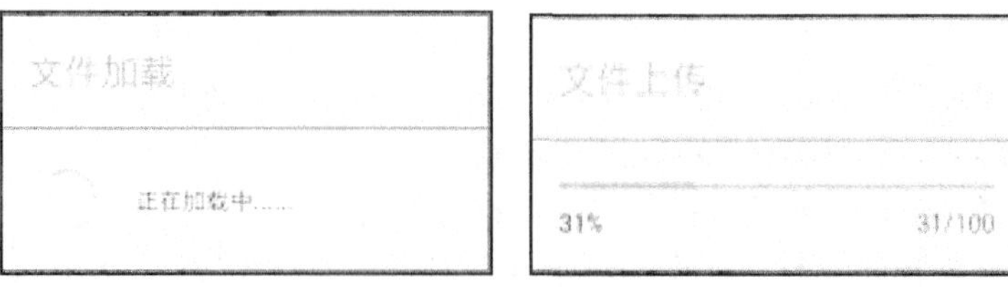

图 4.3.12　两种常见的进度对话框的运行界面

【设计步骤】

(1) 新建名为 Dialog 的 Android 应用工程，勾选创建 Activity，工程文件结构(部分)如图 4.3.13 所示。

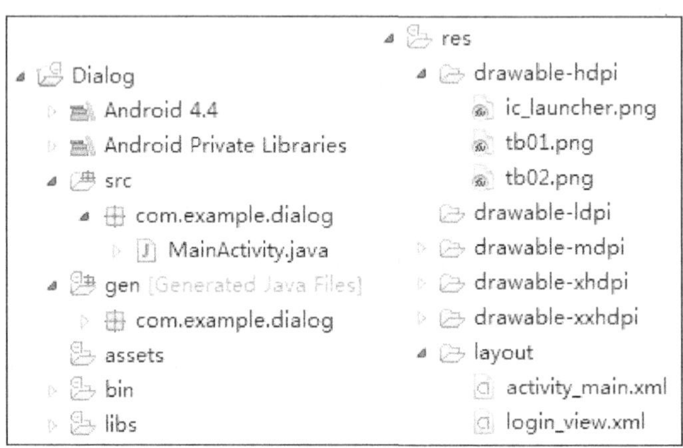

图 4.3.13　Dialog 工程文件结构 (部分)

(2) 编写项目的主 Activity，对应的 MainActivity.java 文件的完整代码如下：

```
package com.example.dialog;
/*
 * 本程序演示了多种对话框(提醒、是否型、单选、多选、进度与进度条)的使用
 */
import android.app.Activity;
import android.os.Bundle;
import android.app.AlertDialog;   //主类
import android.app.AlertDialog.Builder;
import android.content.DialogInterface;
import android.content.DialogInterface.OnClickListener;
import android.content.DialogInterface.OnMultiChoiceClickListener;
import android.app.ProgressDialog;   //主类
import android.widget.Button;
import android.view.LayoutInflater;
import android.view.View;
import android.widget.Toast;
```

```
import com.example.dialog.R;
import android.annotation.SuppressLint;
public class MainActivity extends Activity {
    private Button btn1a,btn1b,btn2,btn3,btn4,btn5;
    private View.OnClickListener listener = new View.OnClickListener()        {
        @Override
        public void onClick(View arg0) {
            // TODO Auto-generated method stub
            if(arg0 == btn1a)
                click1a();      //提醒对话框(只有一个“确定”按钮)
            else if(arg0 == btn1b)
                    click1b();   //是否型对话框
                else if(arg0 == btn2)
                        click2();   //单选
                    else if(arg0 == btn3)
                            click3();   //复选
                        else if(arg0 == btn4)
                                click4();   //进度对话框
                            else if(arg0 == btn5)
                                click5();    //水平进度条对话框
        }
    };    //在重写的监听器的onClick()方法中，需要判断按钮对象
    @Override
    protected void onCreate(Bundle savedInstanceState) {
        super.onCreate(savedInstanceState);
        setContentView(R.layout.activity_main);
        btn1a = (Button)findViewById(R.id.click1a);   //找控件
        btn1b = (Button)findViewById(R.id.click1b);
        btn2 = (Button)findViewById(R.id.click2);
        btn3 = (Button)findViewById(R.id.click3);
        btn4 = (Button)findViewById(R.id.click4);
        btn5 = (Button)findViewById(R.id.click5);
        btn1a.setOnClickListener( listener);   //设置监听器
        btn1b.setOnClickListener( listener);   //一组按钮共用一个监听器
        btn2.setOnClickListener( listener);
        btn3.setOnClickListener( listener);
        btn4.setOnClickListener( listener);
        btn5.setOnClickListener(listener);
    }
```

```
public void click1a() {
    AlertDialog.Builder builder = new Builder(this);
    builder.setIcon(R.drawable.tb01);
    builder .setTitle("消息提醒");
    builder.setMessage("你单击了提醒对话框按钮！");
    builder .setPositiveButton("确定", null);
    //builder.create();
    builder.show();
}
public void click1b() {
    AlertDialog.Builder builder = new Builder(this);
    builder.setIcon(R.drawable.ic_launcher);
    builder.setTitle("对话框标题");
    builder.setMessage("是否升级应用程序？");
    builder.setPositiveButton("确定", new OnClickListener() {
        @Override
        public void onClick(DialogInterface arg0, int arg1) {
            // TODO Auto-generated method stub
            Toast.makeText(MainActivity.this, "确定被点击了",
                                        Toast.LENGTH_SHORT).show();
        }
    });
    builder.setNegativeButton("取消", new OnClickListener() {
        @Override
        public void onClick(DialogInterface arg0, int arg1) {
            Toast.makeText(MainActivity.this, "取消被点击了",
                                        Toast.LENGTH_SHORT).show();
        }
    });
    builder.show();
}
public void click2() {
    AlertDialog.Builder builder = new Builder(this);
    builder.setTitle("单选对话框");
    final String[] items = new String[] { "条目1", "条目2", "条目3" };
    builder.setSingleChoiceItems(items, -1, new OnClickListener() {
        @Override
        public void onClick(DialogInterface dialog, int which) {
            // TODO Auto-generated method stub
```

```
                Toast.makeText(MainActivity.this, items[which] +
                        "(序号为"+which+")被点击了", Toast.LENGTH_SHORT).show();
            }
        });
        builder.setNegativeButton("确定", new OnClickListener() {
            @Override
            public void onClick(DialogInterface arg0, int arg1) {
                //确定的业务逻辑
            }
        });
        builder.show();
    }
    @SuppressLint("ShowToast")
    public void click3() {
        AlertDialog.Builder builder = new Builder(this);
        builder.setTitle("多选对话框");
        final String[] items = new String[] { "条目1", "条目2", "条目3", "条目4" };
        builder.setMultiChoiceItems(items, new boolean[] { true, false, true,
                false }, new OnMultiChoiceClickListener() {
            public void onClick(DialogInterface dialog, int which,boolean isChecked) {
                Toast.makeText(MainActivity.this, items[which] + isChecked,
                                                Toast.LENGTH_SHORT).show();
            }
        });
        builder.setNegativeButton("确定", new OnClickListener() {
            @Override
            public void onClick(DialogInterface arg0, int arg1) {
                //确定的业务逻辑
            }
        });
        builder.show();
    }
    public void click4() {     //进度对话框
        ProgressDialog pd = new ProgressDialog(this);
        pd.setTitle("文件加载");
        pd.setMessage("正在加载中......");
        pd.setCanceledOnTouchOutside(true); //允许
        pd.show();
        //pd.dismiss();     //在事务处理完毕后取消进度显示
```

```
    }
    public void click5() {      //水平进度条对话框
        final ProgressDialog pd = new ProgressDialog(this);
        pd.setProgressStyle(ProgressDialog.STYLE_HORIZONTAL);
        pd.setTitle("文件上传");
        //pd.setMessage("文件上传中......");
        pd.setMax(100);
        pd.setCanceledOnTouchOutside(false); //禁止
        pd.show();
        //Activity是主进程，创建它的一个子线程并运行(Java用法)
        new Thread() {
            @Override
            public void run() {
                for (int i = 0; i < 100; i++) {
                    pd.setProgress(i);   //进度(数字)
                    try {
                        Thread.sleep(50);//控制进度快慢
                    }
                    catch (InterruptedException e) {
                        // TODO Auto-generated catch block
                        e.printStackTrace();
                    }
                }
                pd.dismiss();     //在事务处理完毕后取消进度显示
            };
        }.start();   //
        // 创建子线程的另一种写法(Java用法)
/*          new Thread(new Runnable(){   //传对象
            @Override
            public void run() {
                // TODO Auto-generated method stub
                for (int i = 0; i < 100; i++) {
                    pd.setProgress(i);   //进度(数字)
                    try {
                        Thread.sleep(50);//控制进度快慢
                    }
                    catch (InterruptedException e) {
                        // TODO Auto-generated catch block
                        e.printStackTrace();
```

```
                }
            }
            pd.dismiss();    //在事务处理完毕后取消进度显示
        }

    }).start();*/
  }
}
```

inflate()直观意义是填充，用于实现视图的动态载入，其主要代码如下：

```
LayoutInflater inflater = this.getLayoutInflater();   //获取LayoutInflater对象
View layout = inflater.inflate(R.layout.login_view, null); //加载布局文件
```

【例 4.3.3】动态加载布局的使用——弹窗效果。

【程序运行】程序运行时，单击命令按钮后的界面如图 4.3.14 所示。

【设计步骤】

(1) 新建名为 DynamicLayout 的 Android 应用工程，勾选创建 Activity。

(2) 新建名为 login_view.xml 的布局文件，工程文件结构(部分)如图 4.3.15 所示。

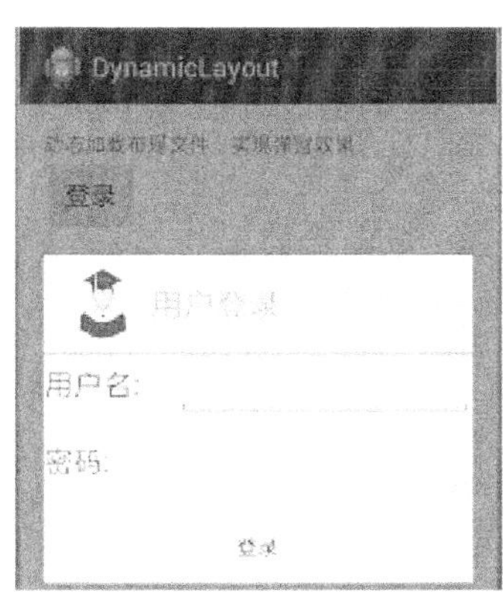

图 4.3.14　动态加载布局程序运行界面

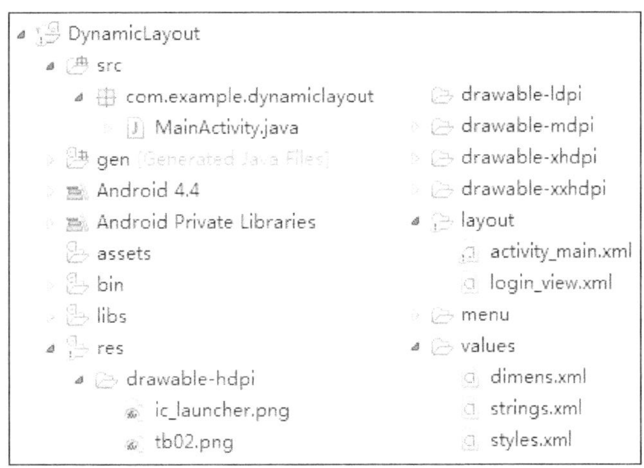

图 4.3.15　DynamicLayout 工程文件结构(部分)

(3) 布局文件 login_view.xml 的完整代码如下：

```
<?xml version="1.0" encoding="utf-8"?>
<LinearLayout xmlns:android="http://schemas.android.com/apk/res/android"
    android:layout_width="match_parent"
    android:layout_height="match_parent"
    android:orientation="vertical" >
    <LinearLayout android:layout_width="match_parent"
        android:layout_height="50dp"
        android:orientation="horizontal">
```

```
        <TextView
        android:text="用户名:"
        android:layout_width="90dp"
        android:layout_height="wrap_content"
        android:textSize="20sp" />
    <EditText
        android:layout_width="fill_parent"
        android:layout_height="wrap_content"
        android:id="@+id/user" />
    </LinearLayout>
    <LinearLayout
        android:layout_width="match_parent"
        android:layout_height="50dp"
        android:orientation="horizontal">
        <TextView
        android:layout_width="90dp"
        android:layout_height="wrap_content"
        android:text="密码:"
        android:textSize="20sp" />
       <EditText
        android:id="@+id/password"
        android:layout_width="fill_parent"
        android:layout_height="wrap_content"
        android:password="true"    />
    </LinearLayout>
</LinearLayout>
```

(4) 编写工程的主 Activity，对应的 MainActivity.java 文件的完整代码如下：

```
package com.example.dynamiclayout;
/*
 * 本程序演示在单击按钮后动态加载布局文件到对话框
*/
import android.app.Activity;
import android.os.Bundle;
import android.view.LayoutInflater;     //主类：布局填充
import android.view.View;
import android.widget.Button;
import android.app.AlertDialog;
```

```
import android.app.AlertDialog.Builder;
import android.content.DialogInterface;
public class MainActivity extends Activity {
    @Override
    protected void onCreate(Bundle savedInstanceState) {
        super.onCreate(savedInstanceState);
        setContentView(R.layout.activity_main);
        Button btn1 = (Button)findViewById(R.id.click1);
        btn1.setOnClickListener( listener);
    }
    private View.OnClickListener listener = new View.OnClickListener()
    {
        @Override
        public void onClick(View arg0) {
            // TODO Auto-generated method stub
            AlertDialog.Builder builder = new Builder(MainActivity.this);
        LayoutInflater inflater = MainActivity.this.getLayoutInflater();
                                                        //获取LayoutInflater对象
            View layout = inflater.inflate(R.layout.login_view, null);  //创建布局
            builder.setIcon(R.drawable.tb02);
            builder.setTitle("用户登录");
            builder.setView(layout);  //动态加载布局
            builder.setPositiveButton("登录", new DialogInterface.OnClickListener(){
    @Override
                public void onClick(DialogInterface arg0, int arg1) {
                    // TODO Auto-generated method stub
                    //单击“登录”按钮后的业务逻辑
                }
            });
            builder.create().show();
        }
    };
}
```

4.3.7　列表控件 ListView 与列表数据适配器、列表项选择监听器

1. 列表控件 ListView 及其数据适配器

ListView 是列表数据显示控件，以列表形式显示。Android 软件包 android.widget，在提供类 ListView 的同时，还提供了用于显示列表数据的数据适配器接口 ListAdapter 及其相关

类，如图 4.3.16 所示。

图 4.3.16　列表控件及其数据适配器接口(类)

注意：列表控件使用广泛但较复杂，因为它的数据项会涉及 Java 泛型。

其中，抽象类 BaseAdapter 实现了 ListAdapter 接口，同时又是类 ArrayAdapter 和 SimpleAdapter 的超类。定义一个继承 BaseAdapter 类的子类 MyAdapter 时，需要重写的方法如下：

```
public class MyAdapter extends BaseAdapter {
    @Override
    public int getCount() {
        // TODO Auto-generated method stub
        return 0;
    }
    @Override
    public Object getItem(int arg0) {
        // TODO Auto-generated method stub
        return null;
    }
    @Override
    public long getItemId(int arg0) {
        // TODO Auto-generated method stub
        return 0;
    }
    @Override
    public View getView(int arg0, View arg1, ViewGroup arg2) {
        // TODO Auto-generated method stub
        return null;
    }
```

ListView 控件的数据适配器都继承抽象类 BaseAdapter，可根据需要选用。例如，单列数据的显示可采用 ArrayAdapter，该类的定义如图 4.3.17 所示。

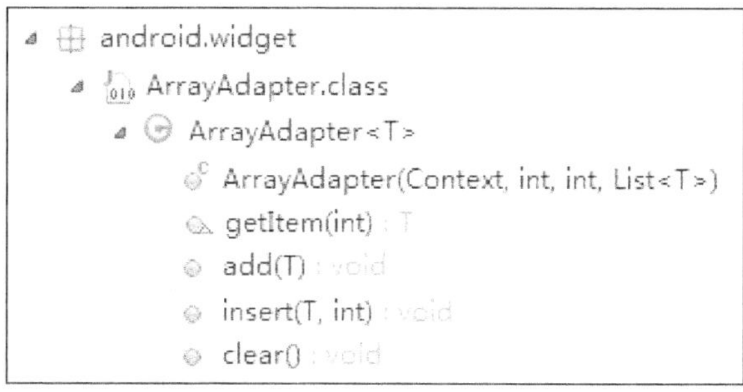

图 4.3.17　数据适配器 ArrayAdapter 的定义

通常使用的构造方法是 ArrayAdapter(Context,int,int,List<T>)，其中第一参数是上下文，第二参数是列表控件的列表项的布局文件，第三参数是列表布局文件中显示数据的控件，第四参数是列表数据。

如果列表项的数据项有多个，则可采用 SimpleAdapter。类 SimpleAdapter 的定义与使用参见第 7.3.4 小节。

2. 列表项选择监听器

在 Android 系统类库中，抽象类 AdapterView 是 ListView 的超类(但不是直接父类)，提供了设置列表项单击事件监听器的重要方法 setOnItemClickListener()和内部接口 OnItemClickListener，用以定义用户选择了某个列表项后的动作。抽象类 AdapterView 的定义如图 4.3.18 所示。

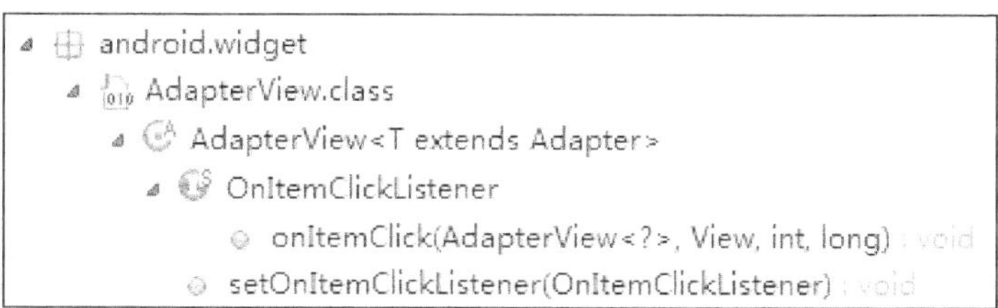

图 4.3.18　数据适配器视图类的定义

【例 4.3.4】使用 ListView 控件和 ArrayAdapter 数据适配器实现列表方式的选择。

【程序运行】程序运行界面如图 4.3.19 所示。

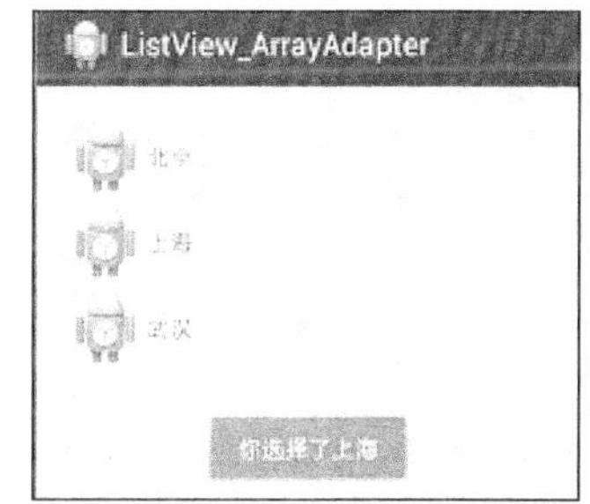

图 4.3.19　例 4.3.4 工程的运行界面

【设计步骤】

(1) 新建名为 ListView_ArrayAdapter 的 Android 应用工程，勾选创建 Activity，工程文件结构(主要部分)如图 4.3.20 所示。

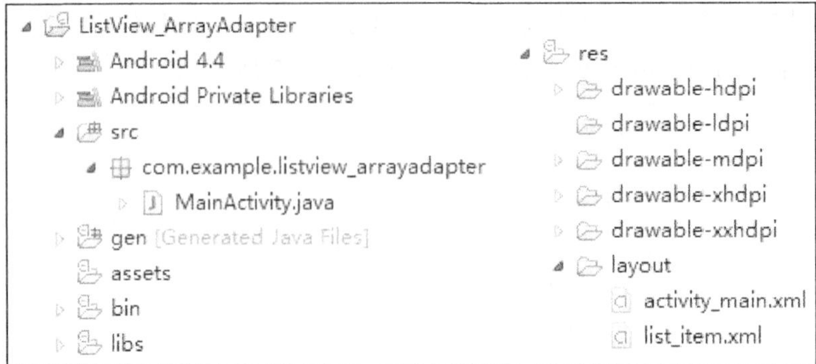

图 4.3.20 ListView_ArrayAdapter 工程文件结构(主要部分)

(2) 创建 MainActivity 对应的布局文件 activity_main.xml，只包含一个 ListView 控件，与该控件相关的列表布局文件 list_item.xml 包含一个 ImageView 控件和一个 TextView 控件，其代码如下：

```
<?xml version="1.0" encoding="utf-8"?>
<LinearLayout xmlns:android="http://schemas.android.com/apk/res/android"
    android:layout_width="match_parent"
    android:layout_height="match_parent"
    android:orientation="horizontal" >
    <ImageView
        android:layout_width="wrap_content"
        android:layout_height="wrap_content"
        android:src="@drawable/ic_launcher"/>
    <TextView
        android:id="@+id/tv"
        android:layout_width="wrap_content"
        android:layout_height="wrap_content"
        android:layout_gravity="center_vertical"/>
</LinearLayout>
```

(3) 编写程序 MainActivity.java，文件代码如下：

```
package com.example.listview_arrayadapter;
import android.app.Activity;
import android.os.Bundle;
import android.widget.ListView; //主类
import android.widget.ArrayAdapter;     //相关类
import android.view.View;   //相关类
import android.widget.AdapterView;   //
import android.widget.AdapterView.OnItemClickListener;   //
import android.widget.Toast;
import android.view.Gravity;
```

```
import java.util.ArrayList;     //相关类
import java.util.List;      //相关类
public class MainActivity extends Activity {
    //List换成ArrayList也可以！习惯上针对接口编程
    private List<String> listData=new ArrayList<String>();
    private ListView listView;
    private   ArrayAdapter<String> adapter; //指定接口参数的类型为String型
    @Override
    protected void onCreate(Bundle savedInstanceState) {
        super.onCreate(savedInstanceState);
        setContentView(R.layout.activity_main);
        listView=(ListView)findViewById(R.id.listView);   //找控件
        listData.add("北京"); listData.add("上海");
        listData.add("武汉");      //赋值列表项
        //构造函数共有4个参数，第二参数为列表布局文件名
        adapter=new ArrayAdapter<String>(this,R.layout.list_item,R.id.tv,listData);
        listView.setAdapter(adapter);
        //      选择项监听器
        listView.setOnItemClickListener(new OnItemClickListener()
        {
            @Override
            public void onItemClick(AdapterView<?> arg0,
                                        View arg1, int arg2,long arg3) {
                // TODO Auto-generated method stub
                String str = "你选择了"+adapter.getItem(arg2);   //获取选项
                Toast toast = Toast.makeText(MainActivity.this,
                                        str, Toast.LENGTH_SHORT);
                toast.setGravity(Gravity.CENTER, 0, 0);   //
                toast.show();
            }
        });
    }
}
```

【例 4.3.5】在 ListView 控件中，使用 BaseAdapter 适配器。

【程序运行】程序运行界面如图 4.3.21 所示。

图 4.3.21　例 4.3.5 工程的运行界面

【设计步骤】

(1) 新建名为 ListView_BaseAdapter 的 Android 应用工程，勾选创建 Activity，工程文件结构(主要部分)如图 4.3.22 所示。

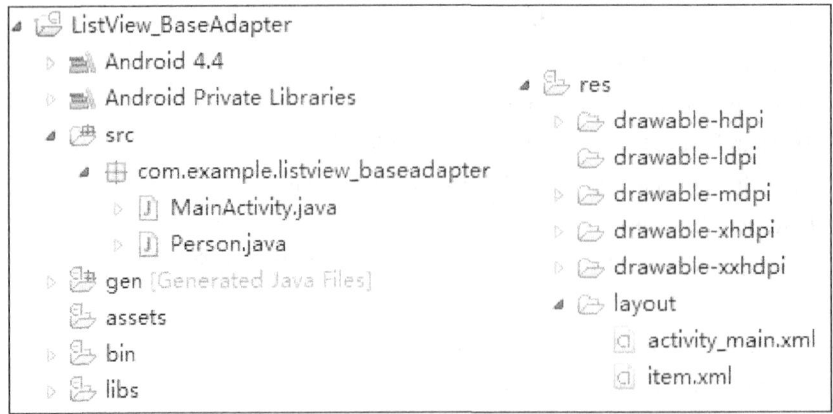

图 4.3.22 ListView_BaseAdapter 工程文件结构(主要部分)

(2) 创建 MainActivity 对应的布局文件 activity_main.xml，它包含一个 TextView 控件和一个 ListView 控件，该控件的列表布局文件 list_item.xml 包含一个 ImageView 控件和一个 TextView 控件，其代码如下：

```
<?xml version="1.0" encoding="utf-8"?>
<LinearLayout xmlns:android="http://schemas.android.com/apk/res/android"
    android:layout_width="match_parent"
    android:layout_height="match_parent"
    android:orientation="horizontal" >
    <ImageView
        android:layout_width="wrap_content"
        android:layout_height="wrap_content"
        android:src="@drawable/ic_launcher"/>
    <TextView
        android:id="@+id/tv"
        android:layout_width="wrap_content"
        android:layout_height="wrap_content"
        android:layout_gravity="center_vertical"/>
</LinearLayout>
```

(3) 创建列表控件使用的布局文件 item.xml，它包含三个 TextView 控件，通过使用布局属性 android:layout_weight 控制控件之间的间隔，其代码如下：

```
<?xml version="1.0" encoding="utf-8"?>
<LinearLayout xmlns:android="http://schemas.android.com/apk/res/android"
    android:layout_width="match_parent"
    android:layout_height="match_parent"
    android:orientation="horizontal"
    android:padding="10dp" >
    <TextView
```

```
        android:id="@+id/idTV"
        android:layout_width="0dp"
        android:layout_height="wrap_content"
        android:layout_weight="1"
        android:text="1"
        android:textSize="20sp" />
    <TextView
        android:id="@+id/nameTV"
        android:layout_width="0dp"
        android:layout_height="wrap_content"
        android:layout_weight="2"
        android:text="张三"
        android:textSize="20sp" />
    <TextView
        android:id="@+id/balanceTV"
        android:layout_width="0dp"
        android:layout_height="wrap_content"
        android:layout_weight="2"
        android:text="50000"
        android:textSize="20sp" />
</LinearLayout>
```

(4) 编写程序 MainActivity.java，文件代码如下：

```
package com.example.listview_baseadapter;
/*
 * 对ListView控件应用数据适配器BaseAdapter
 * 定义了内部类MyBaseAdapter
 * 适配器中的数据是固定的，实际应用中一般来源于SQLite数据库
 * 自定义了表示个人信息的实体类Person
 */
import android.app.Activity;
import android.os.Bundle;
import android.widget.BaseAdapter;    //主类
import java.util.ArrayList;
import java.util.List;
import android.view.View;
import android.view.ViewGroup;
import android.widget.ListView;
import android.widget.TextView;
```

```
public class MainActivity extends Activity {
    private ListView personLV;
    private List<Person> persons;
    public void onCreate(Bundle savedInstanceState) {
        super.onCreate(savedInstanceState);
        setContentView(R.layout.activity_main);
        //创建集合实例
        persons = new ArrayList<Person>();
        // 创建Person实例并添加数据集合persons里
        Person p1 = new Person(1,"张三",5000);
        persons.add(p1);      //
        Person p2 = new Person(2,"李四",5600);
        persons.add(p2);
        Person p3 = new Person(3,"王五",5800);
        persons.add(p3);
        // 获取ListView
        personLV = (ListView) findViewById(R.id.personLV);
        // 给ListView添加Adapter, 按照Adapter中的方法对ListView添加条目
        personLV.setAdapter(new MyBaseAdapter());          //
    }
    // 定义Adapter, 把每个Person对象生成一个条目, 将所有条目装入ListView
    private class MyBaseAdapter extends BaseAdapter {
        // 返回ListView中要装入的条目的数量
        public int getCount() {
            return persons.size();
        }
        // 返回指定位置上的对象
        public Object getItem(int position) {
            return persons.get(position);
        }
        // 返回条目的id
        public long getItemId(int position) {
            return position;
        }
        // 返回指定位置上的条目, 条目会被自动添加到ListView中
        public View getView(int position, View convertView, ViewGroup parent) {
            // 根据布局文件创建View
            View item = View.inflate(getApplicationContext(), R.layout.item, null);
            // 获取这个新生成的View中的TextView
```

```
            TextView idTV = (TextView) item.findViewById(R.id.idTV);

            TextView nameTV = (TextView) item.findViewById(R.id.nameTV);
            TextView balanceTV = (TextView) item.findViewById(R.id.balanceTV);
            // 根据位置获取Person对象
            Person p = persons.get(position);
            // 给TextView设置文本
            idTV.setText(p.getId() + "");

            nameTV.setText(p.getName());
            balanceTV.setText(p.getBalance() + "");
            return item;
        }
    }
}
```

4.3.8　在 ListActivity 中使用 ListView

ListActivity 是 Activity 的子类，它提供了获取 ListView 对象的方法 getListView()，该类的定义如图 4.3.23 所示。

图 4.3.23　ListActivity 类的定义

注意：在 ListActivity 中使用 ListView 参见例 8.2.2。

4.3.9　下拉列表控件 Spinner

Spinner 控件提供下拉列表式的输入方式，可有效地节省手机屏幕的显示空间。

与 ListView 类似，Spinner 控件也有设置数据适配器方法 setAdapter()、监听选择项方法

setOnItemSelectedListener()。

Spinner 比 ListView 多使用的一个方法是 setOnTouchListener()，用以实现触屏处理：当单击 Spinner 对象时才出现列表。

4.4 其他 Widget 控件介绍

4.4.1 日期和时间选择器(DatePicker 和 TimePicker)

Android 软件包 android.widget 提供了两个实用控件 DatePicker 和 TimePicker，以方便用户以图形化的方式输入日期和时间。其中，日期信息(年、月、日)通过 DatePicker 提供的 init()方法获取，时间信息(时、分)通过 TimePicker 的内部接口 OnTimeChangedListener 提供的 onTimeChanged()方法获取。

控件 DatePicker 和 TimePicker 均包含了内部监听器接口，其定义如图 4.4.1 所示。

```
android.widget
  DatePicker.class
    DatePicker
      OnDateChangedListener
        onDateChanged(DatePicker, int, int, int) : void
      init(int, int, int, OnDateChangedListener) : void
  TimePicker.class
    TimePicker
      OnTimeChangedListener
        onTimeChanged(TimePicker, int, int) : void
      setOnTimeChangedListener(OnTimeChangedListener) : void
```

图 4.4.1 日期和时间选择器类的定义

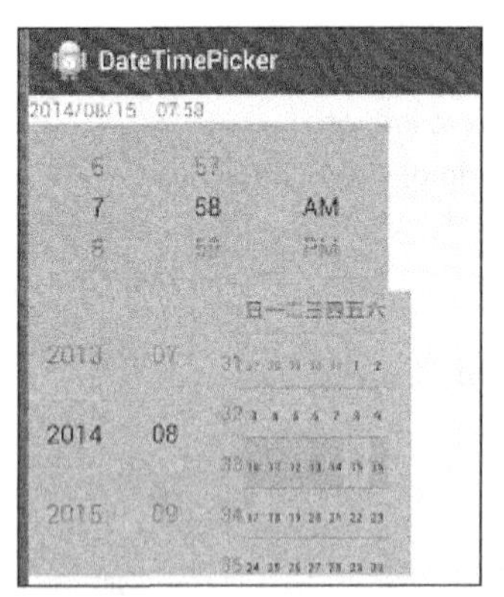

图 4.4.2 日期与时间选择器程序运行界面

注意：实际使用这两个控件时，需要使用标准的 Java 日历类 java.util.Calendar 且月份值是 0～11。

【例 4.4.1】日期与时间选择器(控件)的使用。

【程序运行】用户界面中包含了一个用于显示日期与时间信息的标签控件、一个用于获取系统日期信息的 DatePicker 控件和一个用于获取时间信息的 TimePicker 控件。程序运行时，显示为当前的日期和时间。在用户调整日期或时间后，文本框里的内容会立即进行相应的改变。程序运行界面如图 4.4.2 所示。

【设计步骤】

(1) 新建名为 DateTimePicker 的 Android 应用工程，勾选创建 Activity，主 Activity 和布局文件采用默认名称。创建后工程文件结构如图 4.4.3 所示。

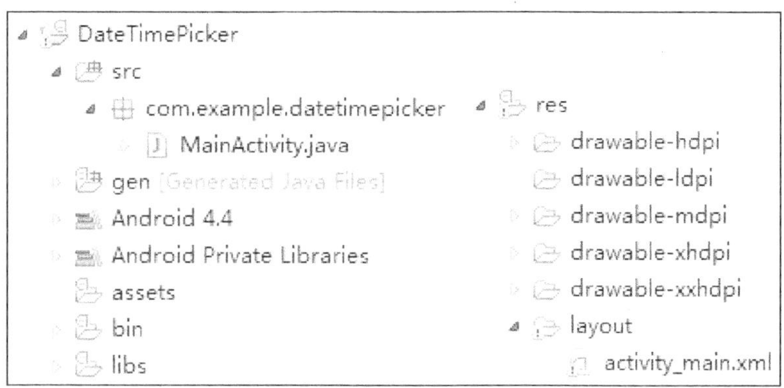

图 4.4.3　DateTimePicker 工程文件结构

(2) 修改默认的布局文件 activity_main.xml，使之包含一个 TextView 控件、一个 DatePicker 控件和一个 TimePicker 控件，采用垂直线性布局。

(3) 编写程序 MainActivity.java，文件代码如下：

```
package com.example.datetimepicker;
import android.app.Activity;
import android.os.Bundle;
import android.widget.DatePicker;   //主类
import android.widget.TimePicker;   //主类
import java.util.Calendar; //相关类
import android.widget.TextView;
public class MainActivity extends Activity {
    private TextView textview;
    private TimePicker timepicker;
    private DatePicker datepicker;
    /* 声明日期及时间变量 */
    private int year;
    private int month;
    private int day;
    private int hour;
    private int minute;
    @Override
    public void onCreate(Bundle savedInstanceState) {
        super.onCreate(savedInstanceState);
        setContentView(R.layout.activity_main);
        Calendar calendar = Calendar.getInstance();   //实例
        year = calendar.get(Calendar.YEAR);
```

```
month = calendar.get(Calendar.MONTH);
day = calendar.get(Calendar.DAY_OF_MONTH);
hour = calendar.get(Calendar. HOUR_OF_DAY);   //24小时制
minute = calendar.get(Calendar.MINUTE);
datepicker = (DatePicker) findViewById(R.id.datepicker);
timepicker = (TimePicker) findViewById(R.id.timepicker);
textview = (TextView) findViewById(R.id.timeview);
textview.setText(new StringBuilder().append(year).append("/")
        .append(format(month +1)).append("/")
        .append(format(day))
        .append("   ").append(format(hour)).append(":")
        .append(format(minute)));
datepicker.init(year, month, day,new DatePicker.OnDateChangedListener() {
        @Override
        public void onDateChanged(DatePicker view, int year,
                int monthOfYear, int dayOfMonth) {
            // TODO Auto-generated method stub
            MainActivity.this.year = year;
            month = monthOfYear;
            day = dayOfMonth;
            textview.setText(new StringBuilder().append(year)
                    .append("/").append(format(month + 1))
                    .append("/").append(format(day))
                    .append("   ")
                    .append(format(hour)).append(":")
                    .append(format(minute)));
        }
    });
    timepicker.setOnTimeChangedListener(
                        new TimePicker.OnTimeChangedListener() {
        @Override
        public void onTimeChanged(TimePicker view,
                                    int hourOfDay,int minute) {
            // TODO Auto-generated method stub
            hour = hourOfDay;
            MainActivity.this.minute = minute;
            textview.setText(new StringBuilder().append(year)
                .append("/").append(format(month + 1))
                .append("/").append(format(day)).append("   ")
                .append(format(hour)).append(":")
```

```
                    .append(format(minute)));
                }
            });
    }
    private String format(int dt) {    //对日期或时间做补0处理
        String str = "" + dt;
        if (str.length() == 1)
            str = "0" + str;
        return str;
    }
}
```

4.4.2　自动完成文本控件 AutoCompleteTextView

自动完成文本控件 AutoCompleteTextView 用于实现文本的快速输入，其原理是事先将用于输入的文本存放在一个字符数组里，根据输入字符前方一致进行匹配(匹配字符个数由控件属性 completionThreshold 进行设置)。作为实用类的 AutoCompleteTextView 控件，其定义如图 4.4.4 所示。

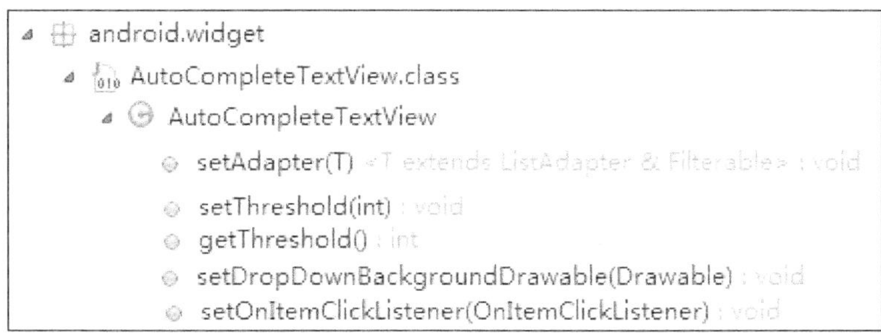

图 4.4.4　自动完成文本控件的定义

【例 4.4.2】自动完成文本控件 AutoCompleteTextView 的使用。

【程序运行】程序运行后，在文本框内输入字母 I 或 i 后，立即弹出一个下拉列表，列表项根据前方一致自动匹配，单击某个选项后的文本自动出现在文本框内。这种快速输入方式实质上是字典输入方式，程序运行界面如图 4.4.5 所示。

图 4.4.5　自动完成文本控件程序运行界面

【设计步骤】

(1) 新建名为 AutoCompleteText 的 Android 应用工程，勾选创建 Activity，主 Activity 和布局文件采用默认名称。创建后的工程文件结构(部分)如图 4.4.6 所示。

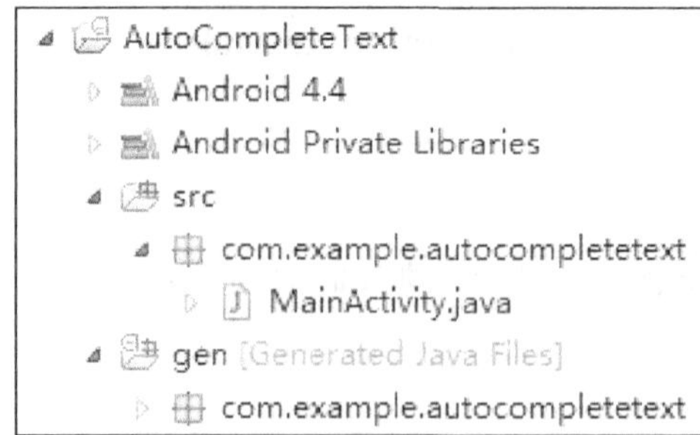

图 4.4.6　AutoCompleteText 工程文件结构(部分)

(2) 修改默认的布局文件 activity_main.xml，添加一个自动完成文本控件 AutoCompleteTextView，并设置自动搜索时匹配字符个数属性为“1”，即 android:completionThreshold="1"。在布局文件中，控件代码如下：

```
<AutoCompleteTextView
    android:id="@+id/autoCompleteTextView1"
    android:layout_width="match_parent"
    android:layout_height="wrap_content"
    android:text=""
    android:completionThreshold="1"/>
```

(3) 编写程序 MainActivity.java，文件代码如下：

```
package com.example.autocompletetext;
/*
 * 将一组国家名称存放在一个字符数组里
 * 使用AutoCompleteTextView控件，根据前方一致自动匹配，实现快速输入
 * 忽略字母大小写。例如，在程序运行时输入字母I或i，结果等效
 */
import android.app.Activity;
import android.os.Bundle;
import android.widget.AutoCompleteTextView;   //
import android.widget.ArrayAdapter;

public class MainActivity extends Activity {
    private AutoCompleteTextView textView;
    private static final String[] autotext = {
                                        "China","Canada","India","Italy","Iran","Iraq"};
    @Override
    protected void onCreate(Bundle savedInstanceState) {
        super.onCreate(savedInstanceState);
        setContentView(R.layout.activity_main);
```

```
        textView = (AutoCompleteTextView )findViewById(
                                                    R.id.autoCompleteTextView1);
        ArrayAdapter<String> adapter=new ArrayAdapter<String>(
                                    this,android.R.layout.simple_dropdown_item_1line,autotext);
        textView.setAdapter(adapter);     //绑定
    }
}
```

4.4.3 菜单 Menu 设计

菜单也是人机交互的重要方式。在前面的应用中，我们没有设置 Activity 的选项菜单。实际上，如果我们在 Activity 中设置了菜单，配合手机上的菜单键，就能使用菜单操作了。

在 Android 4.4 中，新建一个工程时，Activity 文件中的语句“import android.view.Menu;”就是导入用于菜单设计的接口 Menu。菜单接口 Menu 提供了增加菜单项方法 add()，其结果为 MenuItem 接口类型。菜单接口 Menu 提供的主要方法如图 4.4.7 所示。

图 4.4.7　菜单接口 Menu 提供的主要方法

在 Activity 中创建菜单，需要重写方法 onCreateOptionsMenu()，其代码如下：

```
@Override
public boolean onCreateOptionsMenu(Menu menu) {
    // Inflate the menu; this adds items to the action bar if it is present.
    this.getMenuInflater().inflate(R.menu.main, menu);
    return true;
}
```

注意：在 Eclipse 中，可重写某个类的方法的图标是黄色的菱形，而不是普通的绿色小球。

Activity 提供的 getMenuInflater()用于获取 MenuInflater 对象。作为菜单设计必须使用的相关类，类 MenuInflater 提供了从 XML 文件解析(呈现)菜单的主要方法 inflate()，如图 4.4.8 所示。

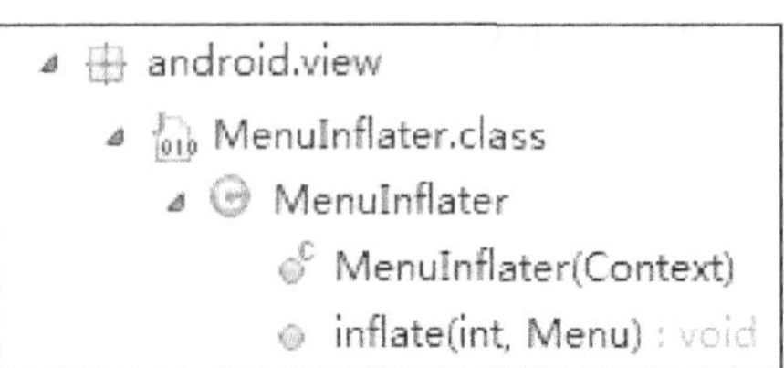

图 4.4.8　类 MenuInflater 的定义

注意：MenuInflater 的作用，与前面介绍的 LayoutInflater 类似。

需要在 Activity 子类中改写的另一个方法是 onOptionsItemSelected()，它表示对用户单击(选择)某个菜单项这个事件的处理，默认时没有加载这个可以重写的方法。

注意：编写 Activity 子类时，在空白处的右键菜单中选择"Source→Override/Implements Methods"，即可添加所有可以改写的父类方法。

与 onOptionsItemSelected()相对应，接口 MenuItem 提供了用于获取用户选择的菜单编号的方法 getItemId()。

设计菜单时，可以将几个菜单项归为一组，形成所谓的子菜单。 Menu 接口提供了增加子菜单的方法 addSubMenu()创建接口 SubMenu 类型的对象。

接口 SubMenu 提供了增加子菜单项的方法 add(int arg0,int arg1,int arg2,CharSequence arg3)。其中，各参数的含义如下：

- 参数 arg0 表示组编号；
- arg1 表示选项的编号，使用接口 MenuItem 提供的方法 getItemId()获取，以判定用户的选择；
- 第三参数 arg2 表示菜单项的显示顺序，默认值为 0，按照增加的顺序来显示；
- 第四参数 arg3 表示选项文本。

注意：子菜单能实现菜单项分类。

上下文菜单注册到 View 对象后，用户长按 View 对象可呼出悬浮于主界面之上的上下文菜单。

【例 4.4.3】菜单(含子菜单和上下文菜单)设计。

【程序运行】在主 Activity 中除了定义一个 TextView 控件和一个 EditText 控件外，还定义了菜单(含子菜单和上下文菜单)。按下手机的 Menu 键，即可出现菜单项，单击某个菜单项，则在 TextView 控件中显示；子菜单作为一个特殊的菜单项，单击它则呈现次级菜单；长按 EditText 控件，则出现"粘贴"的上下文菜单。三种类型的菜单制作如图 4.4.9 所示。

【设计步骤】

(1) 新建名为 MenuDemo 的 Android 应用工程，勾选创建 Activity，设定主 Activity 为 MenusDemoActivity.java，指定相应的主布局文件为 main.xml。工程文件结构如图 4.4.10 所示。

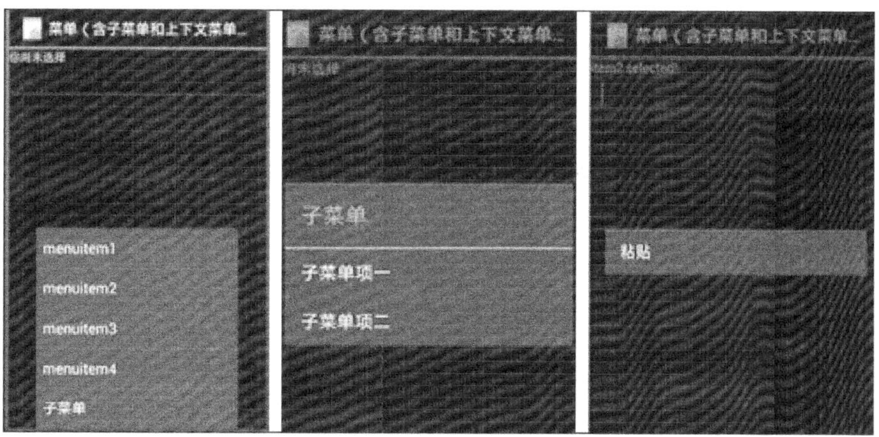

图 4.4.9　三种类型的菜单制作

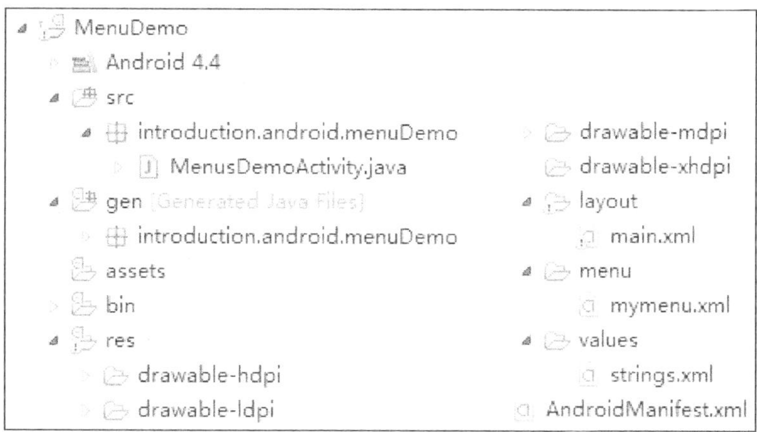

图 4.4.10　MenuDemo 工程文件结构

(2) 修改默认的布局文件，使之包含一个 TextView 控件和一个 EditText 控件。

(3) 在 res\menu 下新建一个名为 mymenu.xml 的文件，其代码如下：

```
<?xml version="1.0" encoding="utf-8"?>
<menu xmlns:android="http://schemas.android.com/apk/res/android" >
    <item
        android:id="@+id/item1"
        android:title="@string/menuitem1"
        android:icon="@drawable/icon01" />
    <item
        android:id="@+id/item2"
        android:title="@string/menuitem2"
        android:icon="@drawable/icon02" />
    <item
        android:id="@+id/item3"
        android:title="@string/menuitem3"
```

```
        android:icon="@drawable/icon03" />
    <item
        android:id="@+id/item4"
        android:title="@string/menuitem4"
        android:icon="@drawable/icon04" />
</menu>
```

(4) 编写程序 MainActivity.java，文件代码如下：

```
package com.example.pendingintent;
package introduction.android.menuDemo;
/**
 * 演示菜单(含子菜单和上下文菜单)的制作
 * 要求在清单文件中指定的Android SDK最低版本为11,否则菜单效果不同
 * 菜单内容存放在menu文件夹里的一个XML文件中
 * */
import android.app.Activity;
import android.os.Bundle;

import android.view.Menu;   //
import android.view.MenuInflater;
import android.view.MenuItem;
import android.view.SubMenu;   //
import android.view.ContextMenu;   //
import android.view.ContextMenu.ContextMenuInfo;

import android.view.View;
import android.widget.TextView;
import android.widget.EditText;

public class MenusDemoActivity extends Activity {
    private TextView textview;
    private EditText edittext;
    private String str="";

    @Override
    public void onCreate(Bundle savedInstanceState) {
        super.onCreate(savedInstanceState);
        setContentView(R.layout.main);

        textview=(TextView)findViewById(R.id.textView1);
```

```
        edittext=(EditText)findViewById(R.id.editText1);

        //注册上下文菜单，长按时弹出
        registerForContextMenu(textview);
        registerForContextMenu(edittext);
    }

    @Override
    public boolean onCreateOptionsMenu(Menu menu) {
        MenuInflater inflater = getMenuInflater();
        inflater.inflate(R.menu.mymenu, menu);
        //下面的两个子菜单项作为主菜单的一个菜单项
        SubMenu submenu=menu.addSubMenu("子菜单");
        submenu.add(0,1,0,"子菜单项一");
        submenu.add(0, 2, 0, "子菜单项二");
        return true;
    }

    @Override
    public boolean onOptionsItemSelected(MenuItem item) {
        // TODO Auto-generated method stub
    switch(item.getItemId()){
            case 1:
                textview.setText("选择了子菜单一");
                break;
            case 2:
                textview.setText("选择了子菜单二");
                break;
            case R.id.item1:
                textview.setText("item1 selected!");
                break;
            case R.id.item2:
                textview.setText("item2 selected!");
                break;
            case R.id.item3:
                textview.setText("item3 selected!");
                break;
            case R.id.item4:
                textview.setText("item4 selected!");
```

```
                break;
            default:
                break;
            }
            return super.onOptionsItemSelected(item);
    }

    @Override
    public void onCreateContextMenu(ContextMenu menu, View v,
            ContextMenuInfo menuInfo) {
        // TODO Auto-generated method stub
        if(v.getId() == R.id.textView1)
            menu.add(0, 1, 0, "复制");
        if(v.getId()==R.id.editText1)
            menu.add(0, 2, 0, "粘贴");
        super.onCreateContextMenu(menu, v, menuInfo);
    }
    @Override
    public boolean onContextItemSelected(MenuItem item) {
        // TODO Auto-generated method stub
        switch(item.getItemId()){
            case 1:
                str = textview.getText().toString();
                break;
            case 2:
                edittext.setText(str);
                break;
            default:
                break;
        }
        return super.onContextItemSelected(item);
    }
}
```

注意：菜单设计的另一个例子，参见例 9.2.1。

4.5　状态栏消息通知 android.app.Notification

4.5.1　通知与通知类 Notification

在 Android 中，通知 Notification 能够通过多种方式提供给用户，例如，发出声音，设备震动，或在位于手机屏幕顶部的状态栏里放置一个持久的图标等。

当手指从状态栏向下滑过时，会展开所有的通知信息，能查看到通知的标题和内容等信息(如未查看的短信)，还能清除通知；当手指从下方向上滑过时将隐藏通知。

当手机短信到来或来电未接时，都会由应用程序自动产生一个通知。系统时间的数字文本、手机网络连接方式的图标、手机充电指示图标等表示系统特定的状态信息，都在手机的状态栏中显示。

注意：通知 Notification 并不属于任何 Activity。

Android 系统提供了对通知进行描述的类 Notification，Android API 16 及以后的版本都推荐使用 Notification 的内部构造器类 Builder 来创建一个 Notification 对象，Notification 类封装了在状态栏里设置发布通知的标题、内容、时间、通知到来时的提示音和标题前的小图标等，如图 4.5.1 所示。

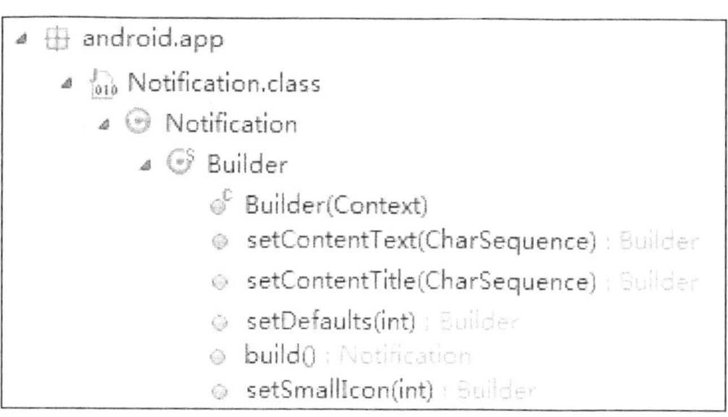

图 4.5.1　状态栏通知类及其内部构造器类

注意：作为消息提醒机制，Notification 与 Toast 类似，但 Toast 位于 android.widget 包；而 Notification 不属于 Activity，它位于 android.app 包，需要使用特定的管理器进行管理。

4.5.2　通知管理器类 NotificationManager

通知是由 NotificationManager 类型的对象进行管理的，而 NotificationManager 对象需要调用系统服务 Context.NOTIFICATION_SERVICE 来创建(参见第 6.1.1 小节)，其代码如下：

```
String ns = Context.NOTIFICATION_SERVICE;
NotificationManager nm=(NotificationManager) getSystemService(ns);
```

通知管理器类 NotificationManager 主要提供了发通知、取消通知两个方法，如图 4.5.2

所示。

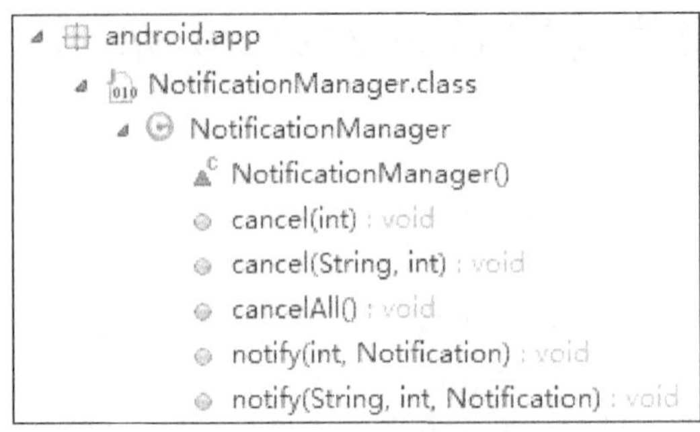

图 4.5.2　通知管理器类的常用方法

【例 4.5.1】通知 Notification 的发布与清除(含快显信息 Toast 的使用)。

【程序运行】单击“产生一个通知”按钮后，在手机状态栏中可以查看到该通知。单击“取消指定的通知”按钮后，有 Toast 方式显示的提示信息，在手机状态栏里该通知被删除。程序运行界面如图 4.5.3 所示。

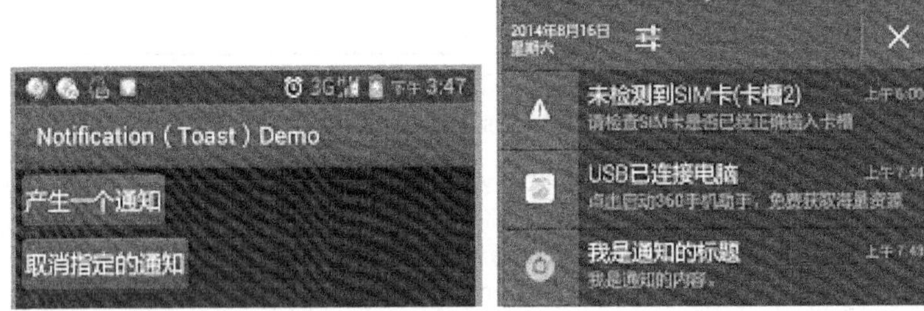

图 4.5.3　例 4.5.1 程序运行界面

【设计步骤】

(1) 新建名为 Notification_Toast 的 Android 应用工程，勾选创建 Activity，指定主 Activity 名为 MainActivity.java，布局文件为 main.xml，工程文件结构(主要部分)如图 4.5.4 所示。

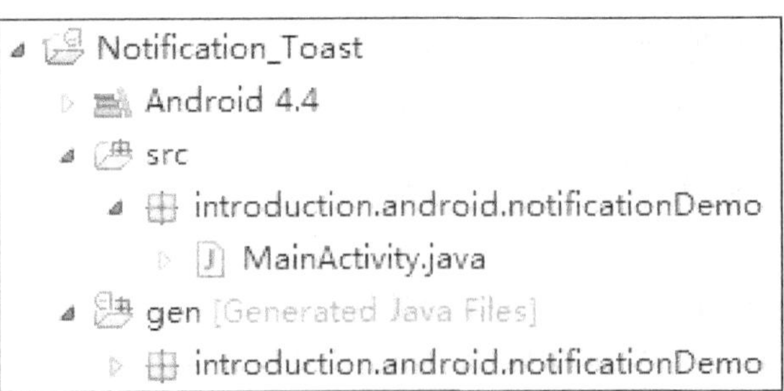

图 4.5.4　Notification_Toast 工程文件结构(主要部分)

(2) 在布局文件 main.xml 中，采用垂直线性布局，依次添加两个命令按钮，布局文件代码如下：

```
<?xml version="1.0" encoding="utf-8"?>
<LinearLayout xmlns:android="http://schemas.android.com/apk/res/android"
    android:layout_width="fill_parent"
    android:layout_height="fill_parent"
    android:orientation="vertical" >
    <Button
        android:id="@+id/button1"
        android:layout_width="wrap_content"
        android:layout_height="wrap_content"
        android:text="产生一个通知" />
    <Button
        android:id="@+id/button2"
        android:layout_width="wrap_content"
        android:layout_height="wrap_content"
        android:text="取消指定的通知" />
</LinearLayout>
```

(3) 编写工程的主 Activity 文件 MainActivity.java，其代码如下：

```
package introduction.android.notificationDemo;
import android.app.Activity;
import android.os.Bundle;

import android.app.NotificationManager;    //
import android.app.Notification;    //
import android.content.Context;
import android.view.View;
import android.widget.Button;
import android.widget.Toast; //

public class MainActivity extends Activity {
    NotificationManager nm;
    Notification notification;
    private static final int NOTIFICATION_ID = 1;
    private Button notifyBtn;
    private Button cancelBtn;
    @Override
    public void onCreate(Bundle savedInstanceState) {
        super.onCreate(savedInstanceState);
        setContentView(R.layout.main);
```

```
        notifyBtn=(Button)this.findViewById(R.id.button1);
        notifyBtn.setOnClickListener(new View.OnClickListener() {
        @Override
        public void onClick(View v) {
                // TODO Auto-generated method stub
                String ns = Context.NOTIFICATION_SERVICE; //
                nm= (NotificationManager) getSystemService(ns);
                int icon = R.drawable.icon01;
                long when = System.currentTimeMillis();

                Notification.Builder builder=new Notification.Builder(MainActivity.this);
                builder.setContentTitle("我是通知的标题")
                        .setContentText("我是通知的内容。")
                        .setSmallIcon(icon)
                        .setWhen(when)
                        .setDefaults(Notification.DEFAULT_SOUND);
                notification=builder.build();
                nm.notify(NOTIFICATION_ID, notification);
            }
        });

        cancelBtn=(Button)this.findViewById(R.id.button2);
        cancelBtn.setOnClickListener(new View.OnClickListener() {
            @Override
            public void onClick(View v) {
                // TODO Auto-generated method stub
                nm.cancel(NOTIFICATION_ID);   //
                Toast.makeText(MainActivity.this,"已经取消了指定的通知，可查验！
                                                        ",1000).show();
            }
        });
    }
}
```

注意：本程序的设计方法要求 Android API 的版本不低于 16，因此在清单文件中需要对使用的 SDK 做如下指定：

```
<uses-sdk android:minSdkVersion="16" />
```

4.6 文件存储

4.6.1 Android 文件读写

文件读写分为内部读写(internal storage)与外部读写(external storage)。

内部存储是将应用程序的数据以文件的方式保存至设备内存中，该文件为其创建的应用程序私有，其他应用程序无权进行操作，当该应用程序被卸载时，其内部存储文件也随之被删除。

外部读写通常是指对 SD 卡上文件的读写操作，需要在清单文件中配置在扩展存储设备里写文件的权限，参见第 4.9 节和例 5.3.2(手机录音程序)。

抽象类 android.content.Context 提供了两个抽象的方法 openFileOutput() 和 openFileInput()，其类型对应于标准 I/O 文件读写类型，如图 4.6.1 所示。

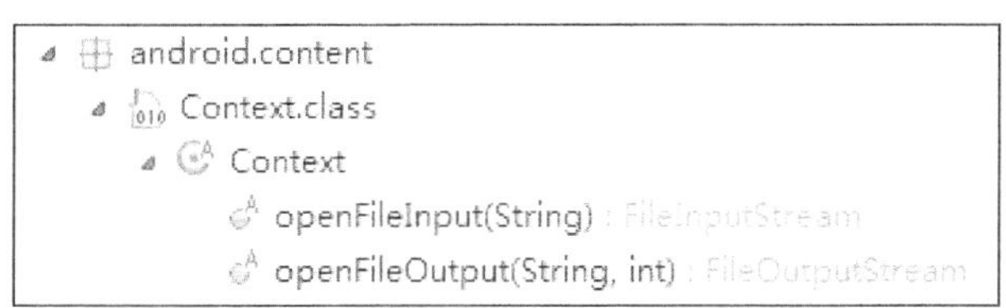

图 4.6.1　抽象类 Context 提供的用于文件读写的两个方法

注意：

(1) 在 Android 应用开发中，还需要配合使用 Java 中的标准 I/O 文件读写方式来实现文件读写。

(2) openFileOutput(String,int)方法中的第一参数为要读取的文件名，第二参数为读取方式。

【例 4.6.1】内部文件读写示例。

【程序运行】在编辑文本框内输入信息后，单击“保存信息”按钮，即可以追加方式保存至文件 text 中。单击“读取信息”按钮，即可将文件内容以 Toast 方式显示。程序运行界面如图 4.6.2 所示。

【设计步骤】

(1) 新建名为 FileDemo 的 Android 应用工程，指定程序的包名为“introduction.android.fileDemo”，勾选创建 Activity，指定主 Activity 名为 FileDemo.java，布局文件名为 main.xml，工程文件结构(主要部分)如图 4.6.3 所示。

图 4.6.2　内部文件读写程序运行界面

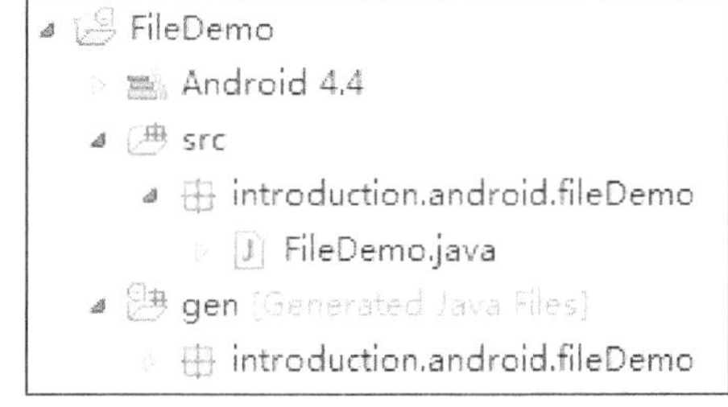

图 4.6.3　FileDemo 工程文件结构(主要部分)

(2) 在布局文件 main.xml 中，采用垂直线性布局，依次添加两个 TextView 控件、一个 EditView 控件和两个命令按钮。其中，两个命令按钮是水平线性布局。布局文件代码如下：

```
<?xml version="1.0" encoding="utf-8"?>
<LinearLayout xmlns:android="http://schemas.android.com/apk/res/android"
    android:layout_width="fill_parent"
    android:layout_height="fill_parent"
    android:orientation="vertical" >
    <TextView
        android:layout_width="fill_parent"
        android:layout_height="wrap_content"
        android:textSize="25sp"
       android:gravity="center_horizontal"
       android:text="文件存储示例" />
    <TextView
        android:layout_width="fill_parent"
        android:layout_height="wrap_content"
        android:text="(追加)输入您存储的信息" />
    <EditText
        android:id="@+id/phone_text"
        android:layout_width="fill_parent"
        android:layout_height="wrap_content"
        android:hint="(追加)输入保存的信息" />
    <LinearLayout
        android:layout_width="wrap_content"
        android:layout_height="wrap_content"
        android:orientation="horizontal" >
        <Button
        android:id="@+id/SaveButton"
        android:layout_width="wrap_content"
        android:layout_height="wrap_content"
        android:text="保存信息"/>
    <Button
        android:id="@+id/LoadButton"
        android:layout_width="wrap_content"
        android:layout_height="wrap_content"
        android:text="读取信息"/>
    </LinearLayout>
</LinearLayout>
```

(3) 编写工程的主 Activity 文件 FileDemo.java，其代码如下：

```
package introduction.android.fileDemo;
import android.app.Activity;
import android.os.Bundle;
import java.io.FileInputStream;   //主类
import java.io.FileOutputStream;   //主类
import android.widget.Button;
import android.view.View;
import android.view.View.OnClickListener;
import android.widget.EditText;
import android.widget.Toast;
public class FileDemo extends Activity {
    private EditText SaveText;
    private Button SaveButton,LoadButton;
    @Override
    public void onCreate(Bundle savedInstanceState) {
        super.onCreate(savedInstanceState);
        setContentView(R.layout.main);
        SaveText=(EditText)findViewById(R.id.phone_text);
        SaveButton = (Button)findViewById(R.id.SaveButton);
        LoadButton = (Button)findViewById(R.id.LoadButton);
        SaveButton.setOnClickListener(new ButtonListener());
        LoadButton.setOnClickListener(new ButtonListener());
    }
    private class ButtonListener implements OnClickListener{
    @Override
    public void onClick(View v) {
        switch (v.getId()) {
        case R.id.SaveButton:   //保存数据
        String saveinfo = SaveText.getText().toString().trim();
        FileOutputStream fos;
        try {
            fos = openFileOutput("text", MODE_APPEND);   //
            fos.write(saveinfo.getBytes());   //
            fos.close();
        } catch (Exception e) {
            e.printStackTrace();
```

```
                }
                Toast.makeText(FileDemo.this,"数据保存成功",Toast.LENGTH_LONG).show();
                break;
                case R.id.LoadButton:   //读取数据
                String get="";
                try {
                    //文件无扩展名且为应用程序私有
                    FileInputStream fis=openFileInput("text");
                    byte [] buffer=new byte[fis.available()];
                    //available()返回一次可以读取到的数据长度
                    fis.read(buffer);
                    get = new String(buffer);
                } catch (Exception e) {
                    e.printStackTrace();
                }
                Toast.makeText(FileDemo.this,"保存的数据是: "+get,
                                                        Toast.LENGTH_LONG).show();
                break;
                default:
                break;
                }
        }
    }
}
```

注意：

(1) 本方式的文件读写，并不需要在清单文件里注册任何权限。

(2) 文件 text 为应用程序私有，保存在手机(或模拟器)的\data\data\introduction.android.fileDemo\files 文件夹里。

4.6.2 Android 系统中文件(目录)的导入/导出

在 DDMS 模式下，可以对模拟器里的文件使用工具"[图标]"和"[图标]"实现 Android 系统与 Windows 系统中文件的导出和导入。例如，工程 FileDemo 部署到模拟器运行时，会在 data/data/introduction.android.fileDemo 包的 files 文件夹里建立一个名为 text 的文本文件，选中该文件时，可以进行导出/导入操作，如图 4.6.4 所示。

注意：由于权限问题，在 DDMS 中使用 File Explorer 一般不能对手机里的 data 文件夹里的文件进行导入和导出操作，但可以对模拟器里的 data 文件夹里的文件进行导入和导出操作。

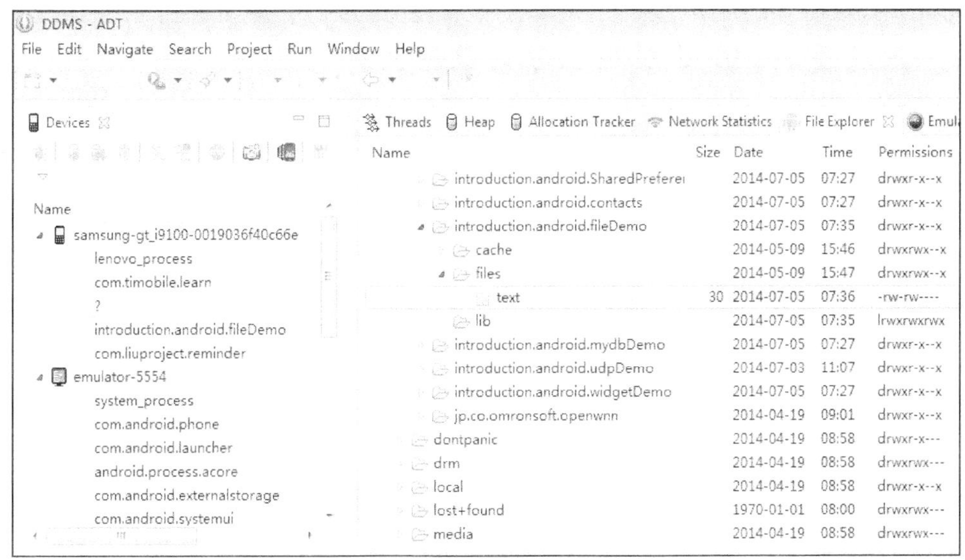

图 4.6.4　文件的导入和导出

4.7　使用 SharedPreferences 进行偏好设定

4.7.1　SharedPreferences 接口

为了方便用户操作，可以将用户名及密码保存至某个文件里，供下一次登录时使用，节省用户输入时间，这就需要使用 Android 提供的 SharedPreferences 接口。用户信息保存在工程包名下的 shared_prefs 文件夹里的一个.xml 文件里，该文件保存在 data/data/包名/shared_prefs 里，以键值对的形式存放 Activity 里的文本框等的历史输入值。

SharedPreferences 接口位于软件包 android.content 里，每个 Activity 都有一个 SharedPreferences 接口类型的对象，该对象通过使用 Activity 父类(Context)提供的方法 getSharedPreferences()得到。

SharedPreferences 接口提供了获取用户信息的方法 getString()及编辑用户信息的 edit()方法。edit()方法的返回值为 SharedPreferences 接口的内部接口类型 Editor。内部接口 Editor 提供了以键值对形式保存用户信息的方法 putString()和提交用户信息方法 commit()。

SharedPreferences 接口的内部接口及其方法，如图 4.7.1 所示。

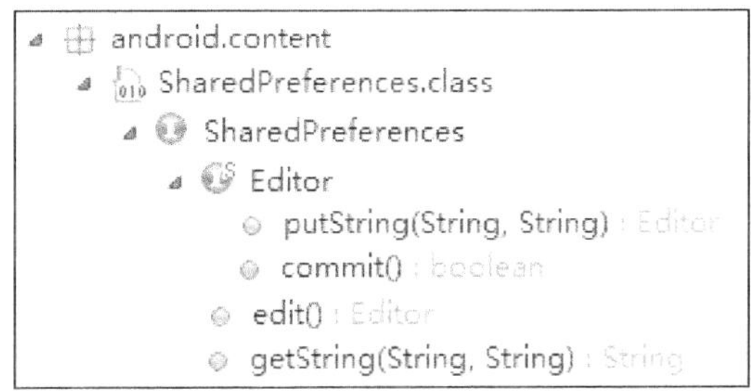

图 4.7.1　SharedPreferences 接口的内部接口及其方法

4.7.2 隶属于 Android 应用程序的数据文件

Android 应用程序默认安装至手机内存里(即手机 ROM，而不是 RAM 或 SD 卡)，项目 SharedPreferencesDemo 会生成隶属于该应用的.xml 文件，存放在系统文件夹 data/data/pn/shared_prefs 里，通过 DDMS 的 File Explorer(文件浏览器)可以查看。图 4.7.2 所示为隶属于工程的私有文件。

Name	Size	Date	Time	Permissions
introduction.android.SharedPreferencesDemo		2014-05-09	05:48	drwxr-x--x
cache		2014-04-18	21:12	drwxrwx--x
lib		2014-05-09	05:48	lrwxrwxrwx
shared_prefs		2014-04-21	07:45	drwxrwx--x
SharedPreferencesDemo.xml	139	2014-04-21	07:45	-rw-rw----

图 4.7.2 隶属于工程的私有文件

注意：Android 应用程序安装至手机(不是 AVD)，生成的数据文件由于权限不够而不能查看，除非手机已经 Root 了。

【例 4.7.1】使用 SharedPreferences 保存用户登录信息。

【程序运行】程序首次运行时输入电话号码(15527643858)和城市名称(武汉)并退出后，再次进入时便能显示上次输入的用户信息，界面如图 4.7.3 所示。

图 4.7.3 例 4.7.1 程序运行界面

【设计步骤】

(1) 新建名为 SharedPreferencesDemo 的 Android 应用工程，指定程序包名为"introduction.android.SharedPreferencesDemo"，勾选并指定主 Activity 名为 SharedPreferencesDemo.java，指定主布局文件名为 main.xml。创建后的工程文件结构(部分)如图 4.7.4 所示。

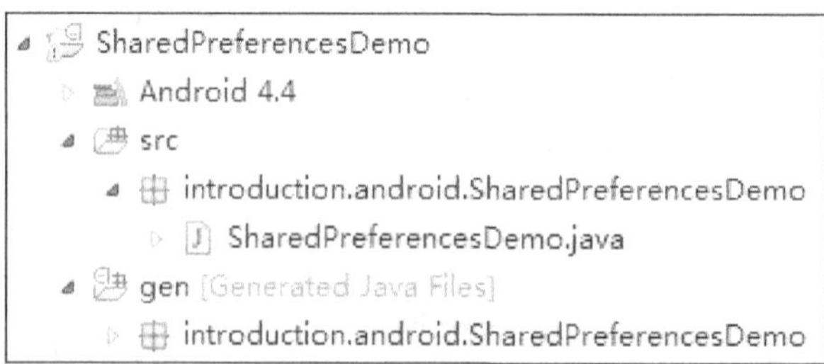

图 4.7.4 SharedPreferencesDemo 工程文件结构(部分)

(2) 主布局文件采用水平线性布局，包含三个 TextView 控件和两个 EditText 控件，代码如下：

```
<?xml version="1.0" encoding="utf-8"?>
<LinearLayout xmlns:android="http://schemas.android.com/apk/res/android"
    android:layout_width="fill_parent"
    android:layout_height="fill_parent"
    android:orientation="vertical" >
    <TextView
        android:layout_width="fill_parent"
        android:layout_height="wrap_content"
        android:text="使用Shared Preferences存储程序信息" />
    <TextView
        android:layout_width="fill_parent"
        android:layout_height="wrap_content"
        android:text="您的电话号码: "/>
    <EditText
        android:id="@+id/phone_text"
        android:layout_width="fill_parent"
        android:layout_height="wrap_content"
        android:hint="输入电话号码"/>
    <TextView
        android:layout_width="fill_parent"
        android:layout_height="wrap_content"
        android:text="您所在的城市"/>
    <EditText
        android:id="@+id/city_text"
        android:layout_width="fill_parent"
        android:layout_height="wrap_content"
        android:hint="输入城市名称"/>
</LinearLayout>
```

(3) 新建类文件 SharedPreferencesDemo.java，建立的 SharedPreferences 接口对象和获取用户信息的代码含于 onCreate()方法内，保存用户输入信息的代码含于 onStop()方法内，整个程序的源代码如下：

```
package introduction.android.SharedPreferencesDemo;
import android.app.Activity;
import android.os.Bundle;
import android.content.SharedPreferences;   //主类
import android.widget.EditText;
public class SharedPreferencesDemo extends Activity {
```

```
    private SharedPreferences sp;      //
    private EditText phoneText,cityText;
    private String phone,city;
    public static final String PHONE = "PHONE";
    public static final String CITY = "CITY";
    @Override
    public void onCreate(Bundle savedInstanceState) {
        super.onCreate(savedInstanceState);
        setContentView(R.layout.main);
        phoneText=(EditText)findViewById(R.id.phone_text);
        cityText=(EditText)findViewById(R.id.city_text);
        sp =this.getPreferences(Activity.MODE_PRIVATE);     //
        //也可以自定义文件名data.xml
        //sp=this.getSharedPreferences("data",0);
        //取出保存的电话号码和地址信息
        phone = sp.getString(PHONE, null);     //接口方法
        city = sp.getString(CITY, null);   //如果有值就使用

        //将取出的信息分别放在对应的EditText控件中
        phoneText.setText(phone);
        cityText.setText(city);
    }
    @Override
    protected void onStop() {
        sp.edit()
        .putString(PHONE,phoneText.getText().toString())
        .putString(CITY,cityText.getText().toString())
        .commit();  //提交事务
        super.onStop();  //必须
    }
}
```

4.8 意图类 android.content.Intent

Intent 的中文含义是意图、打算，用于对某个操作的抽象描述，包括动作名称、操作数据以及附加数据的描述。在 Android 中，通过 Intent 对象实现组件之间相互调用的相关信息。意图类 Intent 在 Android 程序设计中起纽带的作用，负责传递信息和数据。

Android 组件调用时，Intent 除了封装了意图名称外，还封装了一些方法，主要包括

构造方法、存放附加数据方法 putExtra()、获取数据方法 get×××Extra()和设置属性方法 set××× ()等几大类，如图 4.8.1 所示。

```
android.content
  Intent.class
    Intent
      Intent()
      Intent(Intent)
      setAction(String) : Intent
      putExtra(String, char) : Intent
      getStringExtra(String) : String
      setComponent(ComponentName) : Intent
      setClass(Context, Class<?>) : Intent
      setClassName(Context, String) : Intent
      setClassName(String, String) : Intent
      setClipData(ClipData) : void
```

图 4.8.1　Intent 类的常用方法

4.8.1　使用 Intent 对象调用系统应用程序

在前面的应用中，只涉及了一个 Activity。在实际应用中，通常会从一个窗口跳转至另一个窗口，这就需要通过 Intent 对象来传递这个调用意图。

注意：Context 是一个抽象类且为 Activity 的超类，提供了 startActivity()方法，并以 Intent 对象作为参数，用于实现窗体的跳转。

在 Android 中，调用其他应用的动作名称由众多的类提供，也可以自定义。Android 系统提供的常用意图动作名称，如表 4.8.1 所示。

表 4.8.1　Android 中常用的意图动作名称

方　法　名	功 能 描 述
Intent.ACTION_MAIN	标识 Activity 为一个程序的开始，参见清单文件中对主 Activity 的定义
Intent.ACTION_DIAL	调用系统提供的拨号程序，参见例 4.8.1
Intent.ACTION_CALL	呼叫指定的电话，参见例 5.1.1
Intent.ACTION_SENDTO	发短信、E-Mail 等，参见例 5.2.1
Intent.ACTION_VIEW	浏览网页、地图，播放多媒体等
Intent.ACTION_WEB_SEARCH	网络搜索
Intent.ACTION_BATTERY_CHANGED	检测手机电量情况
BluetoothAdapter.ACTION_REQUEST_ENABLE	蓝牙当前是否可用

Android 提供了处理组件名的类 ComponentName，其定义如图 4.8.2 所示。

【例 4.8.1】调用系统提供的拨号程序。

【程序运行】可以将系统拨号程序理解为系统预先定义的一个 Activity。当程序运行时，单击主界面中的命令按钮，即可调用系统的拨号程序；当按手机上的返回键时，就会回退到应用程序的主 Activity。程序运行界面如图 4.8.3 所示。

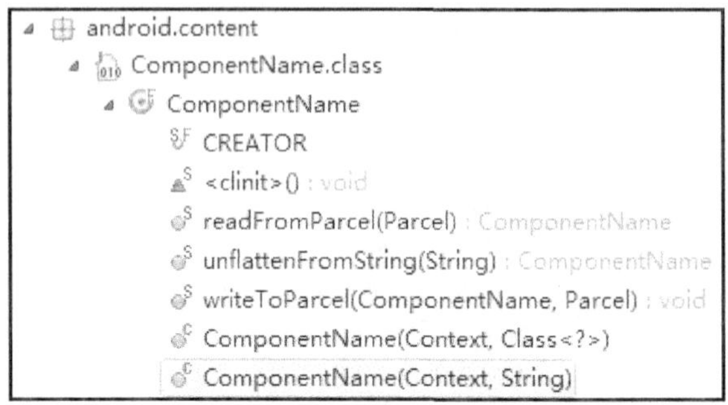

图 4.8.2　组件名类 ComponentName 的定义

图 4.8.3　简单的 Activity 调用与返回程序运行界面

【设计步骤】

(1) 新建名为 Dialing 的 Android 应用工程，勾选创建 Activity。

(2) 修改默认的布局文件。将文本框改写为 Button，并设置其 android:text 属性值为“使用系统提供的拨号程序”，添加属性 android:onClick="dialing"。

(3) 编写程序 MainActivity.java，文件代码如下：

```
ipackage introduce.android.dialing;
import android.os.Bundle;
import android.app.Activity;
import android.content.Intent;   //意图类
import android.view.View;
import android.net.Uri;
public class MainActivity extends Activity {
    @Override
    protected void onCreate(Bundle savedInstanceState) {
        super.onCreate(savedInstanceState);
        setContentView(R.layout.activity_main);
    }
```

```
    public void dialing(View view){        //命令按钮单击事件的响应方法
        Intent intent=new Intent(Intent.ACTION_DIAL); //拨号意图，创建实例
        //Intent intent =new Intent(Intent.ACTION_DIAL,Uri.parse("tel:10086"));
        startActivity(intent);   //调用系统自带的拨号程序
    }
}
```

【例 4.8.2】显示手机的所有应用程序列表。

【程序运行】程序运行界面如图 4.8.4 所示。

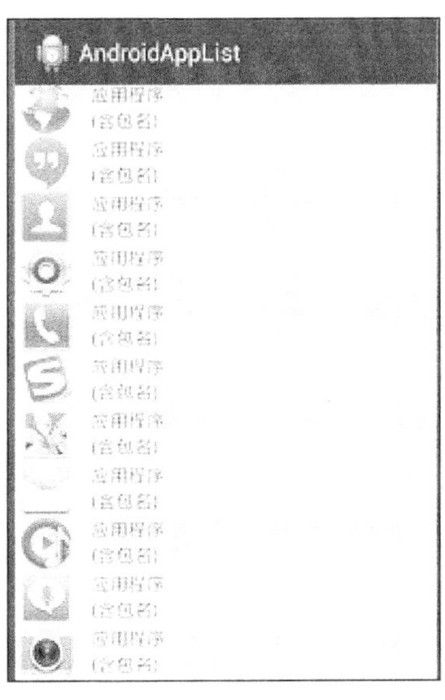

图 4.8.4　显示系统自带的应用程序列表程序运行界面

【设计步骤】

(1) 新建名为 AndroidAppList 的 Android 应用工程，勾选创建 Activity，指定主 Activity 对应的布局文件为 browse_app_list.xml，新建主 Activity 中 ListView 控件对应的布局文件为 xmlbrowse_app_item.xml，工程文件结构(主要部分)如图 4.8.5 所示。

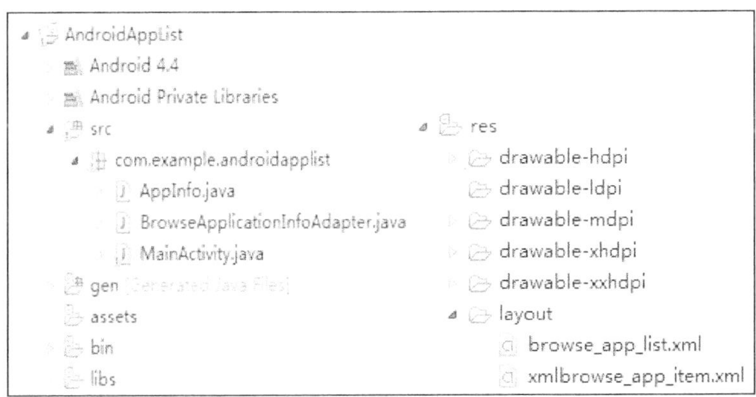

图 4.8.5　AndroidAppList 工程文件结构(主要部分)

(2) 编写程序 MainActivity.java，文件代码如下：

```
package com.example.androidapplist;
/*
 * 显示手机上已安装的应用程序的相关信息(图标、应用程序名及包名)
 * ListView控件显示
 * AppInfo.java是自定义的应用程序实体类
 * BrowseApplicationInfoAdapter.java是继承BaseAdapter的数据适配器类
 * 单击程序名可执行相应的程序
 */
import android.app.Activity;
import android.os.Bundle;
import android.content.ComponentName;   //
import android.content.Intent;
import android.content.pm.PackageManager;   //主类
import android.content.pm.ResolveInfo;    //
import android.graphics.drawable.Drawable;
import android.view.View;
import android.widget.AdapterView;
import android.widget.AdapterView.OnItemClickListener;
import android.widget.ListView;
import android.widget.Toast;
import java.util.ArrayList;
import java.util.Collections;
import java.util.List;
public class MainActivity extends Activity implements OnItemClickListener {
    private ListView listview = null;
    private List<AppInfo> mlistAppInfo = null;
    @Override
    public void onCreate(Bundle savedInstanceState) {
        super.onCreate(savedInstanceState);
        setContentView(R.layout.browse_app_list);
        listview = (ListView) findViewById(R.id.listviewApp);
        mlistAppInfo = new ArrayList<AppInfo>();
        queryAppInfo(); // 查询所有应用程序信息
        BrowseApplicationInfoAdapter browseAppAdapter = new
                                    BrowseApplicationInfoAdapter(this, mlistAppInfo);
        listview.setAdapter(browseAppAdapter);
        listview.setOnItemClickListener(this);
    }
```

```
// 单击跳转至该应用程序
public void onItemClick(AdapterView<?> arg0, View view, int position, long arg3) {
    // TODO Auto-generated method stub
    Intent intent = mlistAppInfo.get(position).getIntent();
    startActivity(intent);
}

// 获得所有启动Activity的信息，类似于Launch界面
public void queryAppInfo() {
    PackageManager pm = this.getPackageManager(); //获得包管理对象
    Intent mainIntent = new Intent(Intent.ACTION_MAIN, null);
    mainIntent.addCategory(Intent.CATEGORY_LAUNCHER);
    // 通过查询，获得所有ResolveInfo对象
    List<ResolveInfo> resolveInfos = pm.queryIntentActivities(mainIntent,
                                          PackageManager.MATCH_DEFAULT_ONLY);
    // 根据应用程序名排序
    Collections.sort(resolveInfos,new ResolveInfo.DisplayNameComparator(pm));
    if (mlistAppInfo != null) {
        mlistAppInfo.clear();
        for (ResolveInfo reInfo : resolveInfos) {
            // 获得该应用程序的启动Activity的name
            String activityName = reInfo.activityInfo.name;
            // 获得应用程序的包名
            String pkgName = reInfo.activityInfo.packageName;
            // 获得应用程序的Label
            String appLabel = (String) reInfo.loadLabel(pm);
            // 获得应用程序图标
            Drawable icon = reInfo.loadIcon(pm);
            // 为应用程序的启动Activity 准备Intent
            Intent intent = new Intent();
            intent.setComponent(new ComponentName(pkgName,
                                                    activityName));
            // 创建一个AppInfo对象，并赋值
            AppInfo appInfo = new AppInfo();
            appInfo.setAppLabel(appLabel+activityName);
            appInfo.setPkgName(pkgName);
            appInfo.setAppIcon(icon);
            appInfo.setIntent(launchIntent);
            mlistAppInfo.add(appInfo); // 添加至列表中
            System.out.println(appLabel + " activityName---" + activityName
```

```
                    + " pkgName---" + pkgName);
            //Toast.makeText(MainActivity.this,activityName,1000).show();
        }
    }
  }
}
```

4.8.2 使用 Intent 显式调用自定义的 Activity 组件

Intent 显式调用是通过使用显式意图明确要激活的组件，其方法是通过 Intent 提供的方法 setClass()、setClassName()或 setComponent()来指定的。

注意：Activity 调用时，若存在数据传递，则需要使用类 android.os.Bundle 捆绑数据，并通过 Intent 提供的 putExtra()方法传递。最后，在目录 Activity 里通过 Intent 的 getBundleExtra()方法获取 Bundle 对象。

【例 4.8.3】带数据传递的 Activity 显式调用。

【程序运行】在主 Activity 中输入用户名和密码后，单击“登录”按钮，其登录信息在另一个 Activity 中显示，程序运行界面如图 4.8.6 所示。

图 4.8.6　例 4.8.3 程序运行界面

【设计步骤】

(1) 新建名为 CallActivity 的 Android 应用工程，勾选创建 Activity，使用默认的布局文件。

(2) 新建表示被调用的 Activity 文件 ShowActivity.java，指定其布局文件为 activity_show.xml，工程文件结构(主要部分)如图 4.8.7 所示。

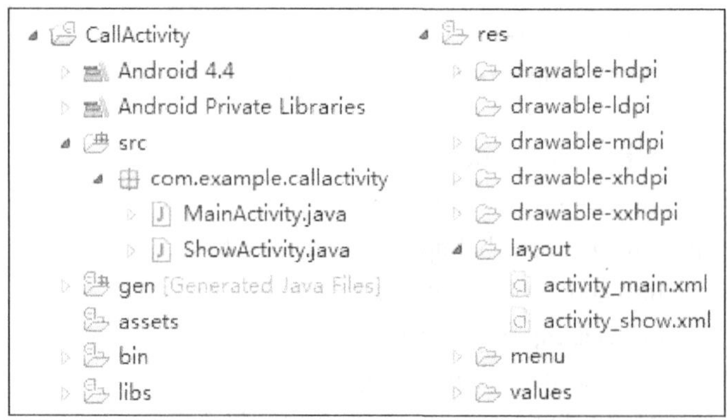

图 4.8.7　CallActivity 工程文件结构(主要部分)

(3) 主 Activity 文件 MainActivity.java 的代码如下：

```
package com.example.callactivity;
/*
 * 本程序演示了带数据传递的Activity显式调用
 * 通过Bundle实现数据的捆绑，且使用键值对形式
 */
import android.app.Activity;
import android.os.Bundle;   //
import android.content.Intent;
import android.widget.EditText;
import android.widget.Button;
import android.view.View;
import android.content.ComponentName;
public class MainActivity extends Activity {
    private Button mybutton ;
    private EditText myet1,myet2;
    @Override
    protected void onCreate(Bundle savedInstanceState) {
        super.onCreate(savedInstanceState);
        setContentView(R.layout.activity_main);
        mybutton = (Button)findViewById(R.id.logButton);
        myet1 = (EditText)findViewById(R.id.edit1);
        myet2 = (EditText)findViewById(R.id.edit2);
        mybutton.setOnClickListener(new View.OnClickListener() {
            @Override
            public void onClick(View arg0) {
                // TODO Auto-generated method stub
                //显式意图的多种表达形式
                //Intent intent = newIntent(MainActivity.this,ShowActivity.class);
                //Intent intent = new Intent(MainActivity.this,ShowActivity.class);
                Intent intent=new Intent();
                //intent.setClass(MainActivity.this,ShowActivity.class);
                intent.setComponent(new ComponentName(
                                        MainActivity.this,ShowActivity.class));
                Bundle b = new Bundle();    //准备捆绑数据
                b.putString("name", myet1.getText().toString());
                b.putString("password", myet2.getText().toString());
                intent.putExtra("data", b);
                startActivity(intent);   //显式调用Activity
            }
```

```
        });
    }
}
```

(4) 被调用 Activity 文件 ShowActivity.java 的代码如下：

```
package com.example.callactivity;
/*
 * 被调用的Activity文件
*/
import android.app.Activity;
import android.os.Bundle;
import android.content.Intent;
import android.widget.TextView;
public class ShowActivity extends Activity {
    private TextView tv;
    private Intent intent;
    private Bundle b1 = null;
    @Override
    protected void onCreate(Bundle savedInstanceState) {
        super.onCreate(savedInstanceState);
        setContentView(R.layout.activity_show);
        tv = (TextView)findViewById(R.id.et);
        intent = getIntent();
        b1 = intent.getBundleExtra("data");
        if(b1 != null){
            tv.setText("用户名: "+b1.getString("name")+
                                    " 密码: "+b1.getString("password"));
        }
    }
}
```

注意：有返回数据的 Activity 调用，参见例 5.5.1。

4.8.3 使用 Intent 隐式调用 Activity 组件

对 Activity 的调用可以分为显式调用和隐式调用，隐式调用需要在清单文件中使用标签<intent-filter>(称之为意图过滤器)来指定其动作名称，这不同于显式调用。

隐式意图就是在意图激活 Activity、Service 或 BroadcastReceiver 这三类组件时，不需要显式指出组件的名称，而是指定 action 及 category，Android 系统会根据其特征找到相应的组件并激活。

注意：组件 Service 和 BroadcastReceiver 的使用，将在第 6 章介绍。

【例 4.8.4】隐式意图的使用。

【程序运行】单击主 Activity 中的命令按钮，即可调用另一个名为 SecondActivity 的程序，程序运行界面如图 4.8.8 所示。

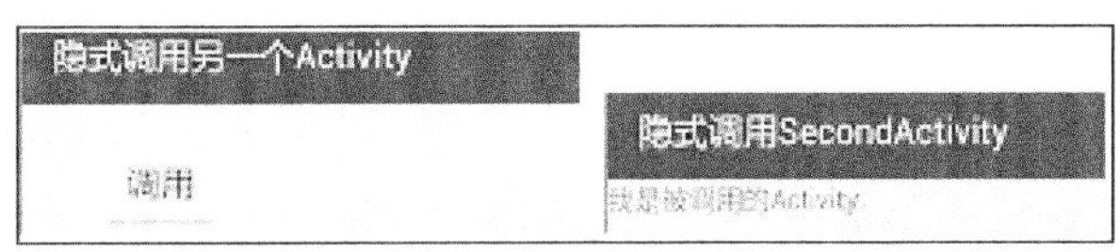

图 4.8.8 Activity 的隐式调用程序运行界面

【设计步骤】

(1) 新建名为 ImplicitIntent 的 Android 应用工程，勾选创建 Activity。

(2) 修改默认的布局文件 activity_main.xml，增加一个 Button 按钮。

(3) 新建名为 second.xml 的布局文件，增加一个 TextView 控件。

(4) 新建名为 SecondActivity.java 且继承 Activity 的类文件，该 Activity 使用 second.xml 布局文件。工程文件结构如图 4.8.9 所示。

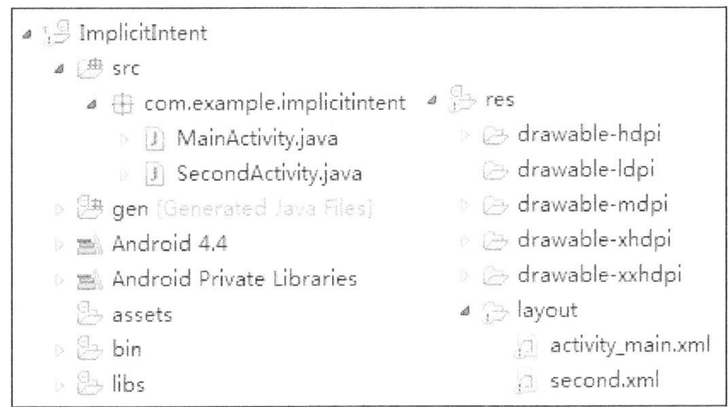

图 4.8.9 ImplicitIntent 工程文件结构

(5) 在清单文件里，增加对 SecondActivity 的配置，其代码如下：

```
<activity
    android:name="com.example.implicitintent.SecondActivity"
    android:label="隐式调用SecondActivity" >
    <intent-filter>
        <action android:name="android.intent.action.yydy" />
        <category android:name="android.intent.category.DEFAULT"/>
    </intent-filter>
</activity>
```

(6) 编写程序 MainActivity.java，文件代码如下：

```
package com.example.implicitintent;
import android.os.Bundle;
```

```
import android.app.Activity;
import android.view.View;
import android.content.Intent;
public class MainActivity extends Activity{
    @Override
    protected void onCreate(Bundle savedInstanceState) {
        super.onCreate(savedInstanceState);
        setContentView(R.layout.activity_main);
    }
    public void dy(View v){
            Intent intent=new Intent();
            //设置已在清单文件中配置的动作名称
            intent.setAction("android.intent.action.yydy" );    //
            startActivity(intent);    //
        }
}
```

(7) 编写程序 SecondActivity.java，文件代码如下：

```
package com.example.implicitintent;
import android.app.Activity;
import android.os.Bundle;
public class SecondActivity extends Activity {
    @Override
    protected void onCreate(Bundle savedInstanceState) {
        super.onCreate(savedInstanceState);
        setContentView(R.layout.second);
    }
}
```

4.8.4 延期意图类 android.app.PendingIntent

Intent 一般用于 Activity、Service、BroadcastReceiver 之间的数据传递，而 PendingIntent 一般用在 Notification 上。可以理解为，PendingIntent 为延期执行的 Intent，是对 Intent 的一个包装。延期意图类的定义，如图 4.8.10 所示。

类 Notification 中使用 PendingIntent 类型参数的方法，如图 4.8.11 所示。

【例 4.8.5】延期意图类 PendingIntent 的使用。

【程序运行】当程序运行时，单击命令按钮，将会产生一个通知。展开通知栏，单击通知内容，就会打开另一个 Activity。程序运行界面如图 4.8.12 所示。

```
◢ android.app
    ◢ PendingIntent.class
        ◢ PendingIntent
            getActivities(Context, int, Intent[], int) : PendingIntent
            getActivities(Context, int, Intent[], int, Bundle) : PendingIntent
            getActivity(Context, int, Intent, int) : PendingIntent
            getActivity(Context, int, Intent, int, Bundle) : PendingIntent
            getBroadcast(Context, int, Intent, int) : PendingIntent
            getService(Context, int, Intent, int) : PendingIntent
            readPendingIntentOrNullFromParcel(Parcel) : PendingIntent
```

图 4.8.10　延期意图类的定义

```
◢ android.app
    ◢ Notification.class
        ◢ Notification
            setLatestEventInfo(Context, CharSequence, CharSequence, PendingIntent) : void
```

图 4.8.11　Notification 类中关于 PendingIntent 类的方法

图 4.8.12　延期意图类使用效果

【设计步骤】

(1) 新建名为 PendingIntent 的 Android 应用工程，勾选创建 Activity，工程文件结构(主要部分)如图 4.8.13 所示。

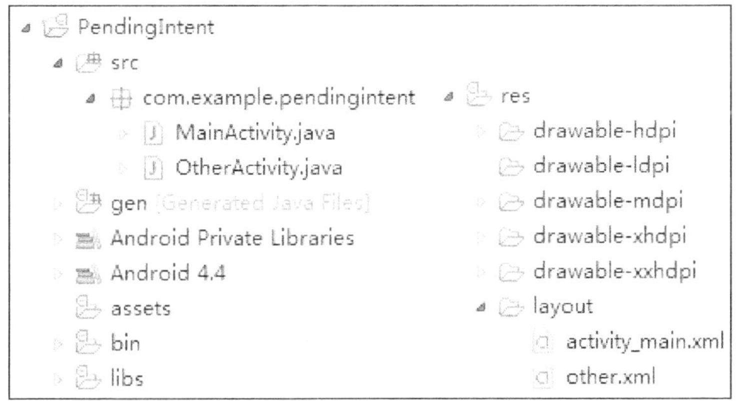

图 4.8.13　PendingIntent 工程文件结构(主要部分)

(2) 修改默认的布局文件。将文本框改写为 Button，并设置其 android:text 属性值为“产生一个延期意图”。

(3) 编写主 Activity 程序文件 MainActivity.java，其代码如下：

```
package com.example.pendingintent;
/*
 * 本程序演示延期意图的使用，包含了通知的使用和Activity的显式调用
 * 要求Android API版本不低于11
 */
import android.app.Activity;
import android.os.Bundle;import android.app.NotificationManager;
import android.app.Notification; import android.content.Intent;
import android.app.PendingIntent;   //import android.content.Context;
import android.view.View;
import android.widget.Button;
public class MainActivity extends Activity {
    NotificationManager nm;
    Notification notification;
    private static final int NOTIFICATION_ID = 1;
    private Button notifyBtn;
    @Override
    public void onCreate(Bundle savedInstanceState) {
        super.onCreate(savedInstanceState);
        setContentView(R.layout.activity_main);
        notifyBtn=(Button)this.findViewById(R.id.button1);
        notifyBtn.setOnClickListener(new View.OnClickListener() {
        @SuppressWarnings("deprecation")
        @Override
        public void onClick(View v) {
                // TODO Auto-generated method stub
                String ns = Context.NOTIFICATION_SERVICE;
                nm= (NotificationManager) getSystemService(ns);
                int icon = R.drawable.icon01;
                long when = System.currentTimeMillis();
                Notification.Builder builder=new Notification.Builder(MainActivity.this);
                builder.setSmallIcon(icon)
                        .setWhen(when)
                        .setDefaults(Notification.DEFAULT_SOUND);
```

```
                notification=builder.build();
                Intent intent = new Intent(MainActivity.this, OtherActivity.class);    //
                PendingIntent p_intent = PendingIntent.getActivity(MainActivity.this, 0,
                                                                        intent, 0);    //
                String temp="单击我后将打开一个新的Activity";
                notification.setLatestEventInfo(MainActivity.this,
                                        "我是通知的标题",temp ,p_intent);    //
                nm.notify(NOTIFICATION_ID, notification);
            }
        });
    }
}
```

4.9　注册应用程序所需要的权限

在程序中使用手机的打电话、发送短信等功能，或者读取手机联系人信息、读取手机 WiFi 状态信息等，都需要在工程的清单文件 AndroidManifest.xml 中注册该应用程序所需要的权限。

例如，拨通某个电话，就要求该应用程序具有使用手机的打电话权限，需要在工程的清单文件中使用<uses-permission>标签进行如下注册：

```
<uses-permission android:name="android.permission.CALL_PHONE"/>
```

在 android.jar 文件中的内部类 android.Manifest.permission 中，定义了所有的 Android 权限(共 106 个)，如图 4.9.1 所示。

Android 4.4
android.jar
android
Manifest.class
Manifest
permission
ACCESS_CHECKIN_PROPERTIES
ACCESS_COARSE_LOCATION
ACCESS_FINE_LOCATION
ACCESS_LOCATION_EXTRA_COMMANDS
ACCESS_MOCK_LOCATION
ACCESS_NETWORK_STATE
……

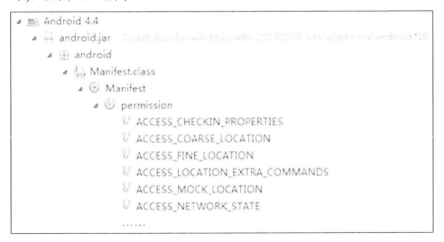

图 4.9.1　Manifest 类及其内部类 permission

在 Android 应用中，应用程序常用的权限如表 4.9.1 所示。

表 4.9.1 Android 常用的手机用户权限

方　法　名	功 能 描 述
CALL_PHONE	允许使用手机的拨打电话功能
SEND_SMS	允许使用手机的发送短信功能
READ_CONTACTS	允许读取手机的联系人信息
INTERNET	允许手机访问 Internet 网络
ACCESS_NETWORK_STATE	允许访问手机的网络连接状态
WRITE_EXTERNAL_STORAGE	允许在 SD 卡中创建文件(夹)
CAMERA	使用摄像头
RECORD_AUDIO	使用录音机
ACCESS_COARSE_LOCATION	使用 GPRS 基站定位或 WiFi 网络定位，参见第 10.2.1 小节
ACCESS_FINE_LOCATION	使用 GPS 定位，参见第 10.2.1 小节

注意：当为其他应用程序提供代码或者数据时，需要在本应用中使用标签<permission>自定义权限。使用标签<permission>还可以定义权限组，不同于标签<uses-permission>。

习　题　4

一、判断题

1. 在 Android 应用中，界面布局有图形和代码两种设计方式。
2. 所有界面元素都是 View 类的子类。
3. 在 Android 中，R.java 只是工程的 res 文件下的资源的索引文件。
4. 在 Android 中，资源文件所在的包名总是与工程的包名一致的。
5. Activity 是一个用来实现与用户交互的 View 容器。
6. Context 是 Activity 的直接父类。
7. 手机通知栏的通知(Notification)，用户只能查看，不能编辑。
8. AlertDialog 和 ProgressDialog 都是 Dialog 的直接子类。
9. AlertDialog 位于软件包 android.widget 内。
10. 普通的 Android 应用程序建立的数据文件存放在手机内存的 dada/data 文件里。
11. 布局标签不能嵌套。
12. 意图类 Intent 含于软件包 android.app 里。
13. 线性布局就是控件对象的水平排列。
14. 任何一个 Activity 都可以实现内部接口 android.view.View.OnClickListener。
15. 隐式调用 Activity 时，必须在清单文件中对被调用的 Activity 使用意图过滤器标签 <intent-filter>。

二、选择题

1. 下列 Android 事件中，其回调函数没有返回值的是____。

   A. OnClick　　B. OnKey　　C. OnTouch　　D. OnFocusChange

2. 在 Eclipse 中使用 Android 4.4 创建 Android 应用工程时，默认使用的布局类型是____。

   A. 相对布局　　B. 帧布局　　C. 表格布局　　D. 绝对布局

3. 在 Android 布局文件中，新建一个资源共享 id 的方法是____。

   A. android:id="id/name"　　B. android:id="@id/name"

   C. android:id="@+id/name"　　D. android:id="@id+/name"

4. 要使线性布局中的 TextView 控件内的文本在水平和垂直两个方向上居中，应设置其属性为____。

   A. android:gravity="center_vertical|center_horizontal"

   B. android:gravity="center "

   C. android: layout _gravity =" center_vertical "

   D. A 或 B

5. 下列选项中，不能使用 UI 的图形化设计方式的是____。

   A. TextView　　B. EditText　　C. Toast　　D. ImageView

6. 手机自带的联系人程序，其 UI 设计时使用的主要控件是____。

A. ListView　　B. TextView　　C. EditText　　D. Spinner

7. 下列方法中，不是 Activity 和 Context 共同具有的是____。

A. startService()　　B. stopService()

C. findViewById()　　D. getSystemService()

8. 关于 Activity 和 Service 两个组件，下列说法中不正确的是____。

A. 通常 Activity 有用户界面，而 Service 在后台运行

B. 使用 Activity 和 Service 前，都必须在清单文件中注册

C. 后台服务程序都继承于抽象类 android.app.Service

D. 都是通过 finish()方法关闭

9. 通过 Notification 的内部类 Builder 设置通知到来时提示音的方法是____。

A. setContentTitle()　　B. setSmallIcon()

C. setWhen()　　D. setDefault()

10. 构造方法 ArrayAdapter(Context,int,int,List<T>)中的第二参数表示____。

A. 上下文　　B. 列表控件的列表项对应的布局文件

C. 列表项控件　　D. 列表项数据

三、填空题

1. 设置垂直或水平线性布局，需要在 LinearLayout 中使用的属性是____。
2. 设置文本框控件中内容的对齐方式，应使用的属性是____。
3. 在布局文件中，设置控件大小通常使用 dp 作为单位，而文字大小单位是____。
4. 对于处于运行态的 Activity，当用户按返回键退出时，将调用方法____。
5. 对 EditText 控件应用____属性后，默认的提示文本在控件获得焦点后自动消失。
6. 使用以前输入过的历史信息，需要使用包 android.content 内的____接口。
7. 意图类 Intent 含于 Android 的____软件包内。
8. 使用 SharedPreferences 建立的文件，其格式为____。
9. 调整线性布局内的各个控件之间的间距，应使用 android: ____属性。
10. 设置控件的单击事件监听器，需要实现的方法是____。
11. 获取使用 ArrayAdapter 作为适配器的列表的选择项，需要该类提供的____方法。

实验 4(A)　Android 应用开发基础(一)

一、实验目的

1. 掌握 Activity 组件作为应用程序的运行窗口。
2. 掌握常用的布局方法。
3. 掌握界面元素的常用属性。
4. 掌握 Button(或 TextView)的单击事件监听器的设计方法。
5. 掌握对 ListView 控件应用数据适配器的设计方法(含列表项的单击事件监听器)。
6. 掌握三种消息提醒方式(Toast、AlertDialog 和 Notification)的用法区别。
7. 掌握应用程序私有的文本文件的读写方法。
8. 掌握使用接口 SharedPreferences 保存用户登录信息的方法。
9. 掌握通过 Intent 调用 Activity 组件的两种用法(显式与隐式)的区别。

二、实验内容及步骤

【预备】访问 http://www.wustwzx.com/ android/index.html，单击“实验 4(A)”超链接，下载本次实验内容的源代码压缩文件并解压(得到文件夹 ch04a)。

1. 在 Android 集成开发环境中，通过双击类名的方法查看 UI 控件之间的继承关系。

(1) View 是 TextView 的直接基类。

(2) TextView 是 Button 的直接基类。

(3) Button 是 CompoundButton 的直接基类。

(4) CompoundButton 是 CheckBox 的直接基类。

(5) CompoundButton 也是 EditText 的直接基类。

(6) TextView 又是 EditText 的直接基类。

(7) Dialog 是 AlertDialog 的直接基类。

(8) AlertDialog 是 ProgressDialog 的直接基类。

2. 掌握各种布局的特点。

(1) 线性布局需要指定方向属性 android:orientation，取值为 horizontal 或 vertical。

(2) 相对布局需要指定相邻控件之间的位置关系，其布局内的控件对象通常使用诸如 layout_ParentLeft(是否与父元素左对齐)、layout_below(放置到指定的控件对象下面)之类的属性。

(3) 表格布局可以通过线性布局的嵌套实现。

(4) 使用帧布局时，控件对象默认固定在屏幕的左上角，通过 layout_gravity 属性可适当修改其位置，后放入的控件对象会在前一控件对象上进行覆盖而形成遮挡。

3. 一个简单的用户登录系统(参见例 4.3.1)。

(1) 再次解压 ch04a 文件夹里的压缩文件 UserLogin.zip。

(2) 导入名为 UserLogin 的 Android 应用工程。

(3) 查看主 Activity 对应的布局文件 res/layout/activity_main.xml，掌握垂直线性布局、水平线性布局以及它们的嵌套用法。

(4) 查看命令按钮单击事件的处理方式(未使用监听器但实现了监听器的功能)：在布局文件中使用了 onClick 属性，指定了单击事件的处理方法名称。

(5) 部署工程到手机并做运行测试。

(6) 对按钮使用单击事件监听器的方式改写源程序后再部署工程并做运行测试。

4. Android 对话框控件和进度控件的使用(参见例 4.3.2)。

(1) 再次解压 ch04a 文件夹里的压缩文件 Dialog.zip。

(2) 导入工程名为 Dialog 的 Android 应用工程。

(3) 打开源程序文件 MainActivity.java。

(4) 查看显示一个简单的提醒对话框的代码，见方法 click1a()。

(5) 查看显示一个确认对话框的代码，见方法 click1b()。

(6) 查看单选对话框的代码，见方法 click2()。

(7) 查看多选对话框的代码，见方法 click3()。

(8) 查看进度对话框的代码，见方法 click4()。

(9) 查看水平进度条对话框的代码，见方法 click5()，验证两种创建子线程方法的等效性。

5. 对 ListView 控件应用 ArrayAdapter 数据适配器(参见例 4.3.4)。

(1) 再次解压 ch04a 文件夹里的压缩文件 ListView_ArrayAdapter.zip。

(2) 导入工程名为 ListView_ArrayAdapter 的 Android 应用工程。

(3) 打开源文件 MainActivity.java，通过按“Ctrl+T”键查看类 ArrayList 和 ArrayAdapter 的继承关系。

(4) 查看 ArrayAdapter 的构造方法中各个参数的含义。

(5) 查看列表控件 ListView 的列表项选择监听器 OnItemClickListener 的实现代码。

(6) 部署工程到手机运行时，单击某个列表项进行测试。

6. 通知 Notification(含 Toast)的设计与使用(参见例 4.5.1)。

(1) 再次解压 ch04a 文件夹里的压缩文件 Notification_Toast.zip。

(2) 导入工程名为 Notification_Toast 的 Android 应用工程。

(3) 打开源文件 MainActivity.java，查看创建 Notification 对象的代码，验证当去掉类型强转时源程序报错。

(4) 查看 Notification 内部类 Builder(构造器)的五个设置方法和一个 build()方法。

(5) 掌握 NotificationManager 提供的 notify()方法的作用。

(6) 部署工程到手机做运行测试(注意：在清单文件中，已经指定 Android SDK 最低版本为 16，因此部署到手机上时，要求手机 Android 版本不低于 4.1.2。否则，部署时找不到手机)。

7. 掌握手机文本文件读写的设计方法(参见例 4.6.1)。

(1) 再次解压 ch04a 文件夹里的压缩文件 FileDemo.zip。

(2) 导入工程名为 FileDemo 的 Android 应用工程。

(3) 打开 FileDemo.java 文件，查看读取文本文件并以 Toast 方式显示的代码。

(4) 查看将输入的文本以追回方式写入到一个文本文件里的代码。

8. Activity 的显式调用：调用系统提供的 Activity(参见例 4.8.1)。

(1) 再次解压 ch04a 文件夹里的压缩文件 Dialing.zip。

(2) 导入工程名为 Dialing 的 Android 应用工程。

(3) 查看 MainActivity.java 文件可知，本例是通过意图对象调用系统自带的拨号程序，在清单文件中并不需要配置拨打电话的权限(与工程 Dialing2 不同，参见例 5.1.1)。

(4) 部署工程到手机做运行测试。

(5) 在拨号界面，当按手机上的返回键时，就会回退到应用程序的主 Activity 界面。

9. 带数据传递的 Activity 的显式调用：调用自定义的 Activity(参见例 4.8.3)。

(1) 再次解压 ch04a 文件夹里的压缩文件 CallActivity.zip。

(2) 导入工程名为 CallActivity 的 Android 应用工程。

(3) 查看 MainActivity.java 文件中捆绑数据时涉及的对象(Bundle 和 Intent)及其方法。

(4) 查看 ShowActivity.java 文件中接收数据时涉及的对象(Intent)及其方法。

(5) 部署工程到手机做运行测试。

三、实验小结及思考

(由学生填写，重点写上机中遇到的问题。)

实验 4(B)　Android 应用开发基础(二)

一、实验目的

1. 掌握日期和时间选择器控件的使用。
2. 掌握自动完成文本控件的使用。
3. 掌握三种类型的菜单设计方法。
4. 掌握对列表控件 ListView 应用数据适配器 BaseAdapter 的方法。
5. 掌握延期意图类 PendingIntent 的使用方法。
6. 掌握动态加载布局至对话框的设计方法。

二、实验内容及步骤

【预备】访问 http://www.wustwzx.com/android/index.html，单击“实验 4(B)”超链接，下载本次实验内容的源代码压缩文件并解压(得到文件夹 ch04b)。

1. 实用控件(日期与时间选择器、自动完成文本)的使用(参见例 4.4.1 和例 4.4.2)。

(1) 再次解压 ch04b 文件夹里的压缩文件 DataTimePicker.zip。

(2) 导入工程名为 DataTimePicker 的 Android 应用工程。

(3) 打开源程序 MainActivity.java，查看通过控件 DataPicker 和 TimePicker 分别获取日期与时间的代码。

(4) 部署工程到手机(或 AVD)做运行测试。

(5) 再次解压 ch04b 文件夹里的压缩文件 AutoCompleteText.zip。

(6) 导入工程名为 AutoCompleteText 的 Android 应用工程。

(7) 打开主 Activity 对应的布局文件，查看 AutoCompleteText 控件的关键属性 completionThreshold 属性(匹配字符的个数)。

(8) 查看源文件 MainActivity.java 中设置 AutoCompleteText 控件的数据适配器的代码。

(9) 部署工程到手机(或 AVD)并做运行测试。

2. 手机菜单设计(参见例 4.4.3)。

(1) 再次解压 ch04b 文件夹里的压缩文件 MenuDemo.zip。

(2) 导入工程名为 MenuDemo 的 Android 应用工程。

(3) 打开工程中菜单文本文件 res/menu/mymenu.xml，查看其内容。

(4) 打开主 Activity 文件 MenusDemoActivity.java，查看创建菜单(含子菜单)的代码。

(5) 查看 MenusDemoActivity.java 中创建上下文菜单的代码。

(6) 部署工程后，使用菜单键做菜单(含子菜单)测试。

(7) 部署工程后，通过长按做上下文菜单测试(长按 TextView 控件对象复制其内容，长按 EditText 控件对象粘贴刚才复制的内容)。

3. 显示手机已安装的应用程序列表(参见例 4.8.2)。

(1) 再次解压 ch04b 文件夹里的压缩文件 AndroidAppList.zip。

(2) 导入工程名为 AndroidAppList 的 Android 应用工程。

(3) 打开封装应用程序相关信息的实体类源文件 AppInfo.java，查看其代码。

(4) 打开继承了 BaseAdapter 的工具类源文件 BrowseApplicationInfoAdapter.java，查看列表项单击事件监听器的代码。

(5) 打开源文件 MainActivity.java，查看获取手机应用程序相关信息(如包名等)的代码。

(6) 部署工程做运行测试。

4. 延期意图类的使用(参见例 4.8.5)。

(1) 再次解压 ch04b 文件夹里的压缩文件 PendingIntent.zip。

(2) 导入工程名为 PendingIntent 的 Android 应用工程。

(3) 查看源程序文件 MainActivity.java 中产生通知的代码。

(4) 查看产生通知的代码内产生延期意图对象的代码。

(5) 部署工程并做运行测试。

5. 带偏好设置的用户登录系统，运行效果如图 4-实-1 所示。

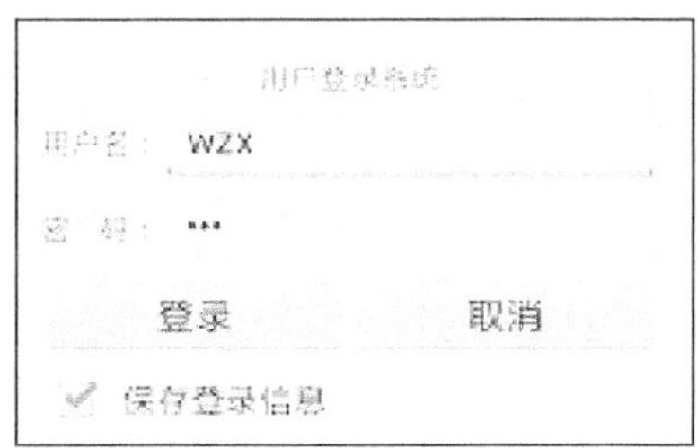

图 4-实-1　带偏好设置的用户登录界面

(1) 新建名为 UserLogin2 的 Android 应用工程。

(2) 在例 4.3.1 的布局文件的基础上增加一个复选框。

(3) 将登录信息作为偏好保存，参见例 4.7.1。

(4) 在程序中增加对复选框的监听器 OnCheckedChangeListener。

(5) 参考文件夹 ch04a 下 UserLogin2 工程里的相关源代码。

6. 在单击命令按钮后动态加载布局至对话框(参见例 4.3.3)。

(1) 再次解压 ch04b 文件夹里的压缩文件 DynamicLayout.zip。

(2) 导入工程名为 DynamicLayout 的 Android 应用工程。

(3) 打开源文件 MainActivity.java，查看动态加载布局文件的代码。

(3) 部署工程并做运行测试。

(4) 修改源程序：使用匿名方式设置命令按钮的监听器，重新部署工程并做运行测试。

三、实验小结及思考

(由学生填写，重点写上机中遇到的问题。)

第5章 手机基本功能程序设计

如今，手机不再是仅仅作为打电话和发短信的工具。事实上，视频(音频)播放与录制、二维码扫描等功能也深受人们欢迎。本章主要介绍了手机常用功能的 Android 程序设计，其学习要点如下：

- 掌握打电话程序的设计方法；
- 掌握收发短信程序的设计方法；
- 掌握(前台)媒体播放程序的设计方法；
- 掌握录音程序和视频拍摄程序的设计方法；
- 掌握手机拍照程序和视频拍摄程序的设计方法；
- 掌握二维码扫描及生成程序的设计方法。

5.1 打 电 话

5.1.1 抽象类 android.net.Uri 及其静态方法 parse()

URI(universal resource identifier)是通用资源标识符，一般由三部分组成：

- 访问资源的命名机制；
- 存放资源的主机名；
- 资源自身的名称，由路径表示。

典型情况下，这种字符串以 scheme(命名 URI 的名字空间的标识符——一组相关的名称)开头，语法如下：

[scheme:] scheme-specific-part

URI 以 scheme 和冒号开头。scheme 使用字母开头，后面为字母、数字、加号、减号和点号。冒号把 scheme 与 scheme-specific-part 分开了，并且 scheme-specific-part 的语法和语义(意思)由 URI 的名字空间决定。

例如，在“http://www.cnn.com”中，“http”就是 scheme，“//www.cnn.com”是 scheme-specific-part，并且它的 scheme 与 scheme-specific-part 被冒号分隔开。

在 android.net 包中，提供了表示 URI 的抽象类 Uri 及其静态方法 parse()，如图 5.1.1 所示。

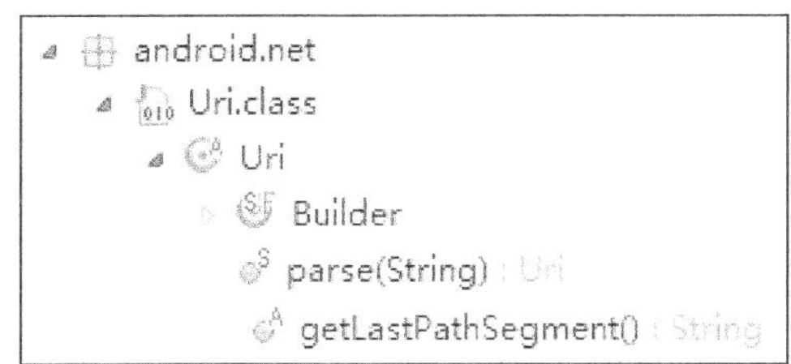

图 5.1.1　抽象类 Uri 及其静态方法

又如：Uri.parse("tel:02751012663")的值就是 Uri 类型，参见例 5.1.1。

注意：

(1) URI 是比 URL(uniform resource location，统一资源定位符)更加广泛的概念，因为 URL 是描述 Internet 上的信息资源的字符串。例如表示 Web 服务器资源的 http://www.peopledaily.com.cn/channel/welcome.htm，又如表示 FTP 服务器资源的 file://ftp.yoyodyne.com/pub/files/foobar.txt。

(2) 在 Android 应用开发中， URI 多用于本地通信，而 URL 用于网络通信。

(3) 表示内容提供者路径的 URI 中，以“content:// ”开头，参见第 8.2.3 小节。

5.1.2　打电话程序设计

在例 4.5.1 中，我们介绍了调用系统自带的拨号程序拨打任意电话的方法。下面介绍立即拨打指定电话的实现方法。

【例 5.1.1】拨打指定电话。

【设计思想】在主 Activity 中分别放置了提示输入的 TextView 控件、用于输入电话号码的 EditView 控件和一个 Button 命令按钮。用户输入电话并单击按钮后就会呼叫刚才输入的电话，程序运行效果如图 5.1.2 所示。

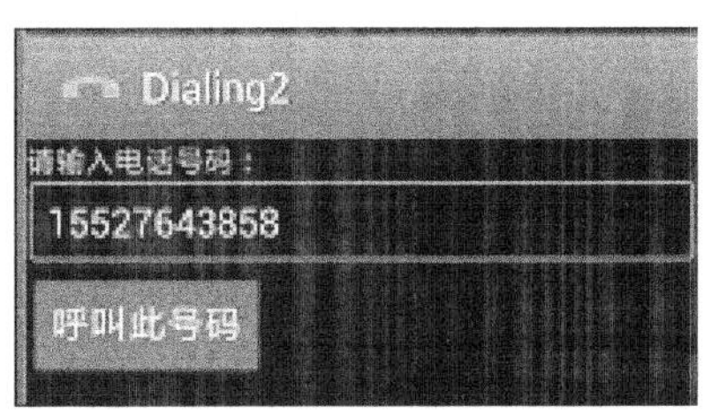

图 5.1.2　打电话程序运行效果

【设计步骤】

(1) 新建名为 Dialing2 的 Android 应用工程，勾选创建 Activity，布局文件命名为 main.xml，工程文件结构如图 5.1.3 所示。

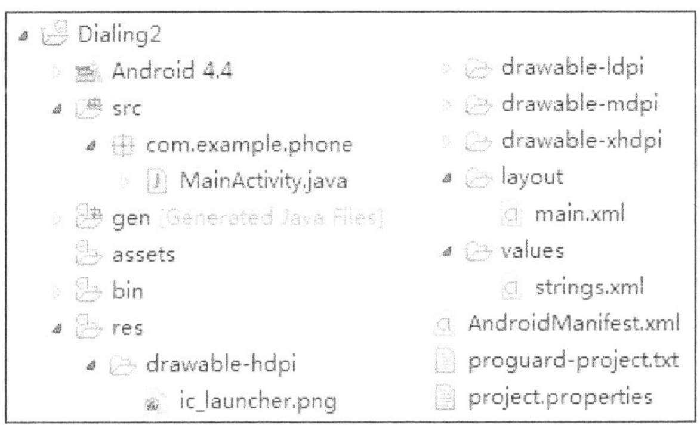

图 5.1.3　Dialing2 工程文件结构

(2) 在清单文件中，注册使用拨打电话的权限：

```
<uses-permission android:name="android.permission.CALL_PHONE" />
```

(3) 布局文件 main.xml 采用垂直线性布局，并采用国际化的设计方法，将布局文件中用到的常量在 res\values\strings.xml 文件中定义。布局文件 main.xml 代码如下：

```
<?xml version="1.0" encoding="utf-8"?>
<LinearLayout xmlns:android="http://schemas.android.com/apk/res/android"
    android:layout_width="fill_parent"
    android:layout_height="fill_parent"
    android:orientation="vertical" >

    <TextView
        android:layout_width="fill_parent"
        android:layout_height="wrap_content"
        android:text="@string/insert_number" />
    <EditText
        android:id="@+id/numET"
        android:layout_width="fill_parent"
        android:layout_height="wrap_content"
        android:inputType="phone" />
    <Button
        android:id="@+id/callBT"
        android:layout_width="wrap_content"
        android:layout_height="wrap_content"
        android:onClick="onClick"
        android:text="@string/call" />

</LinearLayout>
```

(4) 编写工程的主 Activity 对应的 MainActivity.java 文件，其完整代码如下：

```
package com.example.phone;
import android.app.Activity;
import android.os.Bundle;
import android.content.Intent;   //
import android.view.View;   //主类
import android.view.View.OnClickListener;    //View类的内部接口
import android.widget.Button;
import android.widget.EditText;
import android.net.Uri;      //辅助类
public class MainActivity extends Activity {
    private EditText numET;              // 由于多个方法使用，定义为成员变量
```

```
    public void onCreate(Bundle savedInstanceState) {
        super.onCreate(savedInstanceState);
        setContentView(R.layout.main);
        numET = (EditText) findViewById(R.id.numET);              // 找到文本框对象
        Button callBT = (Button) findViewById(R.id.callBT);       // 找到按钮对象
        callBT.setOnClickListener(new MyOnClickListener());       // 添加监听器
    }
    //定义内部的监听器类，实现接口OnClickListener
    private class MyOnClickListener implements OnClickListener {
        @Override
        public void onClick(View arg0) {                  //接口必须实现的方法
            // TODO Auto-generated method stub
            String num = numET.getText().toString();  // 获取其中号码
            Intent intent = new Intent();             // 创建意图对象
            intent.setAction(Intent.ACTION_CALL);     //设置意图动作(拨打电话)
            intent.setData(Uri.parse("tel:" + num));  //设置意图数据(电话号码)
            startActivity(intent);              //使用意图开启一个界面(拨打电话的界面)
        }
    }
}
```

注意：

(1) 拨打固定电话时，需要在电话前加区号。

(2) 例 4.8.1 程序使用的是系统自带的拨号程序，不需要使用打电话权限。本程序必须在清单文件中添加打电话的权限，否则，工程虽然可以正确部署，但运行时在单击按钮后会异常中止。读者可以通过在清单文件中屏蔽或放开权限进行测试。

(3) 通话结束后，按返回键则返回到程序的主界面。

5.2 短信程序

5.2.1 SMS 简介

SMS(short message service)，即短信息服务，是一种存储和转发服务。短信息并不是直接从发信人发送到接收人的，而是通过 SMS 中心进行转发的。如果接收人处于未连接状态，例如手机没有信号或者关机，则短信息在接收人再次连接时发送。

5.2.2 短信管理器 android.telephony.SmsManager

短信管理器类 SmsManager 提供了静态方法 getManager()来获取 SmsManager 实例，也提供了对短信文本进行分块(因为可能内容较多)的方法和对短信发送的方法，如图 5.2.1 所示。

```
android.telephony
  SmsManager.class
    SmsManager
      getDefault() : SmsManager
      divideMessage(String) : ArrayList<String>
      sendTextMessage(String, String, String, PendingIntent, PendingIntent) : void
```

图 5.2.1　SmsManager 类及其常用方法

5.2.3 发送短信程序

要实现短信发送，需要在清单文件中注册如下权限：

<uses-permission android:name="android.permission.SEND_SMS"/>

【例 5.2.1】发送短信程序。

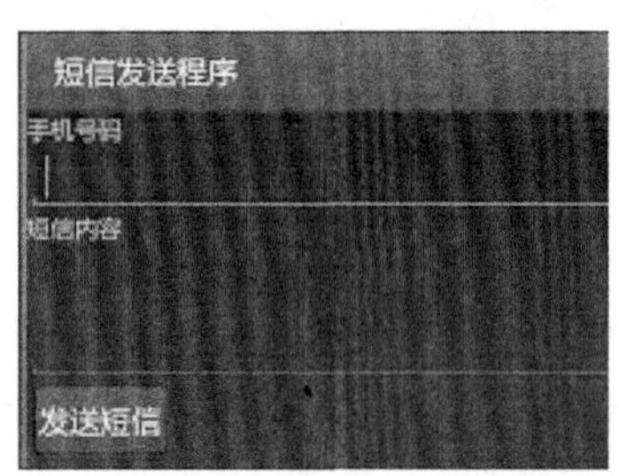

图 5.2.2　发送短信程序运行效果

【设计思想】在主 Activity 中分别放置了提示输入手机号码的 TextView 控件与 EditView 控件、用于输入短信内容的 TextView 控件和 EditView 控件，最后是一个 Button 命令按钮。用户输入电话号码和短信内容后单击按钮就会发送短信，程序运行效果如图 5.2.2 所示。

【设计步骤】

(1) 新建名为 SmsSend 的 Android 应用工程，指定程序包名为“SmsSend”，勾选创建 Activity，布局文件命名为“main.xml”，工程文件结构如图 5.2.3 所示。

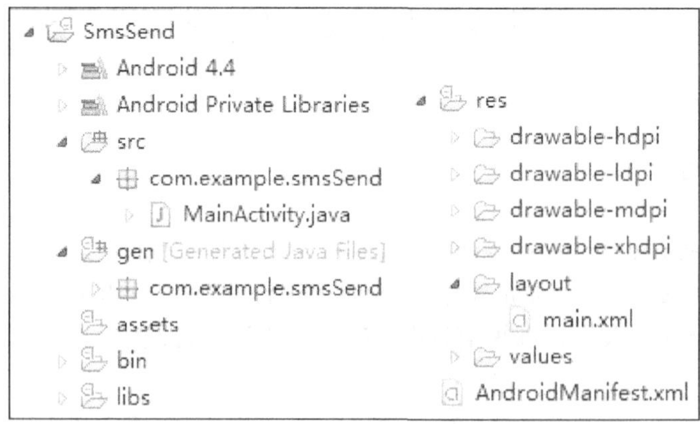

图 5.2.3　SmsSend 工程文件结构

(2) 在清单文件中，注册发短信的用户权限。

(3) 布局文件 main.xml 采用垂直线性布局，其代码如下：

```
<?xml version="1.0" encoding="utf-8"?>
<LinearLayout xmlns:android="http://schemas.android.com/apk/res/android"
    android:layout_width="fill_parent"
    android:layout_height="fill_parent"
    android:orientation="vertical" >
    <TextView
        android:layout_width="fill_parent"
        android:layout_height="wrap_content"
        android:text="手机号码" />
    <EditText
        android:id="@+id/numET"
        android:layout_width="fill_parent"
        android:layout_height="wrap_content"
        android:inputType="phone" />
    <TextView
        android:layout_width="fill_parent"
        android:layout_height="wrap_content"
        android:text="短信内容" />
    <EditText
        android:id="@+id/contentET"
        android:layout_width="fill_parent"
        android:layout_height="wrap_content"
        android:inputType="textMultiLine"
        android:lines="3" />
    <Button
        android:layout_width="wrap_content"
        android:layout_height="wrap_content"
        android:onClick="onClick"
        android:text="发送短信" />
</LinearLayout>
```

(4) 编写工程的主 Activity 对应的 MainActivity.java 文件，其完整代码如下：

```
package com.example.smsSend;
import android.app.Activity;
import android.os.Bundle;
import android.telephony.SmsManager;   //主类
import android.view.View;
```

```
import android.widget.EditText;
import android.widget.Toast;
import java.util.ArrayList;

public class MainActivity extends Activity {
    private EditText numET;
    private EditText contentET;
    public void onCreate(Bundle savedInstanceState) {
        super.onCreate(savedInstanceState);
        setContentView(R.layout.main);
        // 获取2个文本框
        numET = (EditText) findViewById(R.id.numET);
        contentET = (EditText) findViewById(R.id.contentET);
    }
    public void onClick(View view) {   //自定义按钮的单击事件处理方法
        // 获取号码和内容
        String num = numET.getText().toString();
        String content = contentET.getText().toString();
        // 获取短信管理器，静态方法
        SmsManager smsManager = SmsManager.getDefault();
        // 将短信内容分段，装入ArrayList
        ArrayList<String> list = smsManager.divideMessage(content);
        for (String sms : list)              // 分段发送
            smsManager.sendTextMessage(num, null, sms, null, null);   //
        Toast.makeText(getApplicationContext(), "发送成功",
                                                    Toast.LENGTH_SHORT).show();

    }
}
```

注意：短信发送的方法不是唯一的。除了使用短信管理器外，还可使用短信意图。使用短信意图发送短信的主要代码如下：

```
//send sms
Uri uri = Uri.parse("smsto:15527643858");
//创建短信意图对象
Intent intent = new Intent(Intent.ACTION_SENDTO, uri);
intent.putExtra("sms_body", "The SMS text");
startActivity(intent);
//需要手动发送，而不是自动发送
```

5.3　音频播放与录音

5.3.1　媒体播放类 android.media.MediaPlayer

媒体播放类 MediaPlayer 位于 android.media 包内，除了提供创建媒体播放机的构造方法外，还提供创建媒体播放机的静态方法 create()。当然，播放方法 start()和停止方法 stop()是肯定会提供的。类 MediaPlayer 的定义如图 5.3.1 所示。

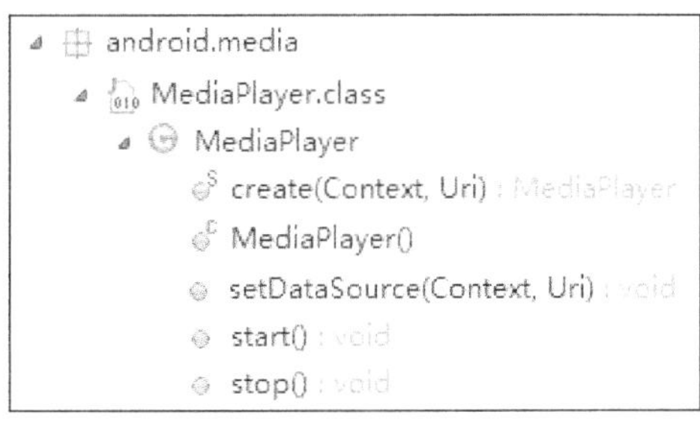

图 5.3.1　媒体播放类 MediaPlayer 的定义

图 5.3.2　前台播放音乐运行界面

5.3.2　前台播放音频

前台播放是指在 Activity 中调用 MediaPlayer 控件，这不同于后台播放(将在第 6.2 节介绍)。下面介绍一个使用 MediaPlayer 播放音频文件的例子。

【例 5.3.1】播放音频文件。

【程序运行】运行程序时，在主窗口中出现一个文本框，同时播放音乐。按手机返回键将关闭窗口，同时音乐停止播放(播放完毕后也会自动停止)。程序运行界面如图 5.3.2 所示。

【设计步骤】

(1) 新建名为 Media_Audio 的 Android 工程，勾选创建 Activity。

(2) 在资源文件夹 res 内建立一个名为 raw 的子文件夹，并复制一个 mp3 音乐文件，创建后的工程文件结构(主要部分)如图 5.3.3 所示。

图 5.3.3　Media_Audio 工程文件结构(主要部分)

(3) 编写源程序文件 MainActivity.java，其代码如下：

```
package com.example.media_audio;
import android.app.Activity;
import android.os.Bundle;
import android.media.MediaPlayer;        //主类
//import android.net.Uri;     //播放SD卡里的音乐时需要
public class MainActivity extends Activity {
    private MediaPlayer mp;   //媒体播放器
    @Override
    protected void onCreate(Bundle savedInstanceState) {
        super.onCreate(savedInstanceState);
        setContentView(R.layout.activity_main);
        mp=MediaPlayer.create(this,R.raw.black);   //不带扩展名
        //mp=MediaPlayer.create(this,Uri.parse("sdcard/music/white.mp3"));   //带扩展名
        mp.start();   //播放
    }
    @Override
    public void finish() {        //关闭窗口
        // TODO Auto-generated method stub
        mp.stop();        //停止音乐播放
        super.finish();
    }
}
```

5.3.3 手机前台录音

```
android.media
  MediaRecorder.class
    MediaRecorder
      MediaRecorder()
      setAudioSource(int) : void
      setOutputFormat(int) : void
      setAudioEncoder(int) : void
      setOutputFile(String) : void
      prepare() : void
      release() : void
      start() : void
      stop() : void
```

图 5.3.4 媒体录制类 MediaRecorder 的常用方法

Android 提供了媒体播放类 MediaPlayer，也提供了媒体录制类 MediaRecorder。类 MediaRecorder 提供的常用方法如图 5.3.4 所示。

下面我们介绍手机录音程序的设计。

【例 5.3.2】手机录音程序设计。

【程序运行】运行程序时，在主窗口中出现“开始录音”和“停止录音”两个按钮，单击“开始录音”按钮后开始录音，单击“停止录音”按钮后停止录音。其中，录音文件保存在 SD 卡的 audioRecords 文件夹里，文件名为 AudioRecord×××.amr。其中，“×××”为若干位随机数

字。程序运行界面如图 5.3.5 所示。

图 5.3.5　录音程序的运行界面

【设计步骤】

(1) 新建名为 Media_Audio 的 Android Application Project，勾选创建 Activity。

(2) 找两个表示录音和停止录音的图片文件 record_start.png 和 record_stop.png，复制到文件夹 res/drawable-hdpi 下，工程文件结构(主要部分)如图 5.3.6 所示。

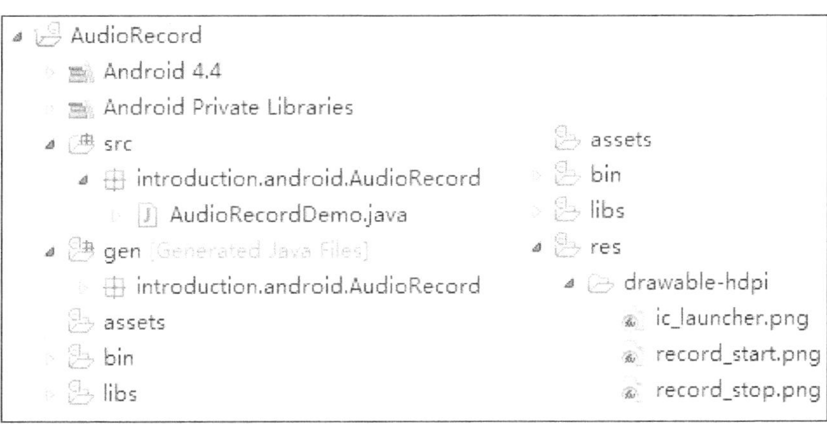

图 5.3.6　Audio Record 工程文件结构(主要部分)

(3) 在清单文件中配置本应用的用户权限，其代码如下：

```
<uses-permission android:name="android.permission.RECORD_AUDIO" />
<uses-permission android:name="android.permission.WRITE_EXTERNAL_STORAGE" />
```

AudioRecordDemo.java 文件代码如下：

```
package introduction.android.AudioRecord;
import android.app.Activity;
import android.os.Bundle;
import android.media.MediaRecorder;   //主类
import android.view.View;
import android.view.View.OnClickListener;
import android.widget.ImageButton;
import android.widget.Toast;
import android.os.Environment;
import java.io.File;   //
import java.io.IOException;
public class AudioRecordDemo extends Activity implements OnClickListener {
    private ImageButton st,stop;
```

```
    private MediaRecorder mRecorder;
    private File recordPath;
    private File recordFile;
    @Override
    public void onCreate(Bundle savedInstanceState) {
        super.onCreate(savedInstanceState);
        setContentView(R.layout.main);
        st = (ImageButton)findViewById(R.id.st);
        stop = (ImageButton)findViewById(R.id.stop);
        st.setOnClickListener(this);
        stop.setOnClickListener(this);
    }
    @Override
    public void onClick(View v) {   //因为主Activity实现了接口OnClickListener
        // TODO Auto-generated method stub
        if(v == st){
            AudioRecordDemo.this.start();
            Toast.makeText(AudioRecordDemo.this, "开始录音......",
                                                    Toast.LENGTH_LONG).show();
        }
        if(v == stop){
            AudioRecordDemo.this.stop();
            Toast.makeText(AudioRecordDemo.this, "录音结束！",
                                                    Toast.LENGTH_SHORT).show();
        }
    }
    public void start() {
        if (checkSDCard()) {
                //设置录音文件的路径及文件名
                recordPath = Environment.getExternalStorageDirectory();
                File path = new File(recordPath.getPath() + File.separator
                                + "audioRecords");      //创建子目录
                if(!path.mkdirs()){
                    Toast.makeText(this, "目录创建失败",Toast.LENGTH_LONG);
                }
                recordPath=path;
        } else {
                Toast.makeText(AudioRecordDemo.this, "SDcard未连接",
                                                    Toast.LENGTH_LONG).show();
                return ;
```

```
        }
        try {
                //使用静态方法File.createTempFile()建立的文件名后多了几个随机的数字
                recordFile = File.createTempFile(String.valueOf("MyAudioRecord"),
                                                                ".amr", recordPath);
        } catch (IOException e) {
            Toast.makeText(this, "文件创建失败",Toast.LENGTH_LONG).show();
        }
        mRecorder = new MediaRecorder();        //
          //设置麦克风
        mRecorder.setAudioSource(MediaRecorder.AudioSource.MIC);
          //文件格式
        mRecorder.setOutputFormat(MediaRecorder.OutputFormat.DEFAULT);
          //音频编码
        mRecorder.setAudioEncoder(MediaRecorder.AudioEncoder.DEFAULT);
          //文件路径
        mRecorder.setOutputFile(recordFile.getAbsolutePath());
        try {
                mRecorder.prepare();   //准备
                mRecorder.start();      // 开始录音
        } catch (IllegalStateException e) {
                e.printStackTrace();
        } catch (IOException e) {
                e.printStackTrace();
        }
}
    public void stop() {
        try{
            if(mRecorder != null) {
                    mRecorder.stop();
                    mRecorder.release();
                    mRecorder = null;
            }
        }
        catch(IllegalStateException e){
            e.printStackTrace();
        }
}
    private boolean checkSDCard() {
```

```
            // TODO Auto-generated method stub
            if(android.os.Environment.getExternalStorageState()
                                .equals(android.os.Environment.MEDIA_MOUNTED))
                return true;
            return false;
        }
}
```

5.4 视频播放

5.4.1 视频播放控件 android.widget.VideoView

控件 MediaPlayer 不仅能播放音频，还可以播放视频，但 Android 也提供了专门播放视频的控件 VideoView。

VideoView 其实也是调用 MediaPlayer 来播放视频的，并提供了一些视频播放的辅助功能。VideoView 位于软件包 android.widget 内，其常用方法如图 5.4.1 所示。

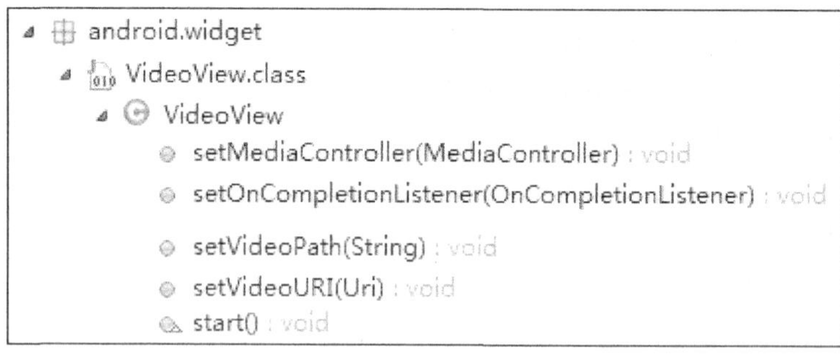

图 5.4.1 视频播放控件 VideoView 的常用方法

注意：

(1) MediaPlayer 与 VideoView 不是处于同一软件包内的，前者处于软件包 android.media 内。

(2) VideoView 的播放完成监听器接口方法的参数为 MediaPlayer 类型，详见例 5.4.1。

5.4.2 媒体播放控制器类 android.widget.MediaController

MediaController 与 VideoView 配套使用，基本能实现播放界面的主要功能,提供的主要方法如图 5.4.2 所示。

5.4.3 使用 VideoView 播放视频

下面通过一个例子完整地介绍使用 VideoView 控件播放视频的设计方法。

【例 5.4.1】使用 VideoView 控件播放手机指定 SD 卡内的视频文件。

【开发步骤】

(1) 新建名为 Media_Video 的 Android Application Project，勾选创建 Activity，工程文件结构如图 5.4.3 所示。

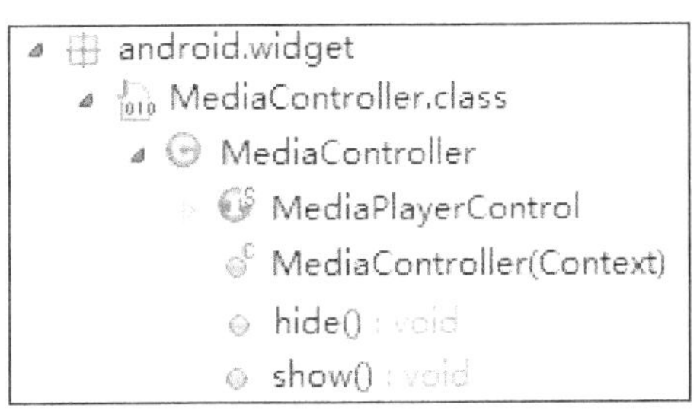

图 5.4.2　MediaController 的主要方法

Media_Video
Android 4.4
Android Private Libraries
src
com.example.media_video
MainActivity.java
gen [Generated Java Files]
com.example.media_video

图 5.4.3　Media_Video 工程文件结构(主要部分)

(2) 编写项目的主 Activity，对应的 MainActivity.java 文件的完整代码如下：

```
package com.example.media_video;
/*
 * 本例演示了使用VideoView控件播放视频
 * 本应用不需要添加任何权限
 * 播放完毕后关闭当前的Activity
 * 从播放完成监听器的参数可以看出VideoView实质上是调用了MediaPlayer
 * 当屏蔽了播放完成监听器后，播放完毕后将静止到最后一帧
 */
import android.os.Bundle;
import android.os.Environment;   //
import android.app.Activity;
import android.view.View;
import android.widget.VideoView;    //主类1
import android.widget.MediaController;   //主类2
import android.net.Uri;
import android.media.MediaPlayer;   //辅类
import android.media.MediaPlayer.OnCompletionListener;
public class MainActivity extends Activity {
    private VideoView videoView1;
    private MediaController mc;
    private String path;
    @Override
    protected void onCreate(Bundle savedInstanceState) {
        super.onCreate(savedInstanceState);
```

```
        setContentView(R.layout.activity_main);
        videoView1=(VideoView)findViewById(R.id.videoView1);  //
        path=Environment.getExternalStorageDirectory().getPath();
        Path+="/Movies/1.3gp";
        videoView1.setVideoURI(Uri.parse(path));  //
        mc=new MediaController(this);  //
        videoView1.setMediaController(mc);  //
        videoView1.start();  //
        videoView1.setOnCompletionListener(new OnCompletionListener(){
            @Override
            public void onCompletion(MediaPlayer arg0) {
                // TODO Auto-generated method stub
                finish();  //播放完毕后关闭当前的Activity
            }
        });
    }
}
```

(3) 在手机内置 SD 卡内建立一个名为 Movies 的文件夹，并将一个名为 1.mp4 的视频文件复制到该文件夹里。

(4) 部署到手机上做运行测试。

注意：

(1) 增加监听器的功能是在视频播放完毕后退出 Activity。如果删除源文件中的监听器代码，则视频播放完毕后画面静止。

(2) 程序设置 path="/sdcard/Movies/1.3gp"也是可以的，但文件保存后会出现感叹号，建议不使用这种方法。

5.5 手机拍照与视频拍摄

5.5.1 有返回值的 Activity 调用

如果想在 Activity 中得到刚打开的 Activity 关闭后返回的数据，需要使用 Activity 具有的方法 startActivityForResult(Intent intent, int requestCode)去打开一个新的 Activity，新的 Activity 关闭后会向前面的 Activity 传回数据，为了得到传回的数据，必须在先前的 Activity 中重写 onActivityResult(int requestCode, int resultCode, Intent data)方法。

5.5.2　手机拍照

目前的 Android 手机都内置了摄像头，这为拍照提供了硬件支持。手机也提供了拍照程序。下面通过两个案例分明说明使用自带的手机拍照程序和使用摄像头 API 实现拍照的方法。其中，这两种方法都提供了对照相机的使用和保存图片至扩展存储设备中，因此，在工程的清单文件中需要注册如下权限：

```
<uses-permission android:name="android.permission.CAMERA"/>
        <uses-feature android:name="android.hardware.camera" />
<uses-permission android:name="android.permission.WRITE_EXTERNAL_STORAGE"/>
```

【例 5.5.1】使用手机自带的相机程序拍照。

【程序运行】程序运行后调用系统自带的相机程序，出现取景界面。按下相机的快门后，出现是否保存的对话框，如图 5.5.1 所示。

单击“储存”按钮(有的手机出现的是图像按钮)后，文件将自动保存至文件夹 SD 卡的 myImage 文件夹里，文件名为 111.jpg。然后，返回至主 Activity，该图片将在 ImageView 控件里显示。

【设计步骤】

(1) 新建名为 Camera1 的 Android 应用工程，勾选创建 Activity，应用的图标从 Clipart 里选定为“ ”，工程文件结构(部分)如图 5.5.2 所示。

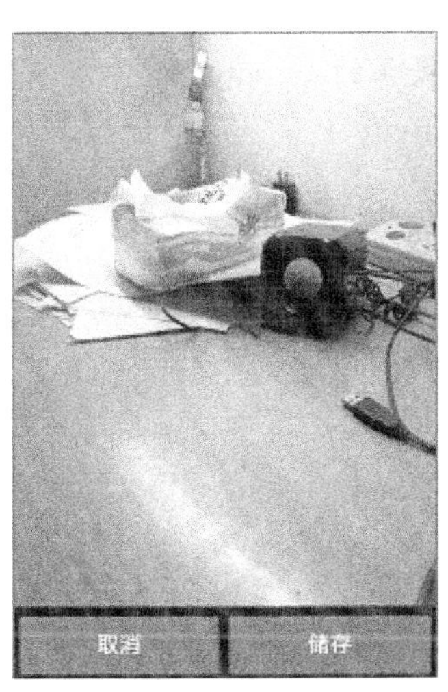

图 5.5.1　是否保存的对话框

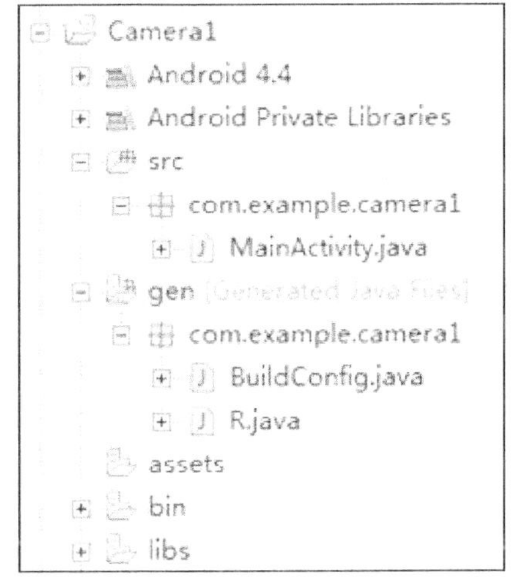

图 5.5.2　Camera1 工程文件结构(部分)

(2) 在清单文件中，注册使用相机和在 SD 卡中创建文件(文件夹)的权限。

```
<uses-permission android:name="android.permission.CAMERA"/>
<uses-permission android:name="android.permission.WRITE_EXTERNAL_STORAGE" />
```

(3) 编写项目的主 Activity 对应的 MainActivity.java 文件，其完整代码如下：

```
package com.example.camera1;
```

```
import android.app.Activity;
import android.os.Bundle;
import android.content.Intent;    //
import android.provider.MediaStore;    //
import android.widget.ImageView;
import android.graphics.Bitmap;    //
import java.io.File;
import java.io.FileOutputStream;
import java.io.IOException;
import java.io.FileNotFoundException;
import android.os.Environment;
import android.util.Log;
public class MainActivity extends Activity {
    private ImageView image;
    @Override
    protected void onCreate(Bundle savedInstanceState) {
        super.onCreate(savedInstanceState);
        setContentView(R.layout.activity_main);
        image = (ImageView)findViewById(R.id.imageView);
        Intent intent=new Intent(MediaStore.ACTION_IMAGE_CAPTURE);
        startActivityForResult(intent,1);    //开启系统提供的Activity
        //第二参数为请求码，用于标识所请求的Activity
    }
    //拍照完成后自动执行的方法
    protected void onActivityResult(int requestCode, int resultCode, Intent data) {
        super.onActivityResult(requestCode, resultCode, data);    //
        if(requestCode == 1)    //
        {
            if(resultCode == RESULT_OK)
            {
                String sdStatus = Environment.getExternalStorageState();
                if (!sdStatus.equals(Environment.MEDIA_MOUNTED)) {
                    Log.v("TestFile", "SD卡无效.");
                    return;
                }
                Bundle bundle = data.getExtras(); //
                // 获取相机返回的数据，并转换为Bitmap图片格式
                Bitmap bitmap = (Bitmap) bundle.get("data");
                FileOutputStream b = null;
```

```
                    File file = new File("/sdcard/myImage/");
                    file.mkdirs();// 创建子目录(文件夹)
                    String fileName = "/sdcard/myImage/111.jpg";
                    try {
                        b = new FileOutputStream(fileName);
                        bitmap.compress(Bitmap.CompressFormat.JPEG, 100, b);
                    } catch (FileNotFoundException e) {
                        e.printStackTrace();
                    } finally {
                        try {
                                b.flush();
                                b.close();
                        } catch (IOException e) {
                                e.printStackTrace();
                        }
                    }
                        image.setImageBitmap(bitmap);    //显示图片
                }
            }
        }
}
```

注意：startActivityForResult()是 Activity 提供的且其父类不具有的方法。

下面介绍的方法使用了相机硬件 API，如图 5.5.3 所示。

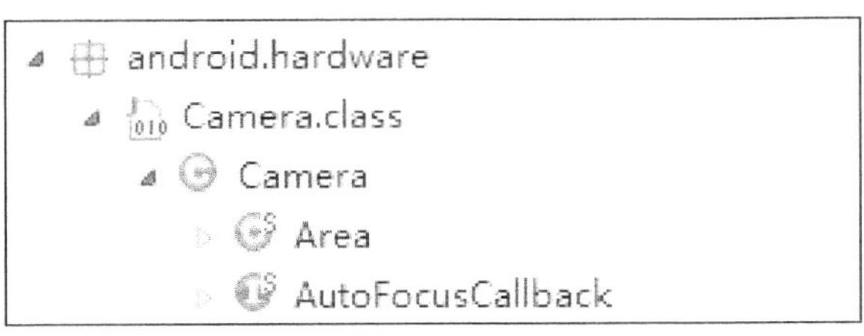

图 5.5.3　Camera 类

通常，在清单文件中注册对相机的使用后面增加一条<uses-feature>标签，表明程序运行时需要使用相机硬件(当没有指定的硬件时程序自动挂断)，代码如下：

```
<uses-permission android:name="android.permission.CAMERA"/>
      <uses-feature android:name="android.hardware.camera" />
```

注意：

(1) 由于目前的 Android 手机都带有摄像头，因此在清单文件中，<uses-feature>标签可以不写。

(2) 在程序中也可以通过代码检测是否存在相机，其代码如下：

```
//检查是否存在相机
```

```
Private Boolean checkCameraHardware(Context context){
  if(context.getPackageManager().hasSystemFeature(PackageManager.FEATURE_CAMERA))
        return true;     //this device has a camera
  else
        return false     //no camera on this device
}
```

【例 5.5.2】使用 Android 提供的 android.hardware.Camera 开发手机拍照程序。

【程序功能】程序运行时，有三个分别用于控制相机操作的三个功能按钮："打开摄像头""拍照"和"关闭摄像头"。单击"拍照"按钮后，生成的照片文件存储在 SD 卡根目录下，文件名为 mypicture.jpg。程序运行界面如图 5.5.4 所示。

【设计思想】使用 SurfaceView 控件并通过 SurfaceHolder 实时显示照相焦点捕捉的图像。在布局文件中使用三个按钮，分别控制摄像头的开启、拍摄和关闭。拍摄后将图像以指定的文件名(mypicture.jpg)和格式(.jpg)保存至指定的位置(SD 卡根目录)。

【开发步骤】

(1) 新建名为 Camera2 的 Android 应用工程，勾选创建 Activity，工程文件结构(部分)如图 5.5.5 所示。

图 5.5.4　手机拍照程序运行界面

图 5.5.5　Camera2 工程文件结构(部分)

(2) 在清单文件中配置与 Camera1 工程相同的权限。

(3) 编写项目的主 Activity，对应的 MainActivity.java 文件的完整代码如下：

```
package com.example.camera2;
import android.app.Activity;
import android.os.Bundle;
import android.os.Environment;
import android.hardware.Camera;   //主类
import android.hardware.Camera.Parameters;
import android.hardware.Camera.PictureCallback;
import android.view.SurfaceHolder;
import android.view.SurfaceView;   //主类
import android.graphics.Bitmap;
import android.graphics.BitmapFactory;
import java.io.FileOutputStream;
```

```
import java.io.IOException;
import java.io.BufferedOutputStream;
import java.io.File;
import java.io.FileNotFoundException;
import android.view.View;
import android.view.View.OnClickListener;
import android.widget.Button;
import android.widget.Toast;
import android.util.Log;
public class MainActivity extends Activity {
    private Button opbtn;
    private Button playbtn;
    private Button clobtn;
    private SurfaceView surfaceView;
    private SurfaceHolder surfaceHolder;
    private Camera camera;
    private int previewWidth = 240;
    private int previewHeight = 320;
    //private String filepath = "/sdcard/mypicture.jpg";  //文件存储路径
    private String filepath =Environment.getExternalStorageDirectory().getPath()
                                                                        +"/mypicture.jpg";
    @Override
    public void onCreate(Bundle savedInstanceState) {
        super.onCreate(savedInstanceState);
        setContentView(R.layout.main);
        opbtn = (Button) this.findViewById(R.id.button1);
        playbtn = (Button) this.findViewById(R.id.button2);
        clobtn = (Button) this.findViewById(R.id.button3);
        surfaceView = (SurfaceView) this.findViewById(R.id.surfaceView1);
        surfaceHolder = surfaceView.getHolder();
        surfaceHolder.addCallback(new SurfaceHolder.Callback() {
            @Override
            public void surfaceDestroyed(SurfaceHolder holder) {
                // TODO Auto-generated method stub
                if (camera != null)
                {
                    camera.release();  //释放资源
                    camera = null;
                }
```

```
            }
            @Override
            public void surfaceCreated(SurfaceHolder holder) {
                // TODO Auto-generated method stub
            }
            @Override
            public void surfaceChanged(SurfaceHolder holder, int format,
                    int width, int height) {
                // TODO Auto-generated method stub
            }
        });
        opbtn.setOnClickListener(new OnClickListener() {
            @Override
            public void onClick(View arg0) {
                // TODO Auto-generated method stub
                openCamera();        //开启摄像头
            }
        });
        playbtn.setOnClickListener(new OnClickListener() {
            @Override
            public void onClick(View v) {
                // TODO Auto-generated method stub
                takePicture();       //照相
            }
        });
        clobtn.setOnClickListener(new OnClickListener() {
            @Override
            public void onClick(View v) {
                // TODO Auto-generated method stub
                closeCamera();         //关闭相机
                MainActivity.this.finish();
            }
        });
    }
    protected void closeCamera() {
        // TODO Auto-generated method stub
        if(camera != null)
        {
            camera.stopPreview();
```

```
                camera.release();
                camera = null;
            }

    }
    protected void takePicture() {
            // TODO Auto-generated method stub
            if (checkSDCard()) {
                camera.takePicture(null, null, jpeg);
                Toast.makeText(MainActivity.this, "拍照成功", Toast.LENGTH_LONG).show();
                try {
                    Thread.sleep(1000);
                } catch (InterruptedException e) {
                    // TODO Auto-generated catch block
                    e.printStackTrace();
                }
                camera.startPreview();
            }
else {
                Log.e("camera", "SD CARD not exist.");
                return;
            }
    }
    private void openCamera() {
            // TODO Auto-generated method stub
            try {
                camera = Camera.open(); // attempt to get a Camera instance
            }
 catch (Exception e) {
                e.printStackTrace();
                return;
            }
            Parameters params = camera.getParameters();
            params.setPreviewSize(previewWidth, previewHeight);
            //params.setPictureFormat(PixelFormat.JPEG);
            params.setPictureSize(previewWidth, previewHeight);
            camera.setParameters(params);
            try {
                camera.setPreviewDisplay(surfaceHolder);
```

```
        } catch (IOException e) {
            // TODO Auto-generated catch block
            e.printStackTrace();
        }
        camera.startPreview();
    }
    private PictureCallback jpeg = new PictureCallback() {
        @Override
        public void onPictureTaken(byte[] data, Camera camera) {
            // TODO Auto-generated method stub
            Bitmap bitmap = BitmapFactory.decodeByteArray(data, 0, data.length);
            File pictureFile = new File(filepath);
            try {
                // 将拍摄的照片写入SD卡中
                FileOutputStream fos = new FileOutputStream(pictureFile);
                BufferedOutputStream bos = new BufferedOutputStream(fos);
                //设置文件格式
                bitmap.compress(Bitmap.CompressFormat.JPEG, 80, bos);
                bos.flush();
                bos.close();
                fos.close();
                Log.i("camera", "jpg file saved.");
            }
catch (FileNotFoundException e) {
                Log.d("camera", "File not found: " + e.getMessage());
            }
catch (IOException e) {
                Log.d("camera", "Error accessing file: " + e.getMessage());
            }
        }
    };
    private boolean checkSDCard() {
        // 判断SD存储卡是否存在
        if (android.os.Environment.getExternalStorageState().equals(
                android.os.Environment.MEDIA_MOUNTED)) {
            return true;
        }
else {
            return false;
```

```
        }
    }
}
```

5.5.3　视频拍摄

视频拍摄本质上是使用系统的摄像头，还需要使用 android.media.MediaRecorder 类的相关方法。此外，由于录制视频时还要使用手机的麦克风，因此，在清单文件中还要注册使用麦克风的权限(与录音程序相同)。

【例 5.5.3】使用手机拍摄视频。

【程序运行】程序运行后的界面，如图 5.5.6 所示。先单击“打开摄像头”按钮，即可取景。单击“拍摄”按钮后，开始拍摄。拍摄过程完成后，单击“关闭摄像头”按钮，视频文件将保存在手机内置存储卡的根目录下，文件名为“myVideo.3gp”，程序终止运行。

【设计步骤】

(1) 新建名为 VideoRecorderDemo 的 Android 应用工程，勾选创建 Activity 并命名为 VideoRecorderDemoActivity。

(2) 使用 Android Tools 修改应用的包名为“com.example.videoRecorderDemo”，工程文件结构(部分)如图 5.5.7 所示。

图 5.5.6　视频拍摄程序运行界面

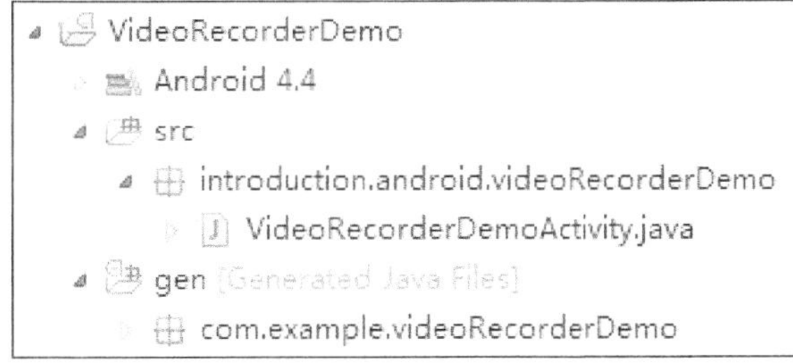

图 5.5.7　VideoRecorderDemo 工程文件结构(部分)

(3) 在清单文件中，增加如下程序使用相机和在扩展存储卡写文件(夹)的权限：

```
<uses-permission android:name="android.permission.CAMERA"/>
<uses-feature android:name="android.hardware.camera" />
<uses-feature android:name="android.hardware.camera.autofocus" />
<uses-permission android:name="android.permission.RECORD_AUDIO"/>
<uses-permission android:name="android.permission.WRITE_EXTERNAL_STORAGE" />
```

(4) 修改文件 res/values/strings.xml 文件的代码，内容如下：

```
<?xml version="1.0" encoding="utf-8"?>
<resources>
    <string name="hello">使用MediaRecorder进行视频录制实例</string>
    <string name="app_name">VideoRecorderDemo</string>
```

```
    <string name="opBtn">打开摄像头</string>
    <string name="play">录制</string>
    <string name="cloBtn">关闭摄像头</string>
</resources>
```

(5) 与主 Activity 对应的布局文件的代码如下：

```
<?xml version="1.0" encoding="utf-8"?>
<LinearLayout xmlns:android="http://schemas.android.com/apk/res/android"
    android:layout_width="fill_parent"
    android:layout_height="fill_parent"
    android:orientation="vertical" >
    <TextView
        android:layout_width="fill_parent"
        android:layout_height="wrap_content"
        android:text="@string/hello" />
    <SurfaceView
        android:id="@+id/surfaceView1"
        android:layout_width="fill_parent"
        android:layout_height="0dip"
        android:layout_weight="0.58" />
    <LinearLayout
        android:id="@+id/linearLayout1"
        android:layout_width="match_parent"
        android:layout_height="wrap_content"
        android:gravity="center">
        <Button
            android:id="@+id/button1"
            android:layout_width="wrap_content"
            android:layout_height="wrap_content"
            android:text="@string/opBtn"/>
        <Button
            android:id="@+id/button2"
            android:layout_width="wrap_content"
            android:layout_height="wrap_content"
            android:text="@string/play"/>
        <Button
            android:id="@+id/button3"
            android:layout_width="wrap_content"
            android:layout_height="wrap_content"
```

```
            android:text="@string/cloBtn" />
    </LinearLayout>
</LinearLayout>
```

(6) 编写主 Activity 对应的文件 MainActivity.java 的完整代码如下：

```
package com.example.videoRecorder;
import android.app.Activity;
import android.os.Bundle;
import android.hardware.Camera;     //主类
import android.view.SurfaceView;    //主类
import android.view.SurfaceHolder;
import android.view.View;
import android.view.View.OnClickListener;
import android.widget.Button;
import android.media.MediaRecorder;    //主类
import android.widget.Toast;
import java.io.IOException;
import android.util.Log;
import com.example.videoRecorderDemo.R;   //应用的包名与本程序的包名不一致时必需
public class VideoRecorderActivity extends Activity {
    private Button opbtn;
    private Button playbtn;
    private Button clobtn;
    private Camera camera;
    private SurfaceView surfaceView;
    private SurfaceHolder surfaceHolder;
    private MediaRecorder videoRecorder;
    private String myVideofilepath = "/sdcard/myVideo.3gp";   //视频文件名
    @Override
    public void onCreate(Bundle savedInstanceState) {
        super.onCreate(savedInstanceState);
        setContentView(R.layout.main);
        opbtn = (Button) this.findViewById(R.id.button1);
        playbtn = (Button) this.findViewById(R.id.button2);
        clobtn = (Button) this.findViewById(R.id.button3);
        videoRecorder=new MediaRecorder();     //
        surfaceView = (SurfaceView) this.findViewById(R.id.surfaceView1);
        surfaceHolder = surfaceView.getHolder();
        opbtn.setOnClickListener(new OnClickListener() {     //
```

```
                @Override
                public void onClick(View arg0) {
                    // TODO Auto-generated method stub
                    openCamera();
                }
            });
            playbtn.setOnClickListener(new OnClickListener() {      //
                @Override
                public void onClick(View v) {
                    // TODO Auto-generated method stub
                    Toast.makeText(VideoRecorderActivity.this,
                                        "开始录像...",Toast.LENGTH_SHORT).show();
                    benginRecording();
                }
            });
            clobtn.setOnClickListener(new OnClickListener() {      //
                @Override
                public void onClick(View v) {
                    // TODO Auto-generated method stub
                    stopRecording();
                    finish();
                }
            });
        }
        @Override
        protected void onPause() {      //
            // TODO Auto-generated method stub
            super.onPause();
            stopRecording();
            releaseCamera();
        }
        protected void stopRecording() {      //
            // TODO Auto-generated method stub
            if(videoRecorder!=null){
            videoRecorder.stop();
            videoRecorder.reset();
            videoRecorder.release();
            videoRecorder=null;
            camera.lock();
```

```
        }
    }
    private void releaseCamera(){        //
        if (camera != null){
            camera.stopPreview();
            camera.release();       //摄像完成后释放Camera对象
            camera = null;
        }
    }
    protected void benginRecording() {
        // TODO Auto-generated method stub
        camera.unlock();  //给摄像头解锁
        //MediaRecorder获取到摄像头的访问权
        videoRecorder.setCamera(camera);
        //设置视频录制过程中所录制的音频来自手机的麦克风
        videoRecorder.setAudioSource(MediaRecorder.AudioSource.CAMCORDER);
        //设置视频源为摄像头
        videoRecorder.setVideoSource(MediaRecorder.VideoSource.CAMERA);
        //设置视频录制的输出文件格式为3gp文件
        videoRecorder.setOutputFormat(MediaRecorder.OutputFormat.THREE_GPP);
        //设置音频编码方式为AAC
        videoRecorder.setAudioEncoder(MediaRecorder.AudioEncoder.AAC);
        // 设置录制的视频编码方式为H.264
        videoRecorder.setVideoEncoder(MediaRecorder.VideoEncoder.H264);
        // 设置视频录制的分辨率，必须放在设置编码和格式的后面，否则报错
        videoRecorder.setVideoSize(176, 144);
        // 设置录制的视频帧率，必须放在设置编码和格式的后面，否则报错
        videoRecorder.setVideoFrameRate(20);
        //videoRecorder.setProfile(CamcorderProfile.get(CamcorderProfile.QUALITY_LOW));
        if(!checkSDCard()){
            Log.e("videoRecorder","未找到SD卡！");
            return;
        }
        videoRecorder.setOutputFile(myVideofilepath);
        videoRecorder.setPreviewDisplay(surfaceHolder.getSurface());
        try {
            videoRecorder.prepare();
        } catch (IllegalStateException e) {
            // TODO Auto-generated catch block
```

```
                e.printStackTrace();
            } catch (IOException e) {
                // TODO Auto-generated catch block
                e.printStackTrace();
            }
            videoRecorder.start();
        }
        private void openCamera() {
            // TODO Auto-generated method stub
            Log.i("videoRecorder","openCamera.");
            try {
                camera = Camera.open(); // attempt to get a Camera instance
            } catch (Exception e) {
                // Camera is not available (in use or does not exist)
                Log.e("camera", "open camera error!");
                e.printStackTrace();
                return;
            }
            try {
                camera.setPreviewDisplay(surfaceHolder);
            } catch (IOException e) {
                // TODO Auto-generated catch block
                Log.e("camera", "preview failed.");
                e.printStackTrace();
            }
            camera.startPreview();
        }
        private boolean checkSDCard() {
            // 判断SD存储卡是否存在
            if (android.os.Environment.getExternalStorageState().equals(
                    android.os.Environment.MEDIA_MOUNTED)) {
                return true;
            } else {
                return false;
            }
        }
}
```

(7) 部署工程到手机后做运行测试。

5.6　二维码(含条码)的扫描与生成

5.6.1　应用概述

二维码(two-dimensional code)，又称二维条码，它是用特定的几何图形按一定规律在平面(二维方向)上分布的黑白相间的图形。

在现代商业活动中，二维码(含条码)的应用十分广泛。二维码的一个典型应用是电子售票。用户通过网络购买车票(或机票)时，输入购票信息，通过电子支付，即可完成车票的预订，稍后手机会收到二维码电子票信息，旅客凭该信息即可到客运站换票或直接检票登车(机)。总之，通过二维码能实现信息的电子化。

越来越多的手机摄像头具备自动对焦的拍摄功能，这也意味着这些手机可以具备条码扫描的功能。

二维码可以存放很多信息。如果信息量大，特别是包含图形、图像等信息时，通常把这些信息制作到一个 Web 页面并存放到 Web 服务器，再制作指向这个 Web 页面的 URL 地址的二维码。

目前，智能手机都具备扫描二维码的功能。微信的“扫一扫”程序，实质上是先得到条码对应的文本，如同图书的 ISBN，通过这个唯一标识在 Internet 上搜索，并以页面形式呈现。使用微信的“扫一扫”程序，扫描《高级 Web 程序设计——JSP 网站开发》(作者吴志祥、王新颖、曹大有)的结果，如图 5.6.1 所示。

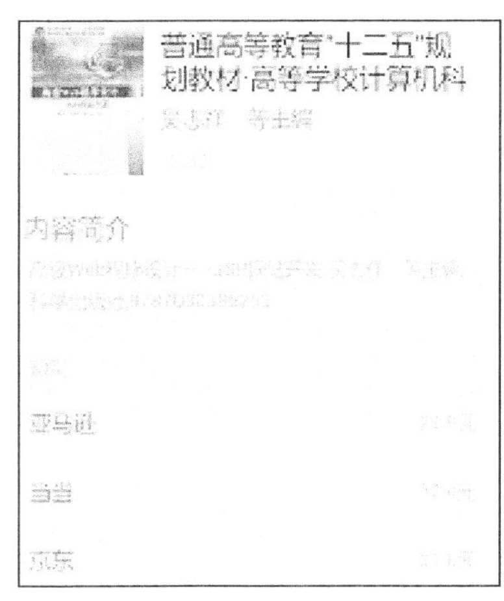

图 5.6.1　微信“扫一扫”程序的扫描结果

注意：

(1) 目前，扫描软件扫描二维码的结果通常是一个表示 URL 的字符串。

(2) 超市里商品的条码一般只是同一种商品的唯一编号，没有 URL 信息。

5.6.2　程序设计

下面介绍一个扫描二维码及生成二维码的 Android 应用程序，它使用了 ZXing 1.6 这个经典的条码/二维码识别的开源类库，实现对二维码的识别和生成。

二维码的 Android 应用工程，需要注册如下权限：

```
<uses-permission android:name="android.permission.CAMERA"/>
<uses-permission android:name="android.permission.WRITE_EXTERNAL_STORAGE"/>
<uses-feature android:name="android.hardware.camera" />
<uses-feature android:name="android.hardware.camera.autofocus" />
```

【例 5.6.1】二维码扫描与生成。

【程序运行】主窗体中的第一个按钮用于扫描二维码或条码，其结果是一个字符串，显示在标签“Scan result:”之后；在文本框内条码对应的文本后，单击第二个按钮，可以生成相应的二维码。程序运行效果，如图 5.6.2 所示。

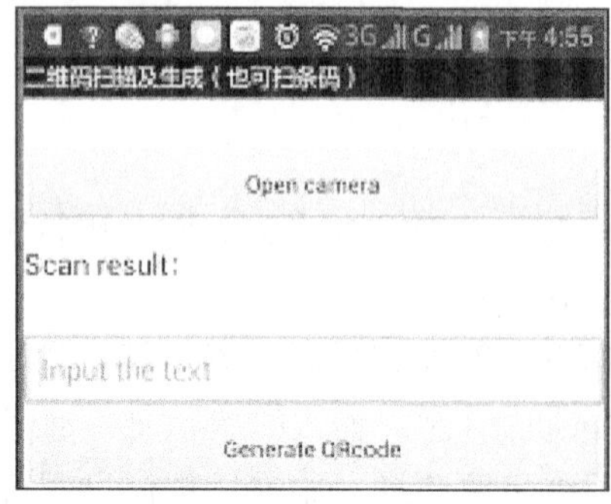

图 5.6.2　二维码扫描与生成程序运行效果

对《高级 Web 程序设计——JSP 网站开发》(作者吴志祥、王新颖、曹大有)的条码应用本程序扫描，其扫描结果就是该书唯一的 ISBN 号，如图 5.6.3 所示。

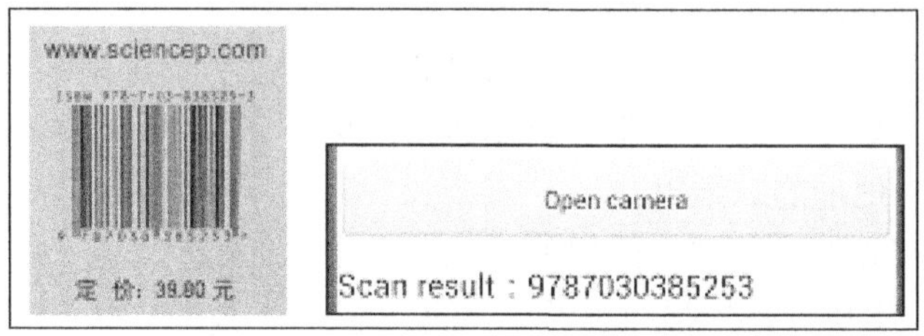

图 5.6.3　条码及其扫描结果

将本书作者的教学网站的网址(http://www.wustwzx.com/)应用本程序生成二维码，如图 5.6.4 所示，扫描生成的二维码也可以得到作者教学网站的网址。

图 5.6.4　作者教学网站的二维码

【设计步骤】

(1) 新建名为 Scan_12Code_Gen2 的 Android 应用工程，勾选创建 Activity，工程文件结构(部分)如图 5.6.5 所示。

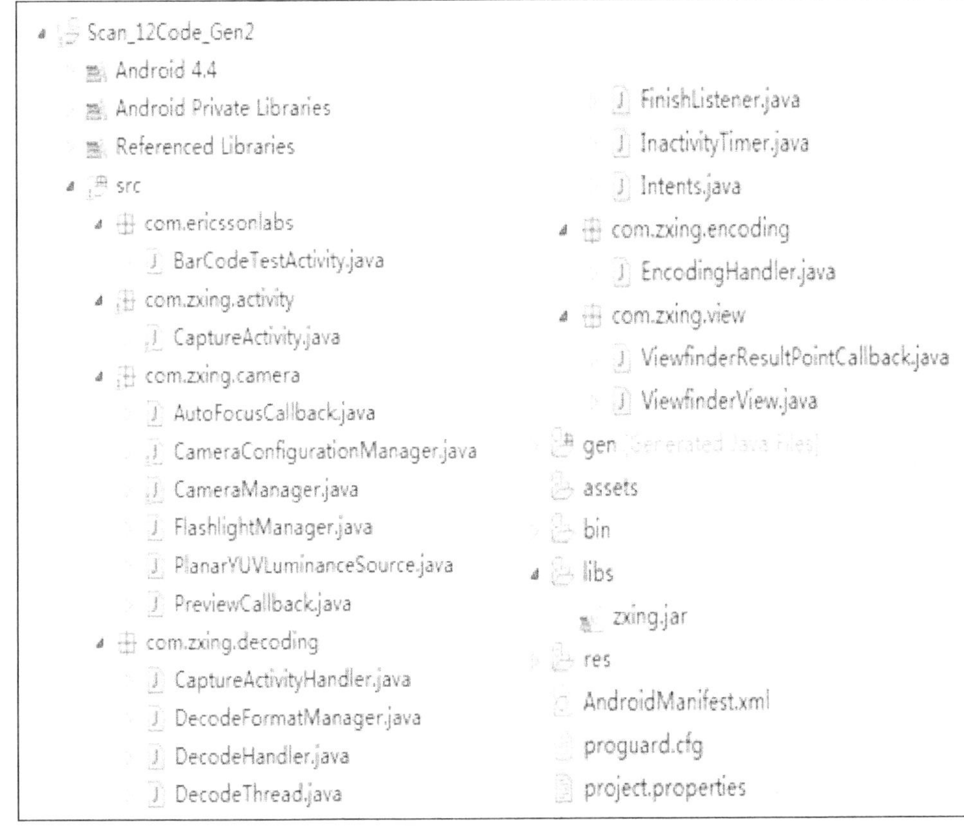

图 5.6.5 Scan_12Code_Gen2 工程文件结构(部分)

(2) 在清单文件中，增加如下两条权限(由标签<uses-permission>指定权限)：

```
<uses-permission android:name="android.permission.VIBRATE" />
<uses-permission android:name="android.permission.CAMERA" />
<uses-feature android:name="android.hardware.camera" />
<uses-feature android:name="android.hardware.camera.autofocus" />
```

(3) 编写项目的主 Activity 对应的文件 BarCodeTestActivity.java 的完整代码如下：

```
/*
 * 扫描二维码，得到商品标识、公司域名等
 * 对任意文本生成二维码
 */
package com.ericssonlabs;
import com.google.zxing.WriterException;
import com.zxing.activity.CaptureActivity;
import com.zxing.encoding.EncodingHandler;
import android.app.Activity;
import android.content.Intent;
import android.graphics.Bitmap;
```

```
import android.os.Bundle;
import android.view.View;
import android.view.View.OnClickListener;
import android.widget.Button;
import android.widget.EditText;
import android.widget.ImageView;
import android.widget.TextView;
import android.widget.Toast;
public class BarCodeTestActivity extends Activity {
    private TextView resultTextView;
    private EditText qrStrEditText;
    private ImageView qrImgImageView;
    @Override
    public void onCreate(Bundle savedInstanceState) {
        super.onCreate(savedInstanceState);
        setContentView(R.layout.main);
        resultTextView = (TextView) this.findViewById(R.id.tv_scan_result);
        qrStrEditText = (EditText) this.findViewById(R.id.et_qr_string);
        qrImgImageView = (ImageView) this.findViewById(R.id.iv_qr_image);
        Button scanBarCodeButton = (Button) this.findViewById(R.id.btn_scan_barcode);
        scanBarCodeButton.setOnClickListener(new OnClickListener() {
            @Override
            public void onClick(View v) {
                Intent openCameraIntent =
                            new Intent(BarCodeTestActivity.this,CaptureActivity.class);
                startActivityForResult(openCameraIntent, 0);
            }
        });
        Button generateQRCodeButton = (Button) this.findViewById(R.id.btn_add_qrcode);
        generateQRCodeButton.setOnClickListener(new OnClickListener() {
            @Override
            public void onClick(View v) {
                try {
                    String contentString = qrStrEditText.getText().toString();
                    if (!contentString.equals("")) {
                        Bitmap qrCodeBitmap
                                = EncodingHandler.createQRCode(contentString, 350);
                        qrImgImageView.setImageBitmap(qrCodeBitmap);
                    }
```

```
                    else {
                        Toast.makeText(BarCodeTestActivity.this,"Text can not be empty",
                                                    Toast.LENGTH_SHORT).show();
                    }
                }
                catch (WriterException e) {
                    // TODO Auto-generated catch block
                    e.printStackTrace();
                }
            }
        });
    }
    @Override
    protected void onActivityResult(int requestCode, int resultCode, Intent data) {
        super.onActivityResult(requestCode, resultCode, data);
        if (resultCode == RESULT_OK) {
            Bundle bundle = data.getExtras();
            String scanResult = bundle.getString("result");
            resultTextView.setText(scanResult);
        }
    }
}
```

习　题　5

一、判断题

1. 意图类 Intent 含于软件包 android.app 内。
2. URL 是比 URI 更广泛的概念。
3. startActivityForResult()是 Activity 提供的且其父类不具有的方法。
4. 实现录音和录像功能的 Android 程序，都需要使用 MediaRecorder 类。
5. MediaPlayer 和 VideoView 是 Android 提供的两个无关的视频播放软件。
6. 扫描二维码的软件本质上不需要网络的支持。

二、选择题

1. 在 Android 中，转换统一资源描述文本为 Uri 类型的方法是____。

   A. convert()　　B. builder()　　C. equals()　　D. parse()

2. 下列控件(类)中，会涉及 Uri 类型的是____。

   A. VideoView　　B. MediaPlayer　　C. MediaRecorder　　D. A 和 B

3. 下列选项中，不是由 androi　d.telephony 包提供的类是____。

   A. SmsMessage　　B. SmsManager

   C. Telephony　　D. TelephonyManager

4. 下列控件(类)中，不需要在清单文件中注册权限的是____。

   A. VideoView　　B. MediaRecorder

   C. SmsManager　　D. B 和 C

5. 下列方法中，不是 Activity 和 Context 共同具有的是____。

   A. startService()　　B. stopService()

   C. findViewById()　　D. getSystemService()

三、填空题

1. 在 Android 应用开发中使用的 Uri 所在的软件包是____。
2. 立即拨打一个输入的电话，需要在工程清单中注册____权限。
3. 拍摄视频并将其保存在 SD 卡里，在工程清单中共需要注册____条权限。
4. 编写手机拍照程序时，为了取得手机内置的相机程序返回的数据，需要使用____方法去启动手机自带的相机程序。
5. 纯扫描软件的扫描结果是一个____。

实验 5　Android 基本功能程序设计

一、实验目的

1. 掌握打电话程序设计的设计方法和监听器的多种用法。
2. 掌握发送短信程序的设计方法。
3. 掌握手机多媒体软件的设计方法。
4. 掌握手机拍照程序的编写，掌握有返回值的 Activity 调用。
5. 掌握手机二维码软件的设计方法。
6. 了解手机二维码扫描及生成的原理，掌握二维码软件的使用方法。

二、实验内容及步骤

【预备】访问 http://www.wustwzx.com/android/index.html，单击“实验 5”超链接，下载本次实验内容的源代码并解压(得到文件夹 ch05)，供研读和调试使用。

1. 打电话程序设计和监听器的多种用法(参见例 5.1.1)。

(1) 再次解压 ch05 内的压缩文件 Dialing2.zip，导入工程 Dialing2。

(2) 在 Eclipse 环境中打开源文件 MainActivity2.java，查看监听器的用法。

(3) 适当修改清单文件，使 MainActivity2 作为应用程序的主界面运行之。

(4) 在 Eclipse 环境中打开源文件 MainActivity3.java，查看监听器的用法。

(5) 适当修改清单文件，使 MainActivity3 作为应用程序的主界面运行之。

(6) 比较本例与例 4.8.1 在创建 Intent 对象时使用参数的差别。

(7) 通过在清单文件中增删打电话权限后部署工程并运行测试，验证本工程必须使用 CALL_PHONE 权限。

2. 短信发送程序设计(参见例 5.2.1)。

(1) 再次解压 ch05 内的压缩文件 SmsSend.zip，导入工程 SmsSend。

(2) 打开源程序 MainActivity.java，查看命令按钮的单击事件过程中发送短信的代码。

(3) 部署工程到手机，做运行测试。

(4) 创建两个 AVD 并启动，部署工程到其中的一个 AVD，然后做运行测试。

(5) 验证当短信内容较少时，可以不使用分段方法 divideMessage()。

3. 前台播放音频程序设计(参见例 5.3.1)。

(1) 再次解压 ch05 内的压缩文件 Media_Audio.zip，导入工程 Media_Audio.zip。

(2) 查看工程文件夹下的音频文件 res/raw/black.mp3。

(3) 部署工程到手机，做运行测试。

(4) 适当修改源程序，复制一个音乐文件至 SD 卡后，部署并做运行测试(当按手机的返回键时，音乐停止播放)。

(5) 屏蔽源文件 MainActivity.java 的 finish()方法内的代码“mp.stop();”后重新部署并做运行测试，验证按手机返回键时，不会停止音乐播放。

4. 掌握手机录音程序的设计方法(参见例 5.3.2)。

(1) 再次解压 ch05 内的压缩文件 AudioRecord.zip，导入工程 AudioRecord。

(2) 打开源程序 AudioRecordDemo.java，查看主 Activity(即 AudioRecordDemo)定义时除了继承 Activity 外，同时实现了内部接口 View.OnClickListener。

(3) 查看命令按钮的单击事件的处理方法。

(4) 部署工程到手机，做运行测试。

5. 使用 VideoView 控件播放视频(参见例 5.4.1)。

(1) 再次解压 ch05 内的压缩文件 MediaVideo.zip，导入工程 MediaVideo。

(2) 打开源程序 MainActivity.java，查看视频文件路径的构造代码。

(3) 准备好视频文件到 SD 卡后，部署工程到手机并做运行测试。

(4) 屏蔽播放完成监听器代码，部署并做运行测试，验证播放完毕后将静止于视频的最后一帧(而不会自动关闭当前 Activity)。

6. 掌握手机拍照程序的设计方法(参见例 5.5.1 和例 5.5.2)。

(1) 再次解压 ch05 内的压缩文件 Camera1.zip 和 Camera2.zip，分别导入工程 Camera1 和 Camera2。

(2) 打开工程 Camera1 中的源程序，查看拍照的实现过程。

(3) 部署工程 Camera1 到手机，做运行测试。

(4) 打开工程 Camera2 中的源程序，查看拍照的实现过程，并与工程 Camera1 比较，找出实现方法的不同点。

(5) 部署工程 Camera2 到手机，做运行测试。

7.掌握视频拍摄程序的设计方法(参见例 5.5.3)。

(1) 再次解压 ch05 内的压缩文件 VideoRecorderDemo.zip，导入工程 VideoRecorderDemo。

(2) 查看清单文件中配置的权限，并与工程 Camera2 比较(多了一个录音的权限)。

(3) 查看源文件中视频的相关参数设置(如音频及视频的编码格式等)。

(4) 查看拍摄完成后关闭摄像头的方法。

(5) 部署工程到手机，做运行测试(打开摄像头→录制→关闭摄像头)。

(6) 播放 SD 卡根目录的视频文件 myVideo.3gp。

8. 了解手机二维码扫描及生成的原理，掌握二维码软件的使用方法(参见例 5.6.1)。

(1) 再次解压 ch05 内的压缩文件 Scan_12Code_Gen2.zip，导入工程 Scan_12Code_Gen2。

(2) 部署工程到手机并运行。

(3) 扫描本教材的条码，核对是否为 ISBN 号码。

(4) 生成武汉科技大学域名的二维码。

(5) 屏蔽清单文件中的两个<uses-feature>标签后，再部署到手机，然后做运行测试，验证对程序运行没有影响。

三、实验小结及思考

(由学生填写，重点写上机中遇到的问题。)

第6章 服务组件与广播组件及其应用

服务组件 Service 和广播接收组件 BroadcastReceiver 是 Android 的两个重要组件。Service 组件是没有运行界面且运行于后台的服务程序，BroadcastReceiver 组件是对广播进行过滤并响应的程序。与 Activity 组件一样，Service 与 BroadcastReceiver 在使用前通常也需要在清单文件里注册(BroadcastReceiver 还可以在程序里动态注册)，并由 Intent 对象去激活。本章学习要点如下：

- 服务组件 Service 介绍，特别是服务的特点及生命周期；
- 服务的多种启动及停止方式；
- 系统提供的常用服务；
- 掌握远程服务的建立与使用；
- 了解 Android 的广播机制；
- 掌握广播接收者的两种注册方式。

6.1 服务组件 Service 的基本用法

6.1.1 服务的概念与 Android 对 Service 的支持

因为手机屏幕尺寸和硬件性能的限制，Android 系统仅允许一个应用程序窗口处于激活状态并显示在手机屏幕上，而暂停其他处于未激活状态的程序(这不同于 Windows 系统可以在屏幕上同时显示多个应用程序窗口)。

Service 本质上是具有一定功能且没有用户界面的程序，它是 Android 系统的服务组件，适用于开发没有用户界面且长时间在后台运行的功能。Service 有利于降低系统资源的开销，而且比 Activity 有更高的优先级(Service 不会在系统资源紧张时被 Android 系统优先终止)。

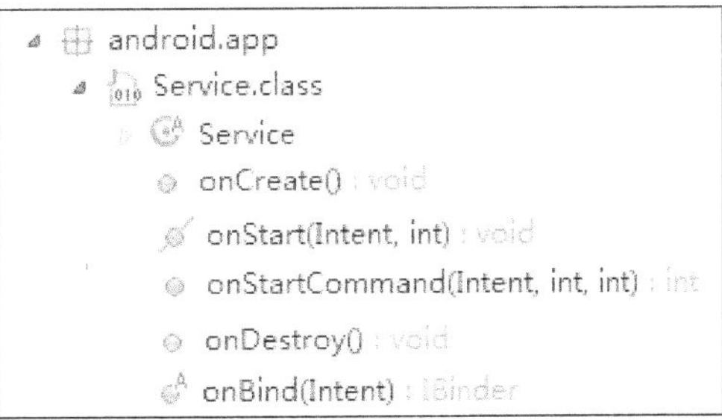

图 6.1.1 Service 类的相关方法

自定义一个 Service，就是实现抽象类 Service 的相关方法，如图 6.1.1 所示。

激活和停止 Service 是由其他组件完成的。例如，组件 Activity 的超类 Context 提供激活和停止 Service 的方法 startService()和 stopService()等，如图 6.1.2 所示。

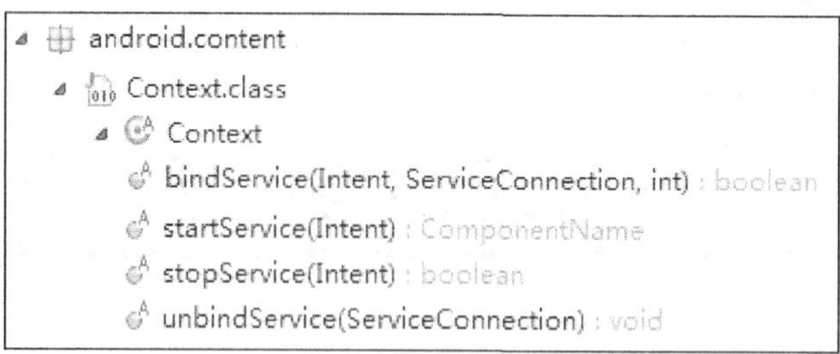

图 6.1.2　Service 的启动与停止方法

注意：

(1) 如同 Activity 组件一样，Service 组件必须在清单文件里使用标签<service>注册，否则，服务无法启动；

(2) Activity 的父类提供的绑定服务方法 bindService()和解除绑定服务方法 unbindService()将在 6.1.5 小节中介绍。

6.1.2　Android 提供的系统服务

在一些 Android 类中，定义了表示系统服务的常量，调用某种系统服务后，就得到相应的服务管理器对象，以实现某种管理服务。

在第 4.5.2 小节，我们知道，通过使用抽象类 Context 提供的 getSystemService()方法调用关于通知的系统服务 Context.NOTIFICATION_SERVICE，可以获取通知管理器 NotificationManager 对象。实际上，Android 类 Context 中定义了许多系统服务常量，如布局填充服务 LAYOUT_INFLATER_SERVICE 等，Android 系统服务常量如图 6.1.3 所示。

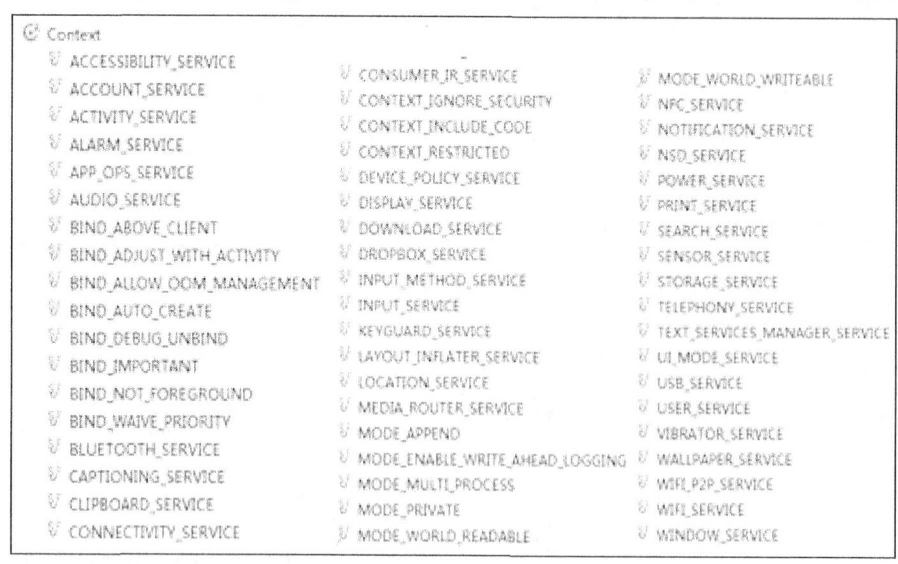

图 6.1.3　抽象类 Context 中定义的系统服务常量

注意： Context 提供的 getSystemService()方法的返回值为 Object 类型，因此，在实际应用时需要进行类型转换(为相应的管理器对象)，参见例 4.5.1。

Android 后台运行的很多 Service 是在系统启动时被开启的，以支持系统的正常工作。比如，MountService 监听是否有 SD 卡安装及移除，ClipboardService 提供剪切板功能，PackageManagerService 提供软件包的安装、移除及查看等。Android 提供的常用服务，如表 6.1.1 所示。

表 6.1.1 Android 中常用的系统服务

服务名称	功能描述
NOTIFICATION_SERVICE	通知服务，得到通知管理器对象，参见第 4.6 节
POWER_SERVICE	电源服务，得到电源管理器对象
AUDIO_SERVICE	实现对音量、音效、声道及铃声等的管理服务
CONNECTIVITY_SERVICE	调用网络连接(GPRS 或 WiFi)服务，获得 ConnectivityManager 对象，参见第 10.2 节
WIFI_SERVICE	WiFi 服务，得到 WiFi 管理器对象，参见第 9.1.3 小节
BLUETOOTH_SERVICE	蓝牙服务，参见第 9.2.3 小节
LOCATION_SERVICE	调用此定位服务，获得位置管理器对象，实现对 GPS 的使用，参见第 10.2 节

下面对 Android 提供的若干系统服务及其相关类加以说明。

1. 应用程序包管理服务 PackageManagerService(PMS)

通过 PackageManagerService 获取已经安装的应用程序的相关信息，参见例 4.8.2。

2. 活动管理服务 ActivityManagerService 及其内部类

调用系统服务 ActivityManagerService(AMS)得到活动管理器 ActivityManager 对象。活动管理器类 ActivityManager 的内部类 RunningServiceInfo 与服务相关，它封装了正在运行的服务程序的信息，如图 6.1.4 所示。

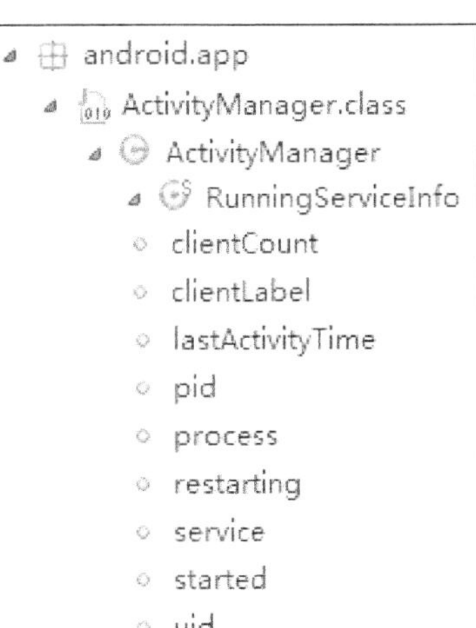

图 6.1.4 内部类 RunningServiceInfo

3. 窗口管理服务 WindowManagerService(WMS)

窗口管理服务 WindowManagerService 是 Android framework 的核心服务，用于对窗口管理。

4. 通知服务 NOTIFICATION_SERVICE

调用系统服务 NOTIFICATION_SERVICE，得到 NotificationManager 对象，进而管理状态栏通知。

5. 网络管理服务 NetworkManagementService

Android 系统网络连接和管理服务由四个系统服务 ConnectivityService、NetworkPolicyManagerService、NetworkManagementService、NetworkStatsService 共同配合完成。

6.WiFi 服务 WIFI_SERVICE

调用系统服务 WIFI_SERVICE，得到 WifiManager 对象，进而管理 WiFi 信息。

7. 蓝牙管理服务类 BluetoothManagerService

BluetoothManagerService 负责蓝牙后台管理和服务。

8. 位置管理器服务 LocationManagerService

LocationManagerService 提供了位置服务、GPS 服务、定位服务等。

9. ContentService

ContentService 即内容服务，主要为数据库等提供解决方法的服务。

10. ClipboardService

ClipboardService 提供了剪贴板服务。

11. 闹铃和定时器类 AlarmManagerService

AlarmManagerService 提供了闹铃和定时器等功能。

12. 音频服务类 AudioService

AudioService 提供了音量、音效、声道及铃声等的管理服务。

13. 电源管理服务 PowerManagerService

PowerManagerService 提供了电源管理服务。

14. 电池管理类 BatteryService

电池管理类 BatteryService 负责监控电池的充电状态、电池电量、电压、温度等信息，当电池信息发生变化时，发出广播通知其他关系电池信息的进程和服务。

15. 屏保服务 DreamManagerService

DreamManagerService 提供了屏幕保护功能。

16. 输入法服务 InputMethodManagerService

输入法服务提供了打开和关闭输入法的功能。

17. 光感应传感服务 LightsService

LightsService 提供了光感应传感器服务。

18. 锁屏设置服务 LockSettingsService

LockSettingsService 提供了锁屏、手势等安全服务。

19. 壁纸管理服务 WallpaperManagerService

WallpaperManagerService 提供了壁纸管理服务。

20. 搜索服务 SearchManagerService

SearchManagerService 提供了搜索服务。

21. 状态栏管理服务 StatusBarManagerService

StatusBarManagerService 提供了对状态栏的管理功能。

22. 振动服务 VibratorService

VibratorService 提供了振动器服务。

6.1.3　自定义服务与服务注册

新建一个服务程序，需要继承抽象类 android.app.Service，并改写 Service 的 onCreate()、onDestroy()等方法。

注意：服务类的设计，与其启动方式有关。启动方式分为 startService(非绑定方式)和 bindService(绑定方式)两种。绑定服务的用法相对复杂些，参见 6.1.5 小节。

自定义的服务，需要在清单文件里使用<service>标签注册，类似于 Activity 组件需要在清单文件中使用<activity>标签注册。

6.1.4　服务的显式启动与隐式启动

创建 Intent 对象时，指明 Service 所在的类，并调用方法 startService(Intent)启动 Service，这种方式称为显式启动。

隐式启动是指在注册 Service 的同时，还内嵌了标签<intent-filter>及<action>。其中，<action>标签的 android:name 属性指定了引用该服务的名称。

隐式启动的好处是不需要指明需要启动哪一个 Activity，而由 Android 系统根据 Intent 的动作和数据来决定，这样有利于降低 Android 组件之间的耦合度，强调 Android 组件的可复用性。

注意：若 Service 与调用它的组件在同一个应用程序中，则既可以使用显式启动，也可以使用隐式启动(显式启动代码简洁)；若服务和调用服务的组件在不同的应用程序中，则服务只能使用隐式启动方式。

【例 6.1.1】服务的创建与使用示例：调用后台音频播放服务。

【程序功能】程序运行时，调用后台服务程序播放音频文件，在主界面中有一个提示正在播放的文本框和一个停止播放的“取消”按钮，当单击“取消”按钮后，停止音乐播放和应用程序的运行。程序运行效果如图 6.1.5 所示。

图 6.1.5　调用后台音频播放服务工程的运行界面

【设计步骤】

(1) 新建名为 Audio_Service1 的 Android 应用工程，源程序包名默认，勾选创建 Activity。

(2) 指向工程名，使用快捷菜单“Android Tools”，修改应用的包名为“com.example.wu_services”。此时，工程文件结构(主要部分)如图 6.1.6 所示。

图 6.1.6　Audio_Service1 工程文件结构(主要部分)

(3) 主布局文件 activity_main.xml 中 TextView 控件与 Button 控件采用垂直线性布局，其代码如下：

```
<LinearLayout xmlns:android="http://schemas.android.com/apk/res/android"
    xmlns:tools="http://schemas.android.com/tools"
    android:layout_width="match_parent"
    android:layout_height="match_parent"
    android:paddingBottom="@dimen/activity_vertical_margin"
    android:paddingLeft="@dimen/activity_horizontal_margin"
    android:paddingRight="@dimen/activity_horizontal_margin"
    android:paddingTop="@dimen/activity_vertical_margin"
    tools:context=".MainActivity" >
    <TextView
        android:id="@+id/textView1"
        android:layout_width="wrap_content"
        android:layout_height="wrap_content"
        android:text="@string/hello_world" />
    <Button
        android:id="@+id/button1"
        android:layout_width="wrap_content"
        android:layout_height="wrap_content"
        android:layout_below="@+id/textView1"
        android:layout_marginLeft="45dp"
        android:layout_marginTop="36dp"
        android:layout_toRightOf="@+id/textView1"
        android:text="取消" />
</LinearLayout>
```

(4) 编写服务程序 MyAudioService.java，文件代码如下：

```
package com.example.media_audio_service1;
/*
```

```
 * 新建Service时，方法 onBind()不可去!
 * 为了防止服务停止了但音乐仍在播放，必须重写Activity的onDestroy()方法
 */
import android.os.IBinder;
import android.app.Service;  //主类
import android.media.MediaPlayer;
import android.content.Intent;
import java.io.IOException;
public class MyAudioService extends Service {
    private MediaPlayer mediaplayer;
    @Override
    public void onCreate() {
        // TODO Auto-generated method stub
        super.onCreate();     //
    }
    @Override
    public void onStart(Intent intent, int startId) {
        // TODO Auto-generated method stub
        String path="sdcard/music/white.mp3";     //
        mediaplayer=new MediaPlayer();
        try {
            mediaplayer.setDataSource(path);
            mediaplayer.prepare();
            mediaplayer.start();
        }catch (IOException e) {
            // TODO Auto-generated catch block
            e.printStackTrace();
        }
    }
    @Override
    //注意：此方法不可去!
    public IBinder onBind(Intent arg0) {
        // TODO Auto-generated method stub
        return null;  // 没有返回代理对象就是非绑定，反之为服务绑定
    }
    @Override
    public void onDestroy() {
        // TODO Auto-generated method stub
        if(mediaplayer!=null){
```

```
                mediaplayer.stop();      //停止播放
                mediaplayer.release();
            }
            super.onDestroy();
        }

}
```

(5) 编写调用服务的主界面程序 MainActivity.java，其文件代码如下：

```
package com.example.media_audio_service1;
/*
 * 本工程中调用的服务，需要在清单文件里注册，而不必配置其过滤器
 * 使用方法startService(intent)启动服务
 * 要求在标准SD卡根目录有音乐文件 music/white.mp3
 * 单击“取消”按钮后，可以停止播放，因为在停止服务时会调用onDestroy()方法
 * 本工程中已屏蔽返回键
 */
import com.example.wu_services.R;  //源程序包名与应用包名不一致时需要导入
import android.app.Activity;
import android.os.Bundle;
import android.content.Intent;
import android.widget.Button;
import android.view.View;
import android.view.View.OnClickListener;
import android.view.KeyEvent;   //键盘事件
//import android.os.Process;     //
public class MainActivity extends Activity {
    Intent intent;
    @Override
    protected void onCreate(Bundle savedInstanceState) {
        super.onCreate(savedInstanceState);
        setContentView(R.layout.activity_main);
        //显式调用服务
        intent=new Intent(this , MyAudioService.class);
        startService(intent);
        //呈现一个停止服务按钮并设置其监听器
        Button button1=(Button)findViewById(R.id.button1);
        button1.setOnClickListener(new buttonListenner());
    }
    class buttonListenner implements OnClickListener{
```

```
        @Override
        public void onClick(View v) {
            // TODO Auto-generated method stub
            intent=new Intent(MainActivity.this , MyAudioService.class);
            stopService(intent);   //停止服务
            //Process.killProcess(Process.myPid());   //清除进程并关闭Activity
            finish();
            }
    }
    @Override
    //重写按键方法onkeyDown()屏蔽按键
    public boolean onKeyDown(int keyCode, KeyEvent event) {
        // TODO Auto-generated method stub
        if(KeyEvent.KEYCODE_BACK == keyCode)     //返回键
            return false;   //return true;
        //return super.onKeyDown(keyCode, event);     //其他按键默认处理
        return MainActivity.this.onKeyDown(keyCode, event);
    }
}
```

6.1.5　绑定服务方式与服务代理

开启一个服务，除了可以选择调用服务方式(显式启动和隐式启动)外，还可以选择启动服务方式(是否绑定服务)。

在绑定服务方法 bindService(Intent,ServiceConnection,int)中，第一个参数表示如果被绑定的 Service 没有启动，则使用 bindService()方法会自动启动该 Service。

bindService()的第一个参数是一个明确指定了要绑定的 Service 的 Intent。

第二个参数是 ServiceConnection 对象。

第三个参数是一个标志，它表明绑定中的操作。它一般应是 BIND_AUTO_CREATE，表示当 Service 不存在时创建一个，其他可选的值是 BIND_DEBUG_UNBIND 和 BIND_NOT_FOREGROUND，不想指定时设为 0 即可。

绑定服务要通过服务连接(ServiceConnection，见图 6.1.7)对象获取 Service 中的状态和数据信息。

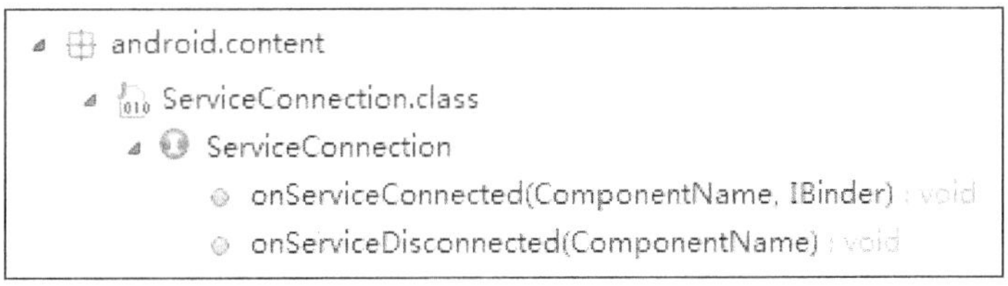

```
android.content
    ServiceConnection.class
        ServiceConnection
            onServiceConnected(ComponentName, IBinder) : void
            onServiceDisconnected(ComponentName) : void
```

图 6.1.7　ServiceConnection 接口

通过绑定方式使用 Service，不仅能够获取 Service 实例，正常启动 Service，还能够调用 Service 中的公有方法和属性。为了使 Service 支持绑定，需要在 Service 中重载 onBind()方法，并在 onBind()方法中返回 Service 实例。这个过程即是通过 Binder(代理)对象获取 Service 对象，进而获取所需的服务。

代理类 Binder 实现了 IBinder 接口，其定义如图 6.1.8 所示。

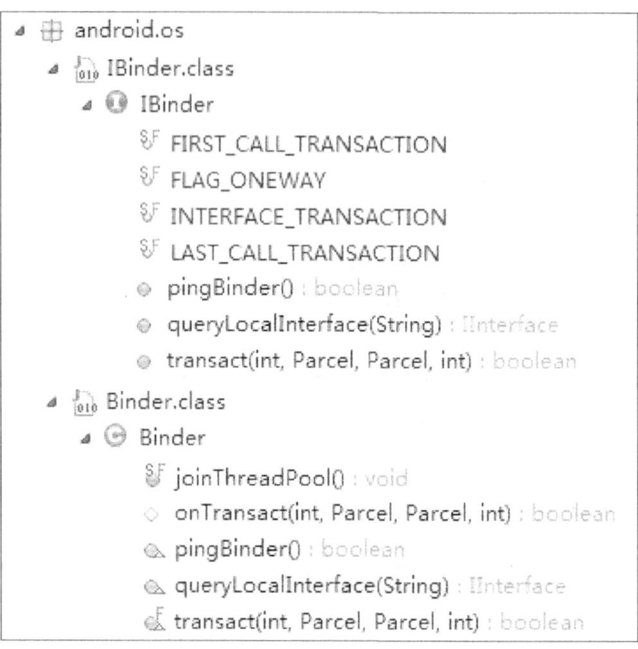

图 6.1.8 服务代理接口 IBinder 与代理类 Binder

使用 bindService()方法启动 Service 的生命周期和 Activity 对象一致，也就是说，使用 bindService()方法启动 Service 后，Service 就与调用 bindService()方法的进程同生共死，当调用 bindService()方法的进程结束了，那么它绑定的 Service 也要跟着被结束。通常，在 Activity 的 onStop()回调函数中加上反绑定 Service 的方法 unbindService()来停止服务连接。

注意：使用 bindService()方法启动服务的方式称为绑定服务方式，而例 6.1.1 中使用 startService()方法启动服务的方式称为非绑定服务方式。

例 6.1.2 中的后台播放，是指在 Activity 中调用一个已经注册的 Service，这不同于前面介绍的前台播放。

【例 6.1.2】以绑定方式调用后台音频播放服务。

【设计步骤】

(1) 新建名为 Audio_Service2 的 Android 应用工程，勾选创建 Activity。工程文件结构(主要部分)如图 6.1.9 所示。

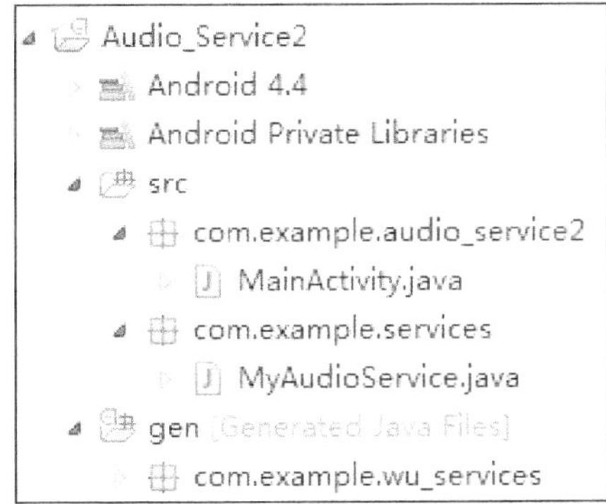

图 6.1.9　Audio_Service2 工程文件结构(主要部分)

(2) 本工程采用显式调用服务方式，如果使用隐式调用服务方式，则需要在工程清单文件中注册服务的同时配置该服务的意图过滤器，其代码如下：

```
<service android:name=".MyAudioService">
 <!--      <intent-filter >
             <action android:name="com.example.audio_service_test.MAS"/>
             <category android:name="android.category.DEFAULT"/>
         </intent-filter>          -->
</service>
```

(3) 编写适合绑定方式的服务程序 MyAudioService.java，文件代码如下：

```
package com.example.services;
/*
 * 本服务程序按照绑定服务方式而写，包含一个继承Binder的内部类PlayBinder;
 * 类PlayBinder定义了播放音乐的方法;
 * 当服务程序被激活时，通过服务类的onBind()方法返回一个PlayBinder对象;
 * 在界面程序中，通过PlayBinder对象调用播放音乐的方法;
 * 通过IBinder对象，还可以跟踪服务的运行状态;
 * 音乐文件存放在SD卡的music文件夹里。
*/
import android.os.IBinder;   //接口
import android.os.Binder;   //类
import android.widget.Toast;
import android.app.Service;
import android.media.MediaPlayer;
import android.content.Intent;
import java.io.IOException;

public class MyAudioService extends Service {
     private MediaPlayer mediaplayer;
```

```
    //定义服务代理类(服务类的内部类)
    public class PlayBinder extends Binder{    //
        public void MyMethod(){
            Toast.makeText(getApplicationContext(), "我是服务里的方法",
                                                    Toast.LENGTH_LONG).show();
            String path="sdcard/Music/white.mp3";     //
            mediaplayer=new MediaPlayer();
            try {
                mediaplayer.setDataSource(path);
                mediaplayer.prepare();
                mediaplayer.start(); }
            catch (IOException e) {
                // TODO Auto-generated catch block
                e.printStackTrace();
            }
        }
    }
    @Override
    public IBinder onBind(Intent arg0) {   //
        // TODO Auto-generated method stub
        return new PlayBinder();       //返回服务代理对象
    }
    @Override
    public void onDestroy() {
        // TODO Auto-generated method stub
        if(mediaplayer!=null){
            mediaplayer.stop();
            mediaplayer.release();
        }
        super.onDestroy();
    }
}
```

(4) 编写界面程序MainActivity.java，文件代码如下：

```
package com.example.audio_service2;
/*
 * 本程序演示了以绑定服务方式启动服务
 * 特点1: 在Activity里创建服务连接对象
 * 特点2: 在服务程序中返回一个服务代理(Binder对象)
```

```
*/
import android.app.Activity;
import android.os.Bundle;
import android.os.IBinder;   //
import android.content.ComponentName;
import android.content.Intent;
import android.content.ServiceConnection; //接口
import android.widget.Button;
import android.view.View;
import android.view.View.OnClickListener;
import android.view.KeyEvent;
import com.example.services.MyAudioService;
import com.example.services.MyAudioService.PlayBinder;
import com.example.wu_services.R;   //
public class MainActivity extends Activity {
    Intent intent;
     // 创建服务连接对象，实现ServiceConnection接口
    private ServiceConnection conn = new ServiceConnection() {
        @Override
        public void onServiceConnected(ComponentName name, IBinder service) {
            // TODO Auto-generated method stub
            PlayBinder myAudioBinder = (MyAudioService.PlayBinder) service; //
            myAudioBinder.MyMethod(); //
        }
        @Override
        public void onServiceDisconnected(ComponentName name) {
            // TODO Auto-generated method stub
        }
    };
    @Override
    protected void onCreate(Bundle savedInstanceState) {
        super.onCreate(savedInstanceState);
        setContentView(R.layout.activity_main);
        intent = new Intent(this, MyAudioService.class); // 显式调用
        // intent=new Intent("com.example.audio_service_test.MAS"); //隐式调用
        bindService(intent, conn, BIND_AUTO_CREATE); // 绑定服务方式
        Button button1 = (Button) findViewById(R.id.button1);
        button1.setOnClickListener(new MyListenner());
    }
```

```
    // 定义内部类——按钮的单击事件监听器
    class MyListenner implements OnClickListener {
        @Override
        public void onClick(View v) {
            // TODO Auto-generated method stub
            unbindService(conn); //
            finish();
        }
    }
    @Override
    public boolean onKeyDown(int keyCode, KeyEvent event) {
        // TODO Auto-generated method stub
        if (KeyEvent.KEYCODE_BACK == keyCode)
            return false; // return true;
        return MainActivity.this.onKeyDown(keyCode, event);
    }
}
```

(5) 部署工程并做运行测试。

6.2 远程服务

6.2.1 本地服务与远程服务

前面介绍的是本地服务，本地服务的特点是使用的服务含于应用的内部，即不存在跨进程调用。

在 Android 系统中，每个应用程序在各自的进程中运行，这些进程之间彼此是隔离的。要完成进程之间数据(对象)的传递，需要使用 Android 支持的进程间通信(inter-process communication，IPC)机制。Android 系统没有使用传统的 IPC 机制(如共享内存、管道、消息队列和 Socket 等)，而是采用了 Intent 和远程服务的方式实现 IPC。

在 Android IPC 中，Binder 是进程通信的媒介，Parcel 是进程通信的内容。

注意：

(1) 本地服务与远程服务不是根据距离划分的，而是根据应用的包名划分的；

(2) 调用 Web 服务，本质上也是调用远程服务，参见例 11.2.1。

6.2.2 Android 跨进程调用与接口定义语言 AIDL

Android 系统中的各进程之间不能直接访问相互的内存空间，为了使数据能够在不同进程间传递，数据必须先转换成能够穿越进程边界的系统级原始语言，同时，在数据完成进程边界穿越后，再转换成原有的格式。

AIDL(android interface definition language)是 Android 自定义的接口描述语言，可以简化进程间数据格式转换和数据交换的代码，通过定义 Service 内部的公有方法，允许在不同进程的调用者和 Service 之间相互传递数据。

AIDL 与 Java 的接口定义非常相似，唯一不同的是 AIDL 允许定义函数参数的传递方向，共有三种方向。标识为 in 的参数将从调用者传递到远程服务中，标识为 out 的参数将从远程服务传递到调用者中，标识为 inout 的参数先从调用者传递到远程服务中，再从远程服务返回调用者。

使用 AIDL 跨进程调用(远程服务)的服务器端开发，其主要步骤如下。

(1) 创建接口描述文件。在 Android 工程的源文件夹的某个包里，创建扩展名为.aidl 的文件(扩展名不是.java)。在该文件中，定义了客户端可用的方法和数据的接口。如果.aidl 文件的内容是正确的，ADT 会在 gen 文件夹里自动生成一个 Java 接口文件(*.java)，该接口有一个继承的命名为 Stub 的内部抽象类(并且实现了一些 IPC 调用的附加方法)，在 Stub 类里又含有一个名为 Proxy(代理)的内部类。

(2) 建立一个服务类(Service 的子类)并实现接口。在服务类中，定义远程服务代理对象，它所在的类重写了抽象类 YourInterface.Stub 里与.aidl 文件中相对应的方法。

(3) 在 Service 的绑定方法 onBind(Intent) 里返回实现了接口的实例对象。

(4) 在清单文件中注册 AIDL 对应的服务。

注意：位于系统文件夹 sdk/build-tools 内的 AIDL 编译器自动编译成接口。

使用 AIDL 跨进程调用(远程服务)的客户端开发，其主要步骤如下。

(1) 创建与服务器端相同的接口描述文件。

(2) 在主 Activity 中，定义远程服务连接对象，在其内创建远程服务对象。

(3) 在主 Activity 中使用绑定方式调用远程服务，通过远程服务对象调用远程服务里的方法。

注意：使用 AIDL 语言建立远程通信接口描述文件，是服务器端与客户端共同拥有的。

6.2.3　远程服务的建立与使用实例

下面通过一个实例，说明远程服务的建立与调用。

【例 6.2.1】以远程服务方式调用自定义的后台音频播放服务程序。

【程序运行】远程服务部署和运行一次后，再运行客户端程序，就能播放远程服务器里的音乐，如图 6.2.1 所示。

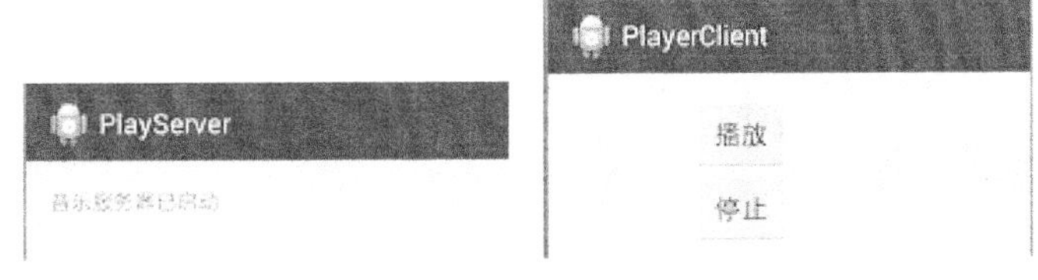

图 6.2.1　以远程服务方式调用后台音频播放服务工程的运行界面

注意：音乐服务器仅需要启动一次。

【设计步骤】

(1) 新建名为 MusicServer 的 Android 应用工程，勾选创建 Activity，程序包名设置为“com.example.playserver”。

(2) 在 res 文件夹里新建文件夹 raw，复制名为 tick.mp3 的音乐文件到该文件夹里。

(3) 在程序包文件夹里，新建服务程序 PlayerService。

(4) 在程序包文件夹里新建使用 AIDL 语言编写的文件 IRemoteService.aidl，其代码如下：

```
package com.example.playserver;
interface IRemoteService {
    void play();
    void stop();
}
```

此时，工程 MusicServer 的文件结构(主要部分)，如图 6.2.2 所示。

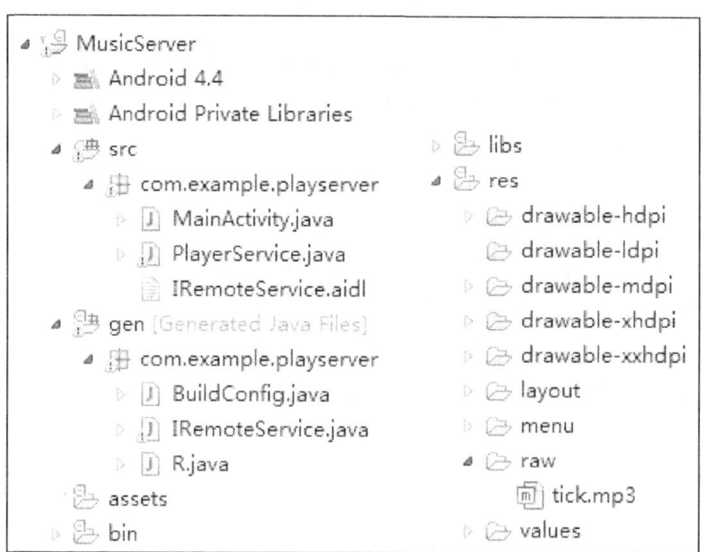

图 6.2.2　MusicServer 工程文件结构(主要部分)

(5) 工程 MusicServer 采用隐式调用服务方式，因此，在工程清单文件中注册服务的同时还需配置该服务的意图过滤器，其代码如下：

```
    <service   android:name="com.example.playserver.PlayerService"
                                    android:process=":remote">   <!-远程服务-->
        <intent-filter>
                <action android:name="com.example.playerserver.PlayerService"/>
        </intent-filter>
</service>
```

(6) 服务程序文件 PlayerService.java 的代码如下：

```
package com.example.playserver;
/*
 *  本服务程序中应用了IRemoteService.aidl对应的接口IRemoteService
 *  音乐播放代码含于服务的onCreate()方法内
 */
import java.io.IOException;
import android.app.Service;   //
import android.content.Intent;
import android.media.MediaPlayer;
import android.os.IBinder; //
import android.os.RemoteException;
import android.widget.Toast;

public class PlayerService extends Service {
    private MediaPlayer mPlayer;
    //定义远程服务代理对象ibinder
    //实现IRemoteService.aidl文件中定义的接口
    //需要重写远程服务接口的内部类Stub中定义的抽象方法
    IBinder ibinder=new   IRemoteService.Stub() {
        @Override
        public void stop() throws RemoteException {
            // TODO Auto-generated method stub
            if(mPlayer.isPlaying()){
                mPlayer.stop();
                try {
                        // stop()后再次start()前需要
                        mPlayer.prepare();
                }
                catch (IOException ex) {
                        ex.printStackTrace();
                }
                mPlayer.seekTo(0);    //从0秒开始
            }
        }
        @Override
        public void play() throws RemoteException {
            // TODO Auto-generated method stub
            if(!mPlayer.isPlaying())
                mPlayer.start();
        }
```

```
    };
    @Override
    public void onCreate() {     //创建服务
        super.onCreate();
        if(mPlayer==null){
            mPlayer=MediaPlayer.create(getApplicationContext(), R.raw.tick);
            mPlayer.setLooping(true);
        }
    }
    @Override
    public IBinder onBind(Intent intent) {       //绑定服务
            Toast.makeText(getApplicationContext(),   intent.getAction(),
                                                  Toast.LENGTH_SHORT).show();
        return ibinder;     //
    }
    @Override
    public boolean onUnbind(Intent intent) {     //解除绑定
        // TODO Auto-generated method stub
        if (mPlayer != null) {
            mPlayer.stop();
            mPlayer.release();   }
        return super.onUnbind(intent);
    }
}
```

MusicClient
- Android 4.4
- Android Private Libraries
- src
  - com.example.playclient
    - MainActivity.java
  - com.example.playserver
    - IRemoteService.aidl
- gen [Generated Java Files]
  - com.example.playclient
  - com.example.playserver
    - IRemoteService.java
      - IRemoteService

图 6.2.3　音乐客户端工程文件结构(主要部分)

(7) 新建名为 MusicClient 的 Android 应用工程，勾选创建 Activity，程序包名设置为 com.example.playclient。

(8) 在程序文件夹 src 里新建一个名为 "com.example.playserver" 的程序包名，并将 MusicServer 工程中的 IRemoteService.aidl 文件复制到包里，此时的工程文件结构(主要部分)如图 6.2.3 所示。

(9) 设计客户端界面及事件过程，其文件 MainActivity.java 的代码如下：

```
package com.example.playclient;
/*
 * 音乐播放客户端，调用远程服务，使用绑定
```

```
服务方式
*/
import com.example.playclient.R;
import com.example.playserver.IRemoteService;   //
import android.os.Bundle;
import android.os.IBinder;
import android.os.RemoteException;
import android.app.Activity;
import android.content.ComponentName;
import android.content.Context;
import android.content.Intent;
import android.content.ServiceConnection;   //
import android.view.Menu;
import android.view.View;
import android.widget.Button;

public class MainActivity extends Activity {
    // 对应服务端 AndroidManifest.xml中的intent-filter action声明的字符串
    public static final String ACTION = "com.example.playerserver.PlayerService";
    //服务是否已绑定
    private boolean isBinded = false;
    //远程服务接口
    private IRemoteService mService;
    //创建服务连接对象
    private ServiceConnection conn=new ServiceConnection(){
        @Override
        public void onServiceConnected(ComponentName arg0, IBinder arg1) {
            mService = IRemoteService.Stub.asInterface(arg1); //实例化
            isBinded = true;
        }
        @Override
        public void onServiceDisconnected(ComponentName arg0) {
            isBinded = false;
            mService = null;
        }
    };
    @Override
    protected void onCreate(Bundle savedInstanceState) {
        super.onCreate(savedInstanceState);
        setContentView(R.layout.activity_main);
```

```
        Intent intent = new Intent(ACTION);
        //创建远程服务连接
        bindService(intent, conn, Context.BIND_AUTO_CREATE);

        Button bt1=(Button)findViewById(R.id.play);
        Button bt2=(Button)findViewById(R.id.stop);

        bt1.setOnClickListener(new View.OnClickListener() {      //播放
            @Override
            public void onClick(View arg0) {
                try {
                    mService.play();
                }
                catch (RemoteException e) {
                    // TODO Auto-generated catch block
                    e.printStackTrace();
                }
            }
        });
        bt2.setOnClickListener(new View.OnClickListener() {      //停止
            @Override
            public void onClick(View v) {
                try {
                    mService.stop();
                } catch (RemoteException e) {
                    // TODO Auto-generated catch block
                    e.printStackTrace();
                }
            }
        });
    }
    @Override
    protected void onDestroy() {
        // TODO Auto-generated method stub
        if (isBinded) {
             unbindService(conn);
             mService = null;
             isBinded = false;
        }
```

```
        super.onDestroy();
    }
    @Override
    public boolean onCreateOptionsMenu(Menu menu) {
        getMenuInflater().inflate(R.menu.main, menu);
        return true;
    }
}
```

(10) 部署工程并做运行测试。

6.3 广播 Broadcast 与广播接收者组件 BroadcastReceiver

6.3.1 Android 的广播机制

在 Android 系统中，广播(Broadcast)是一种广泛运用在应用程序之间传输信息的机制。例如，来电话、来短信、手机没电等系统发送的消息，也可以是一个应用程序自定义的消息，这些消息通过广播方式通知给应用程序，并让相应的广播接收者(BroadcastReceiver)去处理消息。

根据广播的来源，Android 广播可分为系统广播与用户自定义广播两种。

当某种特定的事件发生时，Android 系统都会产生一个 Intent 对象自动进行广播。例如，使用“android.provider.Telephony.SMS_RECEIVED”来表示手机有短信到来时所产生的广播。

广播消息实质上就是将一个 Intent 对象用 sendBroadcast()方法发送出去。在 Android 系统中，上下文类 Context(为抽象类)提供了发送广播的抽象方法，如图 6.3.1 所示。

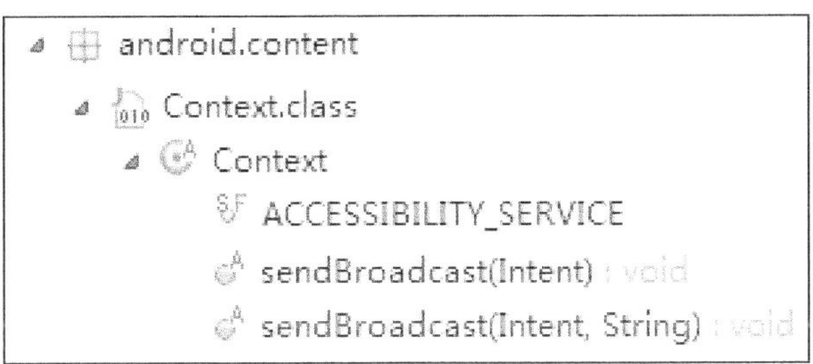

图 6.3.1　发送广播方法

在一个 Activity 里，通过使用继承的方法 sendBroadcast()可以实现对自定义广播的发送(参见表 4.2.1)。

在广播消息前，通过使用 Intent 对象的 putExtras()方法封装广播的数据(Bundle 类型)，在广播接收程序中通过使用 Intent 对象的 getExtras()方法获得 Bundle 类型的数据。

注意：生活中的广播，我们能听到，而 Android 系统中的消息广播，只有与之相应(并非所有)的广播接收者 BroadcastReceiver 才能接收到。此外，有些消息广播还需要有相应的权限。

6.3.2 接收广播的抽象类 android.content.BroadcastReceiver

BroadcastReceiver 是对发送出来的广播进行过滤并响应的一类组件，是 Android 系统的四大组件之一。一个 BroadcastReceiver 对象的生命周期从调用 onReceive(Context,Intent)方法开始，到该方法返回结束。

Android API 提供的 android.content.BroadcastReceiver，是广播接收者类的抽象父类。

在 Android 应用中，每个广播接收者类都是使用标签<receiver>在应用工程的清单文件中注册，只能具有相应权限的广播接收者才能接收广播、获取 Intent 对象中的数据。

除了标准的类继承方式创建广播接收者外，Android 还提供了通过向导创建广播接收者的方法，如图 6.3.2 所示。

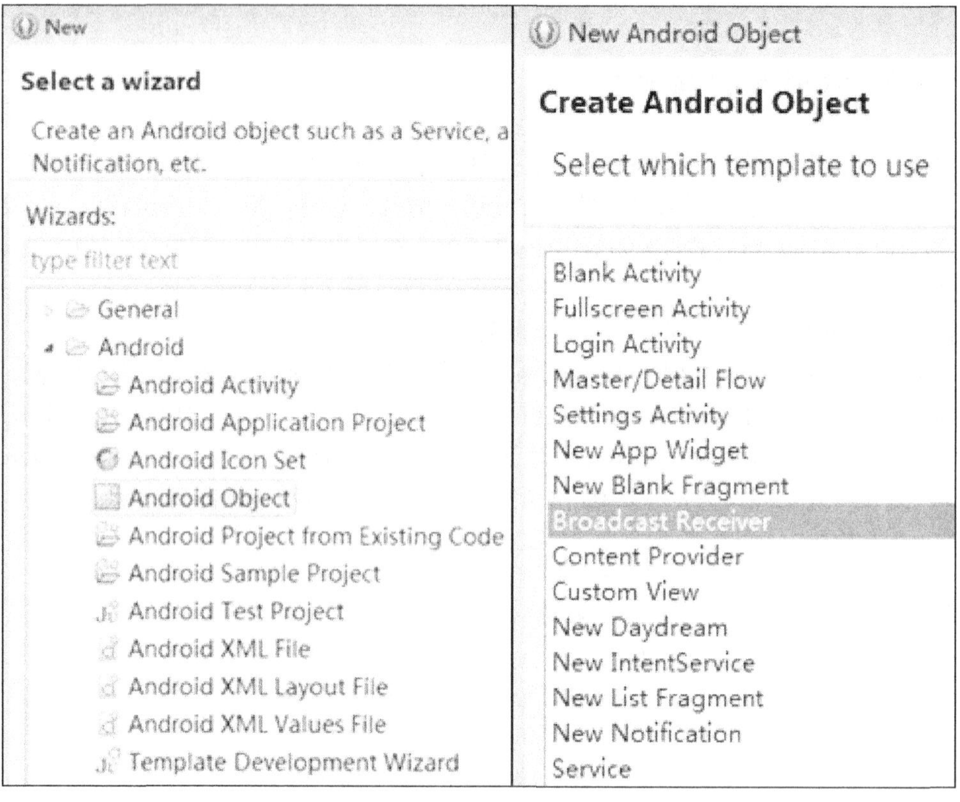

图 6.3.2　使用向导创建广播接收者的方法

下面通过一个简明实例来说明广播接收者的使用方法。

【例 6.3.1】当手机短信到来时，所产生的系统广播触发广播接收者程序——播放音乐。

【开发步骤】

(1) 新建名为 SimpleBroadcastReceiver 的 Android 应用工程，不勾选创建 Activity。

(2) 新建继承 android.content.BroadcastReceiver 的类文件 MyReceiver.java，其代码如下：

```
package com.example.simplebroadcastreceiver;
```

```
import android.content.BroadcastReceiver;     //
import android.content.Context;
import android.content.Intent;
import android.media.MediaPlayer;   //
public class MyReceiver extends BroadcastReceiver {
    public MyReceiver() { }
    @Override
    public void onReceive(Context context, Intent intent) {
        // TODO: This method is called when the BroadcastReceiver is receiving
        // an Intent broadcast.
        //throw new UnsupportedOperationException("Not yet implemented");
        MediaPlayer.create(context,R.raw.tick).start();   //
    }
}
```

(3) 复制音乐文件 tick.mp3 到文件夹 res/raw 里，创建后的工程文件结构(主要部分)如图 6.3.3 所示。

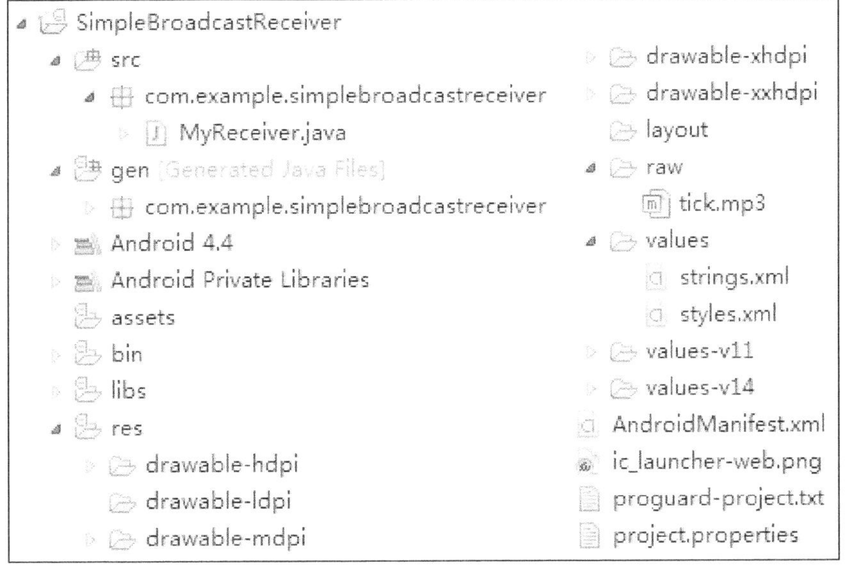

图 6.3.3 SimpleBroadcastReceiver 工程文件结构(主要部分)

(4) 在清单文件中注册接收系统短信广播的权限，代码如下：

```
<uses-permission android:name="android.permission.RECEIVE_SMS"/>
```

(5) 在清单文件中注册接收系统短信广播的广播接收者，代码如下：

```
<receiver
    android:name="com.example.simplebroadcastreceiver.MyReceiver"
```

```
        android:enabled="true"
        android:exported="true" >
        <intent-filter>
            <action android:name="android.provider.Telephony.SMS_RECEIVED"/>
        </intent-filter>
    </receiver>
```

(6) 部署工程到手机成功(观察 Eclipse 控制台)后，向该手机发送一条短信测试。

注意：这是一个没有用户界面的 Android 应用。

【例 6.3.2】当手机短信到来时，所产生的系统广播激活后台音乐播放服务程序和可停止播放的界面程序。

【程序运行】本 Android 应用工程部署到手机后，此时“Stop Music”按钮不可用，按返回键可退出。当手机有短信到来时，所产生的系统广播会同时激活播放后台音乐的服务程序和应用的主 Activity，此时，“Stop Music”按钮可用，单击后停止音乐播放的后台程序并回到手机桌面，效果如图 6.3.4 所示。

图 6.3.4　例 6.3.2 程序运行效果

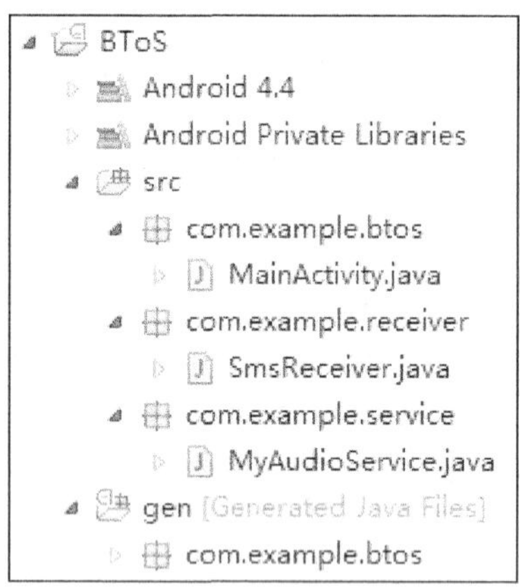

图 6.3.5　BToS 工程文件结构(主要部分)

【开发步骤】

(1) 新建名为 BToS 的 Android 应用工程，勾选创建 Activity。

(2) 新建程序包名 com.example.receiver，在该包下新建一个继承 android.content.BroadcastReceiver 的类文件 SmsReceiver.java。

(3) 新建存放服务程序的包名 com.example.service，在该包下粘贴例 6.1.1 工程里的服务程序文件 MyAudioService.java。此时，工程文件结构(主要部分)如图 6.3.5 所示。

(4) 主 Activity.java 的代码如下：

```
package com.example.btos;
/*
 * 直接运行本应用程序，不会播放音乐
 * 短信(广播)到来时，会自动开启音乐播放的服务程序和本Activity
*/
import com.example.service.MyAudioService;   //
```

```
import android.os.Bundle;
import android.app.Activity;
import android.widget.TextView;
import android.content.Intent;
import android.view.View;
import android.view.View.OnClickListener;

public class MainActivity extends Activity {
    private TextView mText;
    Intent intent;
    private boolean isCast = false; //是否为广播激活
    @Override
    protected void onCreate(Bundle savedInstanceState) {
        super.onCreate(savedInstanceState);
        setContentView(R.layout.activity_main);
        mText=(TextView)this.findViewById(R.id.mText);

        Intent myintent = getIntent();
        isCast = myintent.getBooleanExtra("iscast", false); //广播携带的数据
        mText.setEnabled(isCast);    //
        mText.setOnClickListener(new OnClickListener(){
            @Override
            public void onClick(View v){
                intent=new Intent(MainActivity.this,MyAudioService.class);
                MainActivity.this.stopService(intent);    //
                finish();    //
            }
        });
    }
}
```

(5) 广播接收者程序文件 SmsReceiver.java 的代码如下：

```
package com.example.receiver;
/*
 * 本广播接收者程序分别调用了应用的主Activity程序和播放音乐的服务程序
*/
import com.example.btos.MainActivity;    //
import com.example.service.MyAudioService;    //
import android.content.BroadcastReceiver;    //
```

```
import android.content.Context;
import android.content.Intent;
public class SmsReceiver extends BroadcastReceiver {
    public SmsReceiver() {
    }
    @Override
    public void onReceive(Context context, Intent intent) {
        // TODO: This method is called when the BroadcastReceiver is receiving
        // an Intent broadcast.
        // throw new UnsupportedOperationException("Not yet implemented");
        Intent intent1=new Intent(context,MyAudioService.class);
        context.startService(intent1);     //启动音乐播放服务

        Intent myintent = new Intent(context,MainActivity.class);
        myintent.putExtra("iscast", true);
        //Task(栈)就像一个容器，而Activity就相当于填充这个容器的东西，后进先出
        //如果Activity在Task里存在，移到最顶端,不会启动新的Activity
        //使用Context的startActivity()前，需要开启一个新的Task
        myintent.addFlags(Intent.FLAG_ACTIVITY_NEW_TASK);
        context.startActivity(myintent);
        //Activity是Context的子类，直接使用方法startActivity()
    }
}
```

(6) 部署工程到手机并做运行测试。

6.3.3 自定义广播及广播接收者的两种注册方式

广播接收者在清单文件中的注册是通过使用标签<receiver>完成的，这种注册方式称为静态注册。动态注册是在程序里通过使用 Context 类提供的方法 registerReceiver()完成的。

图 6.3.6 自定义广播与接收(静态注册方式)程序运行界面

【例 6.3.3】自定义广播与接收(静态注册方式)。

【运行效果】程序运行时，单击“发送广播”按钮后，广播接收者程序接收广播消息所携带的数据，然后以 Toast 方式显示，运行界面如图 6.3.6 所示。

【开发步骤】

(1) 新建名为 MyBroadcast(Static) 的 Android Application Project，勾选创建 Activity，指定程序的包名为 com.example.mybroadcast_receiver1。

(2) 新建继承 BroadcastReceiver 的类文件 MyReceiver.java，工程文件结构(主要部分)如

图 6.3.7 所示。

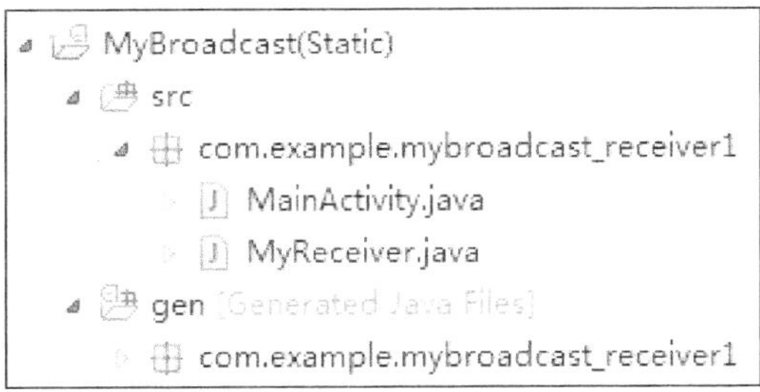

图 6.3.7　MyBroadcast(Static)工程文件结构(主要部分)

(3) 在清单文件中注册广播接收者和广播，其代码如下：

```
<receiver android:name="com.example.mybroadcast_receiver1.MyReceiver">
            <!-- 静态注册时需要使用意图过滤器 -->
            <intent-filter >
                <action android:name="com.example.broadcast.MY_BROADCAST"/>
            </intent-filter>
</receiver>
```

(4) 程序 MainActivity.java 的源代码如下：

```
package com.example.mybroadcast_receiver1;
import com.example.mybroadcast_receiver1.R;
import android.os.Bundle;
import android.app.Activity;
import android.content.Intent;
import android.view.View;
import android.widget.Button;

public class MainActivity extends Activity {
    @Override
    protected void onCreate(Bundle savedInstanceState) {
        super.onCreate(savedInstanceState);
        setContentView(R.layout.activity_main);
        Button myButton = (Button) findViewById(R.id.send);
        myButton.setOnClickListener(new View.OnClickListener() {
            @Override
            public void onClick(View arg0) {
                //发送一个已经在清单文件中注册了广播接收者的广播
                Intent intent = new Intent("com.example.broadcast.MY_BROADCAST");
                Bundle    bundle = new Bundle();    //广播数据封装到Bundle对象里
                bundle.putString("data1", "自定义广播与接收案例");
                bundle.putString("data2", "Writed by WZX");
```

```
                    bundle.putInt("year",2014);
                    intent.putExtras(bundle);   //捆绑数据
                    sendBroadcast(intent);      //发送
                }
            });
        }
}
```

(5) 源程序文件 MyReceiver.java 的代码如下：

```
package com.example.mybroadcast_receiver1;
import android.content.BroadcastReceiver;
import android.content.Context;
import android.content.Intent;
import android.os.Bundle;
import android.widget.Toast;

public class MyReceiver extends BroadcastReceiver {
    public MyReceiver() {
    }

    @Override
    public void onReceive(Context context, Intent intent) {
        // TODO: This method is called when the BroadcastReceiver is receiving
        //从Intent对象里获取广播携带的数据
        Bundle bundle =    intent.getExtras();   //
        String str1 = bundle.getString("data1");
        String str2 = bundle.getString("data2");
        int year=bundle.getInt("year");
        Toast.makeText(context, str1+"\n" +str2+","+year,Toast.LENGTH_LONG).show();
    }
}
```

图 6.3.8　自定义广播与接收(动态注册方式)程序运行界面

(6) 部署工程并做运行测试。

【例 6.3.4】自定义广播与接收(动态注册方式)。

【运行效果】与例 6.3.3 工程相比，本工程没有在清单文件中注册广播接收者，而是在程序中动态注册广播接收者，运行界面如图 6.3.8 所示。

【开发步骤】

(1) 新建名为 MyBroadcast(Dynamic)的 Android 应用工程，指定程序的包名为 com.example.mybroadcast2，勾选创建 Activity，创建

后的工程文件结构(主要部分)如图 6.3.9 所示。

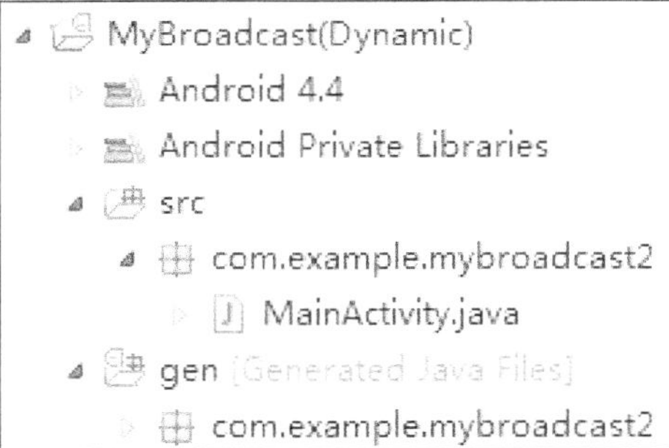

图 6.3.9　MyBroadcast(Dynamic)工程文件结构(主要部分)

(2) 建立与主 Activity 相对应的布局文件。

(3) 文件 MainActivity.java 的代码如下：

```
/*
 * 自定义广播案例：动态注册广播接收者及广播发送
 */
package com.example.mybroadcast2;
import com.example.mybroadcast2.R;
import android.os.Bundle;
import android.app.Activity;
import android.content.BroadcastReceiver;
import android.content.Context;
import android.content.Intent;
import android.content.IntentFilter;//
import android.view.View;
import android.widget.Button;
import android.widget.TextView;
import android.widget.Toast;

public class MainActivity extends Activity {
    private TextView tv;
    private Button myButton;
    private BroadcastReceiver    myReceiver = new BroadcastReceiver() {
        //这里相当于用一个匿名内部类去继承BroadcastReceiver这个类,
        //然后重载父类onReceive(方法), 再新建一个匿名内部类的对象并返回
        //和button.setoncliklistener(new View.onclicklistener(){})一样的原理
        @Override
        public void onReceive(Context context, Intent intent) {
            // TODO Auto-generated method stub
            Bundle mybundle =    intent.getExtras();
```

```
                String str1 = mybundle.getString("data1");
                String str2 = mybundle.getString("data2");
                Toast.makeText(context, str1+" " +str2,Toast.LENGTH_LONG).show();
                tv.setText(str1+" " +str2);
            }
    };
    @Override
    protected void onCreate(Bundle savedInstanceState) {
        super.onCreate(savedInstanceState);
        setContentView(R.layout.activity_main);

        //使用意图过滤器IntentFilter动态注册广播接收者
        IntentFilter myintentfliter = new IntentFilter();
        myintentfliter.addAction("com.example.broadcast.MY_BROADCAST");
        //注册广播接收者
        this.registerReceiver(myReceiver, myintentfliter); //

        tv =(TextView) findViewById(R.id.broadcastshow);
        tv.setText("");
        myButton = (Button) findViewById(R.id.send);
        myButton.setOnClickListener(new View.OnClickListener() {
            @Override
            public void onClick(View arg0) {
                //在程序中动态注册广播接收者
                Intent myintent = new Intent("com.example.broadcast.MY_BROADCAST");
                Bundle    bundle = new Bundle();
                bundle.putString("data1", "自定义广播与接收案例");
                bundle.putString("data2", "张粤~...0v0...");
                myintent.putExtras(bundle);   //捆绑数据
                sendBroadcast(myintent);      //发送广播
            }
        });
    }
}
```

(4) 部署工程并做运行测试。

6.3.4 接收系统广播应用实例——短信接收

在介绍接收系统短信广播程序前，先介绍一下短信的一些知识。

手机接收短信后，其内容将存放到系统的短信数据库内。手机的短信数据库存放在系统文件夹 data/data 的包名 com.android.providers.telephony 内，如图 6.3.10 所示。

com.android.providers.telephony		2014-04-19	09:03	drwxr-x--x
databases		2014-04-19	09:03	drwxrwx--x
mmssms.db	102400	2014-07-13	12:54	-rw-rw----
mmssms.db-journal	16928	2014-07-13	12:54	-rw-------
telephony.db	16384	2014-07-13	06:08	-rw-rw----
telephony.db-journal	8720	2014-07-13	06:08	-rw-------
lib		2014-04-19	09:01	lrwxrwxrwx
shared_prefs		2014-04-19	09:03	drwxrwx--x

图 6.3.10　系统短信数据库

在数据库 mmssms.db 的表 sms 内，存放了手机接收到的短信记录。其中，字段 address 表示短信发送者的电话，body 存放短信的内容。

1. 使用 Bundle 对象处理由 Intent 对象携带的短信内容

Android 设备接收的短信是 PDU(protocol description unit)形式的，是通过 Bundle 类(见图 6.3.11)对象处理由 Intent 对象携带的短信内容。

2. 短消息类 SmsMessage 及其方法

短消息类 SmsMessage 提供了创建该类对象的方法以及解析 PDU 格式信息的数据的方法，如图 6.3.12 所示。

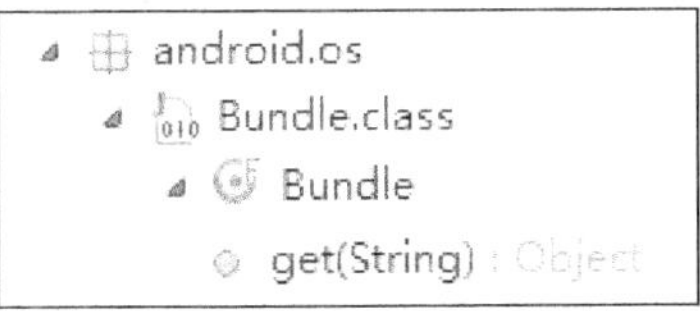

图 6.3.11　Bundle 类

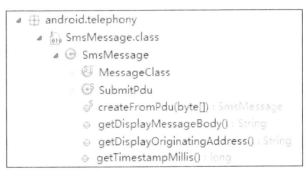

android.telephony
SmsMessage.class
SmsMessage
MessageClass
SubmitPdu
createFromPdu(byte[]) : SmsMessage
getDisplayMessageBody() : String
getDisplayOriginatingAddress() : String
getTimestampMillis() : long

图 6.3.12　SmsMessage 类及其常用方法

当手机有短信到来时，就会产生一个名为 android.provider.Telephony.SMS_RECEIVED 的系统广播，与系统相关的程序不仅会在通知栏里显示该短信，还会把短信的相关信息保存至手机的短信数据库。下面介绍一个处理手机短信的 Android 应用工程。

【例 6.3.5】接收处理 Android 手机的短信广播。

【程序运行】将本应用部署到一部手机上，用另一部手机向第一部手机发送短信(内容为“对接收到的短信广播进行处理。”)，稍后，在第一部手机上出现 Toast 信息(包含了短信发送人的手机号、短信内容和接收时间)，效果如图 6.3.13 所示。

【设计步骤】

(1) 新建名为 ReceiveMessageDemo 的 Android Application Project，勾选创建 Activity。

(2) 新建名为 SmsReceiver.java 且继承 android.content.BroadcastReceiver 的类文件，工程

文件结构(主要部分)如图 6.3.14 所示。

图 6.3.13　接收处理系统短信广播

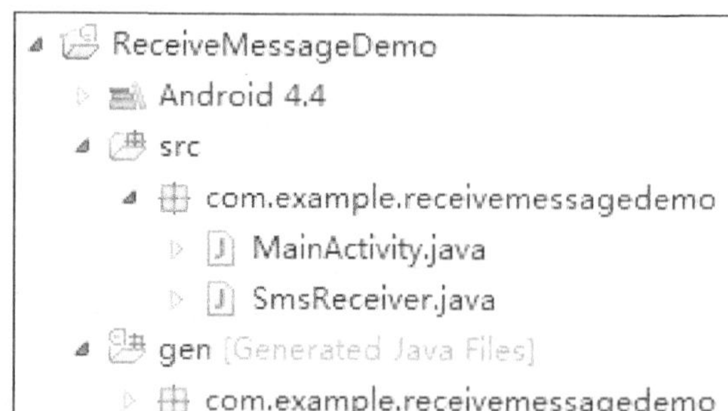

图 6.3.14　ReceiveMessageDemo 工程文件结构(主要部分)

(3) 在清单文件中注册接收系统短信广播的权限 android.permission.RECEIVE_SMS。

(4) 广播接收者文件 SmsReceiver.java 的代码如下：

```
package com.example.receivemessagedemo;
import java.text.SimpleDateFormat;
import android.annotation.SuppressLint;
import android.content.BroadcastReceiver;//广播接收者
import android.telephony.SmsMessage;//短信息类
import android.content.Context;
import android.content.Intent;
import android.os.Bundle;
import android.widget.Toast;

@SuppressLint("SimpleDateFormat")
public class SmsReceiver extends BroadcastReceiver {
    @Override
    public void onReceive(Context arg0, Intent arg1) {
        // TODO Auto-generated method stub
            //得到捆绑数据
            Bundle bundle = arg1.getExtras();
            //得到对象数组
            Object[] objs = (Object[]) bundle.get("pdus");
            //得到SmsMessage对象数组，因为一条长短信是会被切割，分多次发送的
            SmsMessage[] msgs = new SmsMessage[objs.length];
            for (int i = 0; i < objs.length; i++) {
                msgs[i] = SmsMessage.createFromPdu((byte[]) objs[i]);
            }
```

```
            //构造短信相关信息的字符串
            StringBuilder strb=new StringBuilder();
            for(SmsMessage msg : msgs){
                strb.append("发短信人: \n");
                strb.append(msg.getDisplayOriginatingAddress());
                strb.append("\n信息内容: \n");
                strb.append(msg.getDisplayMessageBody());
                //获得当前的日期与时间
                long dt=msg.getTimestampMillis();
                SimpleDateFormat myFormat=new SimpleDateFormat
                                              ("yyyy-MM-dd hh:mm:ss");
                String dtStr=myFormat.format(dt);
                strb.append("\n接收时间: \n");
                strb.append(dtStr);
            }
            Toast.makeText(arg0, strb.toString(), Toast.LENGTH_LONG).show();
    }
}
```

(5) 文件 MainActivity.java 的代码如下：

```
package com.example.receivemessagedemo;
import android.app.Activity;
import android.os.Bundle;
import android.widget.EditText;
import com.example.receivemessagedemo.R;

public class MainActivity extends Activity {
    /** Called when the activity is first created. */
    @Override
    public void onCreate(Bundle savedInstanceState) {
        super.onCreate(savedInstanceState);
        setContentView(R.layout.activity_main);
        EditText text=(EditText) this.findViewById(R.id.editText1);
        text.setText("接收到短信广播后，将由自定义的广播接收者程序SmsReceiver
                          进行处理，本界面可以按返回键关闭啦。");
    }
}
```

注意：

(1) 例 6.3.5 中的 Activity 只是起提示作用，如同例 6.3.1 一样，本应用可以不创建用户界面(Activity)。

(2) 本广播接收者与系统提供的接收者程序并不矛盾，分别起作用。

6.4 组件综合应用实例——自动挂断来电后回复短信

前面我们分别介绍了 Android 的三个重要组件，即 Activity、Service 和 BroadcastReceiver，下面介绍一个关于这三个组件的综合应用实例——防电话打扰应用设计。

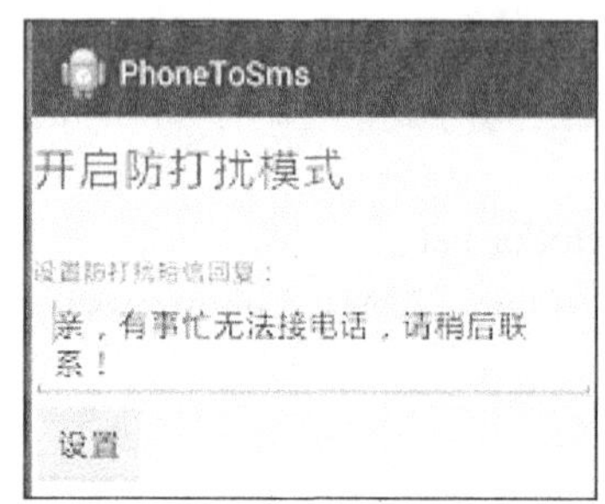

图 6.4.1 设置/取消来电打扰模式及回复短信内容

注意：

(1) 由于挂断电话这个操作不属于手机 App，因此会涉及跨进程通信(远程服务调用)。

(2) 挂断或应答电话，可使用 Android 系统提供的电话服务接口 com.android.internal.ITelephony。

【例 6.4.1】自动挂断来电后回复短信。

【程序运行】将本应用工程部署到一部手机上，单击文本“开启防打扰模式”，可以修改回复短信的内容，效果如图 6.4.1 所示。

在另一部手机上，向第一部手机打电话，则电话自动被挂断，稍后收到来自第一部手机的短信通知。同时，在第一部手机的通知栏里也会出现未接电话的通知，效果如图 6.4.2 所示。

图 6.4.2 出现未接电话的通知

【设计步骤】

(1) 新建名为 PhoneToSms 的 Android 应用工程，程序包名等采用默认值，勾选创建 Activity，完成后的工程文件结构(主要部分)如图 6.4.3 所示。

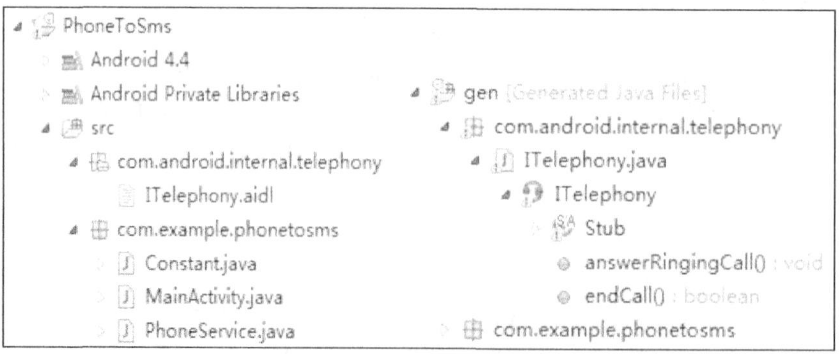

图 6.4.3 PhoneToSms 工程文件结构(主要部分)

(2) 新建名为“com.android.internal.telephony”的包名，用以存放接口文件 ITelephony.aidl 文件，其代码如下：

```
//Android系统提供的电话服务接口ITelephony
package com.android.internal.telephony;
interface ITelephony{
    boolean endCall();
    void answerRingingCall();
}
```

(3) 在源程序包里，新建存放全局变量的类文件 Constant.java，其代码如下：

```
package com.example.phonetosms;
public class Constant {
public static final String ACTION_PLAY = "com.tianjia.service.ACTION_PALY";
public static final String ACTION_PAUSE = "com.tianjia.service.ACTION_PAUSE";
public static final String ACTION_MYSERVICE =
                                        "com.tianjia.service.ACTION_MYSERVICE";
}
```

(4) 在源程序包里，建立服务程序文件 PhoneService.java，其代码如下：

```
package com.example.phonetosms;
/*
 * 本服务程序包含了电话状态监听、通知与延期意图、广播的动态注册、发送短信
* 调用了使用AIDL定义的远程服务接口ITelephony
* 使用了Java的反射机制，得到远程服务对象
*/
import java.lang.reflect.Method;
import com.android.internal.telephony.ITelephony;   //
import com.example.phonetosms.R;
import android.app.Notification;   //
import android.app.NotificationManager;   //
import android.app.PendingIntent;
import android.app.Service;      //
import android.content.BroadcastReceiver;      //
import android.content.Context;
import android.content.Intent;   //
import android.content.IntentFilter;   //
import android.graphics.Color;
import android.os.IBinder;    //
import android.os.RemoteException;
import android.telephony.PhoneStateListener;
import android.telephony.TelephonyManager;
```

```
import android.telephony.gsm.SmsManager;     //
import android.widget.Toast;

@SuppressWarnings("deprecation")
public class PhoneService extends Service {
    private static ITelephony   iTelephony;
    private TelephonyManager manager;
    private String inNumber = null;//记录来电号码
    private String message;//自动回复信息
    private   boolean disturb = false;//防打扰开启标识
    @Override
    public IBinder onBind(Intent arg0) {
        // TODO Auto-generated method stub
        return null;
    }
    public void onCreate()
    {
        try {
            MyReceiver myReceiver = new MyReceiver();
            IntentFilter filter = new IntentFilter();
            filter.addAction(Constant.ACTION_PLAY);
            filter.addAction(Constant.ACTION_PAUSE);
            registerReceiver(myReceiver, filter);
            phonelistener();    //
        }
          catch (Exception e) {
            e.printStackTrace();
        }
    }
    public int onStartCommand(Intent intent, int flags, int startId) {
        return super.onStartCommand(intent, flags, startId);
    }
    public void play() {
        disturb = true;
    }
    public void pause() {
        disturb = false;
    }
    //定义广播接收者，作为内部类
```

```
private class MyReceiver extends BroadcastReceiver {
    @Override
    public void onReceive(Context context, Intent intent) {
        Toast.makeText(context, intent.getAction(),Toast.LENGTH_LONG).show();
        message = intent.getExtras().getString("message");
        try {
            if (Constant.ACTION_PLAY.equals(intent.getAction())) {
                PhoneService.this.play();
            }
                else
                  if (Constant.ACTION_PAUSE.equals(intent.getAction())) {
                PhoneService.this.pause();
            }
        }
          catch (Exception e) {
            e.printStackTrace();
        }
    }
}
//发送短信方法
public void sendSMS(String phonenumber,String msg){
    PendingIntent pi=PendingIntent.getActivity(this, 0, new Intent(), 0);
    SmsManager sms=SmsManager.getDefault();
    //发送信息到指定号码
    sms.sendTextMessage(phonenumber, null, msg, pi, null);
}
//监听并挂断电话方法
public void phonelistener(){
    getphoner();
    manager.listen(new PhoneStateListener(){
    @Override
    public void onCallStateChanged(int state, String incomingNumber) {
    // TODO Auto-generated method stub
    super.onCallStateChanged(state, incomingNumber);
    inNumber = incomingNumber;
    switch(state){
    //判断是否有电话接入
        case 1:
            try {
```

```
                    if(disturb){
                        iTelephony.endCall();//当有电话接入时，自动挂断
                            Toast.makeText(getApplicationContext(),"endcall:+
                                incomingNumber, Toast.LENGTH_SHORT).show();
                        //发送短信
                        sendSMS(inNumber,"<PhoToSms自动回复>\n"+message);
                        //在状态栏提示短信发送成功
                        showNotification();
                    }
                }
                    catch (RemoteException e) {
                    // TODO Auto-generated catch block
                    e.printStackTrace();
                }
                break;
            default :break;
        }
      }
    }, PhoneStateListener.LISTEN_CALL_STATE);
  }

  //使用Java反射机制，程序运行时得到远程服务对象
  public void getphoner(){
      manager = (TelephonyManager)getSystemService(TELEPHONY_SERVICE);
      Class <TelephonyManager> c = TelephonyManager.class;
      Method getITelephonyMethod = null;
      try {
              //从对象得到所在类的方法
          getITelephonyMethod = c.getDeclaredMethod("getITelephony",
                                                          (Class[])null);
              //设置反射的对象在使用时是否应该取消 Java 语言访问检查
              getITelephonyMethod.setAccessible(true);
              //Java反射机制
              iTelephony = (ITelephony) getITelephonyMethod.invoke(manager,
                                                          (Object[])null);
      }
      catch (IllegalArgumentException e) {
              e.printStackTrace();
      }
```

```
        catch (Exception e) {
                e.printStackTrace();
        }
    }

    //通知栏显示
    private void showNotification() {
        // 创建一个NotificationManager的引用
        NotificationManager notificationManager =(NotificationManager)
                    getBaseContext().getSystemService(NOTIFICATION_SERVICE);
        // 定义Notification的各种属性
        Notification notification =new Notification(R.drawable.ic_launcher,
                                        "PhoneToSms",System.currentTimeMillis());
        notification.flags = Notification.FLAG_SHOW_LIGHTS;
        notification.defaults = Notification.DEFAULT_LIGHTS;
        notification.ledARGB = Color.BLUE;
        // 设置通知的事件消息
        String contentText;
        // 通知栏标题
        String contentTitle ="PhoneToSms提醒你："; 
        // 通知栏内容
        contentText ="亲，"+inNumber+"刚才来电话了，系统已自动短信回复了！";
        notification.setLatestEventInfo(getBaseContext(), contentTitle, contentText, null);
        // 把Notification传递给NotificationManager
        notificationManager.notify(0, notification);
    }
    //清理通知
    public void clearNotification(){
        // 启动后，删除之前我们定义的通知
            NotificationManager notificationManager = (NotificationManager)
                        this.getSystemService(NOTIFICATION_SERVICE);
        notificationManager.cancel(0); //取消
    }
}
```

(5) MainActivity.java 的代码如下：

```
package com.example.phonetosms;
/*
 * 本Activity用于设置是否开启免打扰、设置短信回复内容(文本文件读写)
```

```
* 由于挂断电话这个操作不属于主线程，因此涉及跨进程通信(远程服务调用)
*/
import java.io.File;
import java.io.FileInputStream;
import java.io.FileOutputStream;
import java.io.IOException;
import org.apache.http.util.EncodingUtils;
import com.example.phonetosms.R;
import android.app.Activity;
import android.content.Intent;
import android.graphics.Color;
import android.os.Bundle;
import android.os.Environment;
import android.view.View;
import android.view.View.OnClickListener;
import android.widget.Button;
import android.widget.EditText;
import android.widget.TextView;
import android.widget.Toast;

public class MainActivity extends Activity {
    //textview 的状态开启或者取消，初始化为未开启
    private boolean flag = false;
    private TextView tv;
    private String message;//存放回复短信内容
    private EditText et;
    private Button btnOK;
    private String fileName = "test.txt";
    Intent intent;

    @Override
    protected void onCreate(Bundle savedInstanceState) {
        super.onCreate(savedInstanceState);
        setContentView(R.layout.activity_main);

        try{
            tv = (TextView)findViewById(R.id.textView1);
                if(flag)
                    tv.setText(R.string.auto_stop);
                else
```

```
        tv.setText(R.string.auto_start);
    intent = new Intent(Constant.ACTION_MYSERVICE);
    startService(intent);
    et = (EditText)findViewById(R.id.editText1);
    message = readFile(fileName);
    et.setText(readFile(fileName));//读取test.txt到编辑框中

    //设置按钮添加监听
    btnOK = (Button)findViewById(R.id.button1);
    btnOK.setOnClickListener(new OnClickListener(){
        @Override
        public void onClick(View arg0) {
            // TODO Auto-generated method stub
            if("".equals(et.getText().toString()))
                Toast.makeText(getApplicationContext(), "回复为空！请重设！",
                                                Toast.LENGTH_SHORT).show();
            else
            {
                message = et.getText().toString();
                try {
                    writeFile(fileName, message);
                }
                catch (IOException e) {
                    // TODO Auto-generated catch block
                    e.printStackTrace();
                }
            }
     }
    });
    //防打扰设置监听
    tv.setOnClickListener(new OnClickListener() {
        public void onClick(View view) {
            // 单击textview
            if(flag)
            {
                flag = false;
                tv.setText(R.string.auto_start);
             //关闭状态字体为黑色
                tv.setTextColor(Color.parseColor("#000000"));
```

```
                    Intent broadcast = new Intent();
                    broadcast.putExtra("message", message);
                    //取消防打扰
                    broadcast.setAction(Constant.ACTION_PAUSE);
                    sendBroadcast(broadcast);//发送广播
                }
                else
                {
                    flag = true;
                    tv.setText(R.string.auto_stop);
                    //开启状态时将文字颜色改为粉红色
                    tv.setTextColor(Color.parseColor("#ffcccc"));
                    Intent broadcast = new Intent();
                    broadcast.putExtra("message", message);
                    //开启防打扰
                    broadcast.setAction(Constant.ACTION_PLAY);
                    sendBroadcast(broadcast);//发送广播
                }
            }
        });
        }catch(Exception e){
            e.printStackTrace();
        }
}
//读取文件操作test.txt
public String readFile(String fileName) throws IOException
{
    String res = "";
    try{   //判断SD卡是否存在
        if(Environment.getExternalStorageState().equals
                            (android.os.Environment.MEDIA_MOUNTED))
        {
            String SDPATH = Environment.getExternalStorageDirectory().getPath()+"//";
            File file = new File(SDPATH+"/"+fileName);
            if(!file.exists())
            {
                file.createNewFile();
                //首次使用时默认回复的短信内容
```

```
                res = "亲，有事忙无法接电话，请稍后联系！";
                writeFile(fileName,res);
            }
            else    {
              //得到资源中的asset数据流
              FileInputStream fis = new FileInputStream(file);
              int length = fis.available();
              byte [] buffer = new byte[length];
              fis.read(buffer);
              res = EncodingUtils.getString(buffer, "UTF-8");
              fis.close();   }
          }
          else
            Toast.makeText(this, "sd卡不存在,读取失败",
                                                   Toast.LENGTH_SHORT).show();
      }
     catch(Exception e)   {
        e.printStackTrace();
      }
      return res;
  }
  //写文件
  public void writeFile(String fileName ,String write_str) throws IOException
  {
      if(Environment.getExternalStorageState().equals
                                (android.os.Environment.MEDIA_MOUNTED))
      {
          String SDPATH = Environment.getExternalStorageDirectory().getPath()+"//";
          File file = new File(SDPATH+"/"+fileName);
          FileOutputStream fos = new FileOutputStream(file);
          byte[] bytes = write_str.getBytes();
          fos.write(bytes);
          fos.close();
          Toast.makeText(this, "写入成功", Toast.LENGTH_SHORT).show();
      }
      else
          Toast.makeText(this, "sd卡不存在，写入失败", Toast.LENGTH_SHORT).show();
  }
}
```

(6) 部署工程并做运行测试。

习　题　6

一、判断题

1. Android 系统允许应用程序通过 Intent 启动 Activity 和 Service。

2. 使用 BroadcastReceiver 的 Android 应用，可以没有 Activity。

3. Service 的隐式调用与 Activity 类似，在清单文件中注册 Service 的同时还要使用意图过滤器标签<intent-filter>。

4. 如果希望一个应用工程使用另一个应用工程中的服务程序，则在清单文件中注册该服务的同时还要使用意图过滤器。

5. 当手机有短信到来时，只产生通知而不产生广播。

6. 在手机上安装的 Android 系统中，所有广播接收者都会对广播进行各自的处理。

7. 短信接收程序与短信发送程序一样，都要使用短信管理器类 SmsMessage。

8. 在开发 Android 应用的 Eclipse 环境中，BroadcastReceiver 和 Service 等组件都可以通过向导的方式来创建。

二、选择题

1. 非绑定服务的主要业务逻辑代码含于____方法里。

A. onCreate()　　B. onStart()　　C. onDestroy()　　D. onBind()

2. 使用绑定服务时，必须在 Activity 里重写____方法。

A. startActivity()　　B. onServiceConnected()

C. onActivityResult()　　D. onPause()

3.被绑定的服务，需要在服务的____方法里返回一个 Binder 对象。

A. onBind()　　B. onStart ()　　C. onUnbind()　　D. onDestroy()

4. 短信内容是由____对象(组件)传递的。

A. Service　　B. BroadcastReceiver　　C. Intent　　D. ContentProvider

5. 下列方法中，不是 Activity 和 Context 共同具有的是____。

A. startService()　　B. stopService()

C. findViewById()　　D. getSystemService()

三、填空题

1. Service 组件与 Activity 组件一样，含于相同的软件包____里。

2. 服务有显式启动和隐式启动两种。若服务和调用服务的组件在不同的应用程序中，则应使用____启动。

3. 注册短信广播接收者时对所拦截的广播是使用____标签设置的。

4. 定义被绑定的服务时，其 onBind()方法返回的类型是____。

5. 建立远程服务调用，需要使用____语言定义远程服务接口。

6. 发送自定义的广播，携带的数据封装在____对象里。

7. 通过方法 bindService()绑定 Service 时，Service 的方法 onCreate()和____将被调用。

实验6　服务组件与广播组件及其应用

一、实验目的

1. 理解 Service、BroadcastReceiver 与 Activity 的不同点。
2. 掌握 Service 的两种调用方式(显示与隐式)的用法区别。
3. 掌握 Service 的两种启动方式(绑定与非绑定)的用法区别。
4. 掌握本地服务调用与远程服务调用的用法区别。
5. 掌握 Android 的广播机制。
6. 掌握常用的系统广播。
7. 掌握接收系统广播和自定义的用法区别。
8. 掌握注册广播接收者的两种用法(静态注册与动态注册)。

二、实验内容及步骤

【预备】访问 http://www.wustwzx.com/android/index.html，单击“实验 6”超链接，下载本次实验内容的源代码并解压(得到文件夹 ch06)，供研读和调试使用。

1. 服务的创建与使用示例：调用后台音频播放服务(参见例 6.1.1)。

(1) 再次解压文件夹 ch06 内的压缩文件 Audio_Service1.zip，导入工程 Audio_Service1。

(2) 打开清单文件，查看服务的注册代码。

(3) 打开程序文件 MainActivity.java，查看显式调用服务的相关代码。

(4) 部署工程并做运行测试。

(5) 适当修改清单文件和文件 MainActivity.java，使用隐式启动服务方式。

(6) 重新部署工程并做运行测试。

(7) 验证服务程序 MyAudioService.java 里的 onBind()方法不可去。

(8) 打开清单文件，使用“Ctrl+Shift+\”键解除对 Service 过滤器代码的屏蔽。

(9) 打开程序文件 MainActivity.java，屏蔽显式调用并使用隐式调用。

(10) 重新部署文件进行验证。

2. 绑定服务方式的使用。

(1) 再次解压文件夹 ch06 内的压缩文件 Audio_Service2.zip，导入工程 Audio_Service2。

(2) 打开清单文件，查看服务的注册代码。

(3) 打开界面程序文件 MainActivity.java，查看显式调用服务的相关代码。

(4) 部署工程并做运行测试。

(5) 适当修改清单文件和文件 MainActivity.java，使用隐式启动服务方式。

(6) 重新部署工程并做运行测试。

(7) 查看符合绑定服务要求的服务类的内部类 PlayBinder(服务代理)的定义。

(8) 查看服务连接必须重写的 onServiceConnected()方法中对服务代理的使用代码。

(9) 查看创建界面 MainActivity 的 onCreate()方法里以绑定方式启动服务方法里的第 2 个参数的含义。

3. 远程服务的创建与使用。

(1) 再次解压文件夹 ch06 内的压缩文件 MusicServer.zip，导入工程 MusicServer。

(2) 打开清单文件，查看远程服务 PlayerService 的注册代码。

(3) 查看音乐文件 res/raw/tick.mp3。

(4) 查看源程序包文件夹里远程服务接口文件 IRemoteService.aidl 所定义的服务接口。

(5) 查看 PlayerService.java 中创建 IBinder 对象的方法及 onBind()方法。

(6) 部署工程到手机后按返回键退出。

(7) 再次解压文件夹 ch06 内的压缩文件 MusicClient.zip，导入工程 MusicClient。

(8) 查看接口定义文件 IRemoteService.aidl 所在的路径。

(9) 查看主 Activity 中的以绑定方法连接远程服务的相关代码。

(10) 部署工程并做运行测试。

4. 广播组件(含服务组件)用法——手机短信到来时所产生的系统广播启动音乐播放服务。

(1) 再次解压文件夹 ch06 内的压缩文件 SimpleBroadcastReceiver.zip，导入工程 SimpleBroadcastReceiver(参见例 6.3.1)。

(2) 打开工程的清单文件，查验没有注册 Activity 组件，布局文件夹 layout 和值文件内容都为空。

(3) 查看注册接收短信广播的权限的代码。

(4) 查看注册 BroadcastReceiver 组件及其接收的系统短信广播的代码。

(5) 部署工程到手机(注意观察控制台输出的完成信息)，查验没有用户界面和桌面快捷图标，但使用手机助手可以查看本应用。

(6) 从其他手机到本手机发送一条短信，手机接收短信后便开始播放音乐。

(7) 利用手机的应用管理器(可从设置程序进入)卸载本应用。

(8) 再次解压文件夹 ch06 内的压缩文件 BToS.zip，导入工程 BToS(参见例 6.3.2)。

(9) 查看程序 MainActivity 的 onCreate()方法获取意图对象携带的数据的代码。

(10) 查看程序 SmsReceiver 的 onReceive()方法中封装数据到意图对象的代码。

(11) 部署工程到手机，然后做运行测试。

5. 自定义广播及广播接收者的两种注册方式。

(1) 再次解压文件夹 ch06 内的压缩文件 MyBroadcast(Static).zip，导入工程 MyBroadcast(Static)(参见例 6.3.3)。

(2) 查看清单文件中注册广播接收者和静态注册自定义的广播。

(3) 查看 MainActivity 程序的按钮监听器的 onClick()方法内发送广播并封装数据的代码。

(4) 部署工程并做运行测试。

(5) 再次解压文件夹 ch06 内的压缩文件 MyBroadcast(Dynamic).zip，导入工程 MyBroadcast(Dynamic)(参见例 6.3.4)。

(6) 查看程序 MainActivity 的 onCreate()方法获取意图对象携带的数据的代码。

(7) 查看程序 SmsReceiver 的 onReceive()方法中封装数据到意图对象的代码。

(8) 部署工程到手机，然后做运行测试。

6. 广播组件的使用——短信接收程序设计。

(1) 再次解压文件夹 ch06 内的压缩文件 ReceiveMessageDemo.zip，导入工程 ReceiveMessageDemo。

(2) 查看清单文件中注册接收系统短信的权限和接收的广播类型(短信广播)。

(3) 查看广播接收者程序 onReceive()方法内处理短信的代码。

(4) 部署工程并做运行测试。

7. 组件的综合使用——自动挂断来电后回复短信程序设计。

(1) 再次解压文件夹 ch06 内的压缩文件 PhoneToSms.zip，导入工程 PhoneToSms。

(2) 查看源程序包内辅助类(常量)Constant.java 中各个常量的含义。

(3) 查看服务接口定义文件 ITelephony.aidl 与自动生成的接口之间的对应关系。

(4) 查看 MainActivity 程序中监听设置“是否自动挂断来电并回短信”，并发广播的代码。

(5) 查看 MainActivity 程序中读写文本文件的代码。

(6) 查看清单文件中注册服务的代码。

(7) 查看服务程序的 onCreate()方法中动态注册广播及广播接收者的代码。

(8) 查看服务程序监听并挂断来电方法中涉及的相关类(接口)。

(9) 查看服务程序中发送短信的代码。

(10) 查看服务程序中产生通知的代码。

(11) 查看清单文件中需要注册的权限。

(12) 部署工程并做运行测试。

三、实验小结及思考

(由学生填写，重点写上机中遇到的问题。)

第7章 SQLite 数据库编程

在第 4 章，我们介绍了使用 SharedPreferences 和 File 来存取数据。当频繁地存取数据时，需要使用数据库来管理，数据库一直是应用的主要组成部分。在 Android 系统中，通常使用适合于嵌入式系统的 SQLite 数据库，因为它占用资源少、运行高效、可移植性好，并且提供了零配置的运行模式。Android 系统使用 SQLite 来存储、管理、维护数据。对数据库的建立、打开、更新和记录的“增/删/改/查”，是数据库的基本内容。本章学习要点如下：

- 掌握 SQLite 数据库的特点(并与 SharedPreferences 存储和 File 存储比较)；
- 掌握 SQLite 数据库的建立、打开和更新方法；
- 使用 SQLiteDatabase 类实现对 SQLite 数据库的“增/删/改/查”操作；
- 掌握 SQLiteDatabase 提供的用于数据库的“增/删/改/查”操作的多种方法的区别；
- 掌握以 DAO 方式编写数据库访问程序的方法；
- 掌握数据库事务的使用方法。

7.1 SQLite 数据库简介

7.1.1 SQLite 数据库软件的特点

SQLite 是 Android 系统手机自带(即内置)的轻量级数据库软件，类似于 Access 数据库软件，它提供了对数据库的增加、删除、修改和查询等操作。

SQLite 数据库体系结构中最核心的部分是虚拟机，也称虚拟数据库引擎(virtual database engine, VDBE)。与 JVM 类似，VDBE 也用来解释字节代码，以完成对数据库的操作。

注意：当涉及大量的数据存储时，还是需要专门的数据库服务器。

7.1.2 Android 系统对 SQLite 数据库的支持

在标准的 Android 软件包里，提供了与 SQLite 数据库相关的两个软件包，一个是 android.database，另一个是 android.database.sqlite，如图 7.1.1 所示。

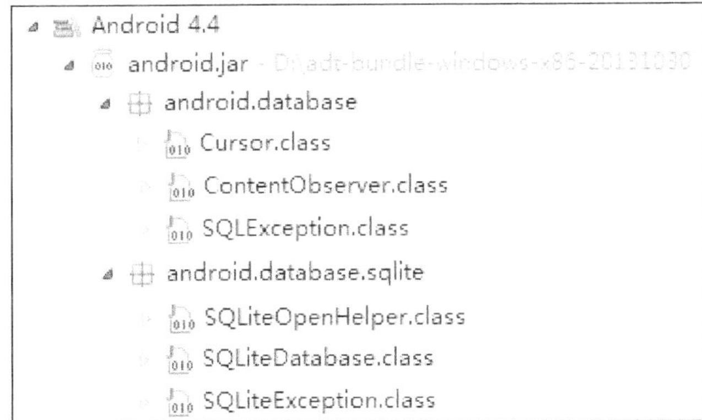

图 7.1.1　Android 提供的 SQLite 软件包及其相关包

注意：内容观察者类 ContentObserver 的使用，参见第 8.2.4 小节。

7.2　使用抽象类 SQLiteOpenHelper 创建、打开或更新数据库

7.2.1　SQLite 数据库及表的创建与打开

抽象类 SQLiteOpenHelper 不仅提供了创建或打开数据库操作的构造方法 SQLiteOpenHelper()，还提供了获取 SQLiteDatabase 对象的方法(getReadableDatabase()和 getWritableDatabase())，如图 7.2.1 所示。

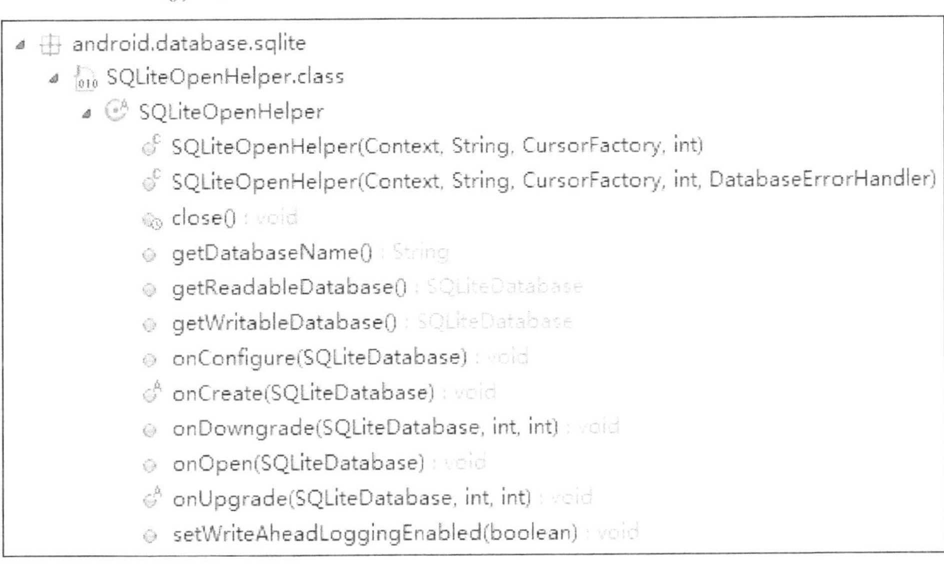

图 7.2.1　抽象类 SQLiteOpenHelper 的主要方法

注意：创建 SQLite 数据库时，以上下文对象作为其中的一个参数。

数据库类 SQLiteDatabase 封装了对数据库操作的方法，如图 7.2.2 所示。

```
android.database.sqlite
  SQLiteDatabase.class
    SQLiteDatabase
      CursorFactory
      create(CursorFactory) : SQLiteDatabase
      deleteDatabase(File) : boolean
      execSQL(String) : void
      execSQL(String, Object[]) : void
      rawQuery(String, String[]) : Cursor
      insert(String, String, ContentValues) : long
      delete(String, String, String[]) : int
      update(String, ContentValues, String, String[]) : int
      updateWithOnConflict(String, ContentValues, String, String[], int) : int
      query(boolean, String, String[], String, String[], String, String, String, String) : Cursor
      query(boolean, String, String[], String, String[], String, String, String, String, CancellationSignal) : Cursor
      query(String, String[], String, String[], String, String, String) : Cursor
      query(String, String[], String, String[], String, String, String, String) : Cursor
```

图 7.2.2　数据库类 SQLiteDatabase 的主要方法

7.2.2　使用 SQLiteSpy 验证创建的数据库

对于部署在 AVD 上的 SQLite 应用，其创建的数据库在 DDMS 视图方式下选择 File Explorer，按照目录顺序“data→data→项目名→包名→databases”，找到应用创建的 SQLite 数据库，其操作如图 7.2.3 所示。

Threads　Heap　Allocation Tracker　Network Statistics　File Explorer　System Information

Name	Size	Date	Time	Permissions
introduction.android.contacts		2014-07-08	07:37	drwxr-x--x
introduction.android.fileDemo		2014-07-08	07:37	drwxr-x--x
introduction.android.mydbDemo		2014-07-08	07:37	drwxr-x--x
cache		2014-04-27	19:33	drwxrwx--x
databases		2014-04-27	19:33	drwxrwx--x
mydb	20480	2014-05-11	16:55	-rw-rw----
mydb-journal	12824	2014-05-11	16:55	-rw-------
lib		2014-07-08	07:37	lrwxrwxrwx
introduction.android.udpDemo		2014-07-03	11:07	drwxr-x--x
introduction.android.widgetDemo		2014-07-08	07:37	drwxr-x--x
jp.co.omronsoft.openwnn		2014-04-19	09:01	drwxr-x--x
dontpanic		2014-04-19	08:58	drwxr-x---
drm		2014-04-19	08:58	drwxrwx---
local		2014-04-19	08:58	drwxr-x--x
lost+found		1970-01-01	08:00	drwxrwx---
media		2014-04-19	08:58	drwxrwx---

图 7.2.3　找到 AVD 里的 SQLite 数据库

使用右上方的“ ”工具，可以将项目里的 SQLite 数据库导出到计算机的硬盘上，如图 7.2.4 所示。

图 7.2.4　导出 AVD 里的 SQLite 数据库

注意：导出时,应指定数据库文件的扩展名为.db。

SQLiteSpy 是一款在 Windows 中运行、以图形方式管理 SQLite 数据库的软件，其运行界面如图 7.2.5 所示。

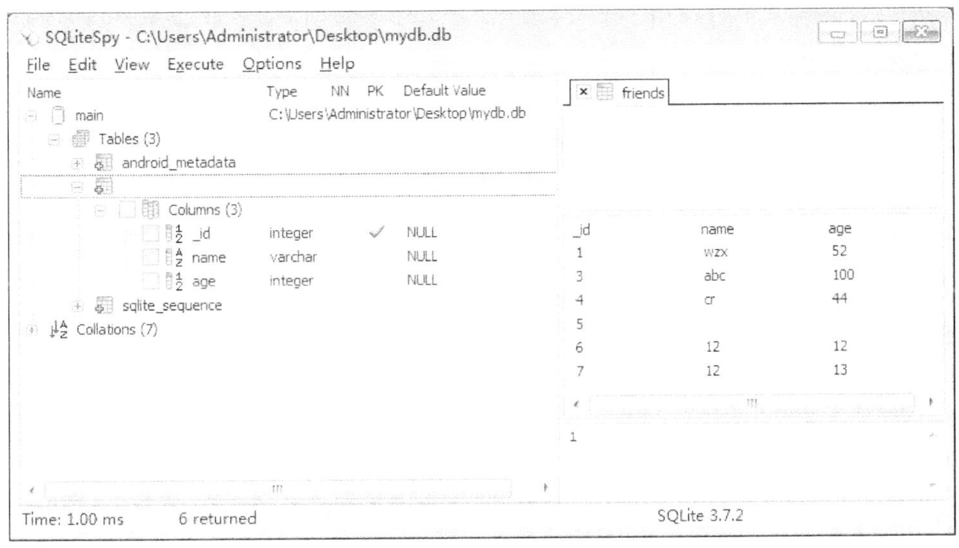

图 7.2.5　SQLiteSpy 的操作界面

注意：当手机 Root 后，使用 Root Explorer 程序只能查看数据库中的记录而不能编辑，而 SQLiteSpy 可以编辑 SQLite 数据库。编辑完成后，在 DDMS 视图方式下，通过使用“ ”工具可导入到 AVD 里。

7.2.3 SQLite 数据库的更新

SQLiteDatabase 类提供了对数据库操作查询的 execSQL(sql)，其中的 sql 参数是对数据库执行操作查询的 SQL 命令(而不是使用 Select 命令的普通查询)，通常是 Insert、Delete、Update 或 Create Table、Alter Table 等命令。

【例 7.2.1】SQLite 数据库及表的创建与更新。

【程序运行】程序首次部署到手机运行后，创建数据库 test.db 及表 person，其中 person 表包含两个字段。当修改源程序，将数据库版本从 1 提升到 2 时，再次部署、运行程序，会执行 onUpgrade()方法，修改 person 表结构，增加一个字段。再次运行后，分别使用 Root Explorer 程序查看数据库表，其效果如图 7.2.6 所示。

图 7.2.6 数据库版本更新前后数据表的变化

【实现步骤】

(1) 新建名为 SQLite_c 的 Android 工程，勾选创建 Activity，文件名采用默认值，工程文件结构(主要部分)如图 7.2.7 所示。

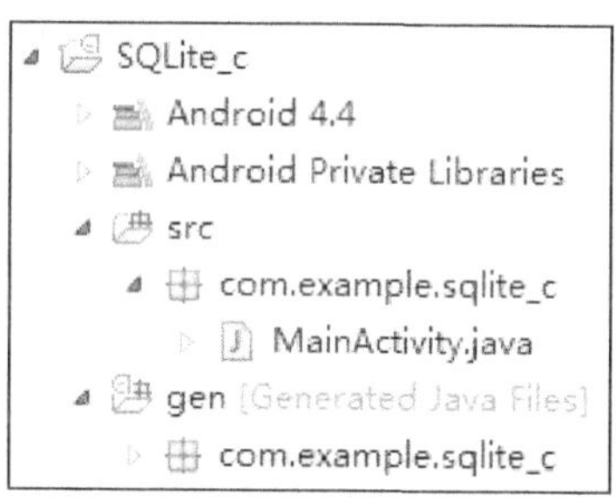

图 7.2.7 工程文件结构(主要部分)

(2) 编写工程的主界面程序 MainActivity.java，其源代码如下：

```
package com.example.sqlite_c;
import android.app.Activity;
import android.os.Bundle;
import android.content.Context;   //
import android.database.sqlite.SQLiteOpenHelper; //主类1
import android.database.sqlite.SQLiteDatabase;   //主类2
public class MainActivity extends Activity {
```

```
@Override
public void onCreate(Bundle savedInstanceState) {
    super.onCreate(savedInstanceState);
    setContentView(R.layout.activity_main);
    MyDbOpenHelper helper = new MyDbOpenHelper(this);
    helper.getWritableDatabase();
}
class MyDbOpenHelper extends SQLiteOpenHelper{  //内部工具类
        public MyDbOpenHelper(Context context) {
        //必须调用抽象父类的构造方法
        super(context, "test.db", null,1);      //建立或打开库
    }
    @Override
    public void onCreate(SQLiteDatabase db) {
        // TODO Auto-generated method stub
        db.execSQL("CREATE TABLE person(id INTEGER PRIMARY KEY
                                AUTOINCREMENT, name VARCHAR(20))");
        db.execSQL("insert into person values(null,'Wu')");
        db.execSQL("insert into person values(null,'Guan')");
    }
    @Override
    public void onUpgrade(SQLiteDatabase db, int oldVersion, int newVersion) {
        // TODO Auto-generated method stub
        //db.execSQL("DROP TABLE IF EXISTS person");
        // 在原表person中增加一个字段tel
        db.execSQL("ALTER TABLE person ADD tel INTEGER");
        db.execSQL("ALTER TABLE person ADD tel INTEGER");
        db.execSQL("update person set tel=15527643858 where name='Wu'");
        db.execSQL("update person set tel=1330862750 where name='Guan'");
    }
}
}
```

7.3　使用 SQLiteDatabase 类实现数据库表的增/删/改/查

7.3.1　使用 execSQL()方法实现记录的“增/删/改”

类 SQLiteDatabase 封装了对 SQLite 数据库的操作方法，如图 7.3.1 所示。

android.database.sqlite
SQLiteDatabase.class
SQLiteDatabase
CursorFactory
newCursor(SQLiteDatabase, SQLiteCursorDriver, String, SQLiteQuery) : Cursor
execSQL(String) : void
execSQL(String, Object[]) : void
insert(String, String, ContentValues) : long
update(String, ContentValues, String, String[]) : int
delete(String, String, String[]) : int

图 7.3.1　SQLiteDatabase 类的常用方法

操作数据库有两种方式：一种是使用 execSQL()，它是操作数据库的简单方法，以 SQL 命令为参数；另一种是使用 insert()、update()和 delete()，这些方法的参数中不包含 SQL 命令，且方法 insert()和 update()还使用了 ContentValues 类型的参数，其用法见第 7.3.2 小节。

注意：

(1) execSQL(String,Object[])方法较 execSQL(String)而言，在 SQL 命令中支持通配符“?”。

(2) 方法 insert()、update()和 delete()中的第一参数都是表名。

(3) 使用 insert()、update()和 delete()方法，本质上也是执行 SQL 命令，只是系统已经组装了 SQL 语句。

7.3.2　使用类 ContentValues 追加或更新记录

前面向数据库中插入一条记录时，通过 db.execSQL(insert 命令)显式使用了 SQL 语句，而使用 SQLiteDatabase 提供的方法 insert(table,null,contentValues)，则不必写 SQL 语句。

注意：当 contentValues 参数为空或者里面没有内容的时候，insert()会失败，因为底层数据库不允许增加一个空行。为了防止这种情况，可以将第二参数设置为一个列名。如果能保证 contentValues 值非空，则可以将第二参数简单地设置为“null”。

在 SQLiteDatabase 提供的两个方法 insert()和 update()中，第一参数都是要操作的数据表的名称，都包含了 ContentValues 类型的参数。ContentValues 类提供了“键-值”对形式的数据，其主要方法如图 7.3.2 所示。

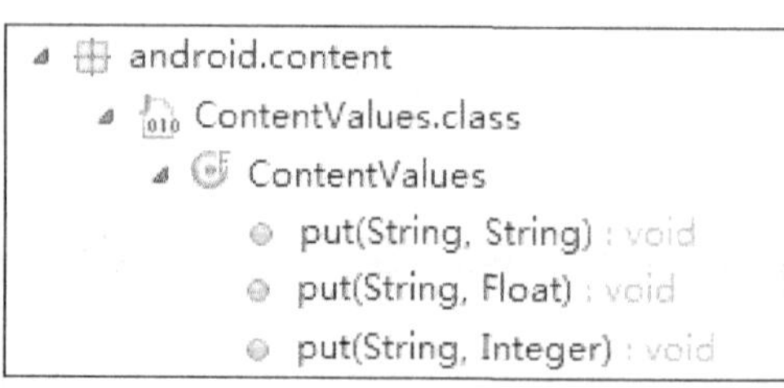

图 7.3.2　ContentValues 类及其主要方法

7.3.3　SQLiteDatabase 类提供的两种查询方法与游标接口 Cursor

使用 SQL 命令查询数据库得到的是内存中的一张虚拟表(记录集)，通过记录指针可以遍历所有记录，游标接口 android.database.Cursor 就是为这一目的设计的。SQLiteDatabase 类提供的两种查询方法 query()和 rawQuery()，如图 7.3.3 所示。

```
query(boolean, String, String[], String, String[], String, String, String, String) : Cursor
query(boolean, String, String[], String, String[], String, String, String, String, CancellationSignal) : Cursor
query(String, String[], String, String[], String, String, String) : Cursor
query(String, String[], String, String[], String, String, String, String) : Cursor
queryWithFactory(CursorFactory, boolean, String, String[], String, String[], String, String, String, String) : Cursor
queryWithFactory(CursorFactory, boolean, String, String[], String, String[], String, String, String, String, CancellationSignal) : Cursor
rawQuery(String, String[]) : Cursor
rawQuery(String, String[], CancellationSignal) : Cursor
rawQueryWithFactory(CursorFactory, String, String[], String) : Cursor
rawQueryWithFactory(CursorFactory, String, String[], String, CancellationSignal) : Cursor
```

图 7.3.3　SQLiteDatabase 类提供的查询方法

含有 2 个参数的 rawQuery()方法，是原始的查询方法，它的第一参数是完整的 SQL 命令，第二参数是条件参数(若 SQL 命令使用了占位符“?”)。例如：

```
Cursor c=db.rawQuery("select * from person where id=?",new String[]{idv+""});
```

注意：由于 id 字段是整型，因此上面的 idv 表达式应为整型。作为 SQL 命令的一部分，它需要转换成字符串。比较标准的做法是使用“Integer(idv).toString()”。

含 7 个参数的 query()方法，不使用完整的 SQL 命令作为参数，各参数的含义在 Eclipse 中时有提示，第一参数为表名，第二参数为字段集，第三参数为选择条件，第四参数为条件参数，最后一个参数为排序依据。例如，查询表 person 中年龄低于 50 者的代码如下：

```
String selectFilter="age<50";
Cursor c=db.query("person", null, selectFilter, null, null, null, "_id ASC");
```

注意：query()方法中的各参数可以拼装成 rawQuery()中的 SQL 命令，因此两种方法本质上是一样的，只是写法不同而已。

SQLiteDatabase 类的两种查询方法都是得到(返回)Cursor 接口类型的对象，该接口提供的常用方法如图 7.3.4 所示。

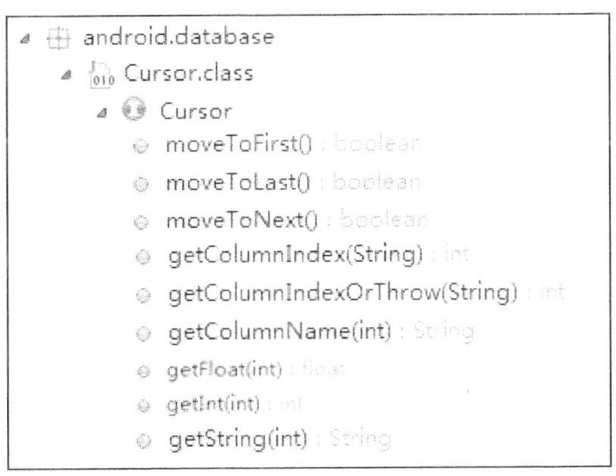

图 7.3.4　Cursor 接口提供的常用方法

7.3.4 使用适配器 SimpleAdapter 显示查询结果

在第 4.3.7 小节，我们介绍了简单的 ArrayAdapter 数据适配器。为了在 ListView 等控件中显示查询得到的结果集，完成一条记录的多项输出，可以使用 SimpleAdapter 数据适配器来完成。例如三星手机的设置程序，就需要使用 SimpleAdapter，其效果如图 7.3.5 所示。

图 7.3.5 三星手机设置程序界面

注意：在第 4.3.7 小节介绍的 ArrayAdapter 只能含有一个数据项，而 SimpleAdapter 可以包含多个数据项。

数据适配器 SimpleAdapter 提供的构造方法用于定义适配器。除此之外，它还有许多方法，如图 7.3.6 所示。

```
android.widget
  SimpleAdapter.class
    SimpleAdapter
      ViewBinder
      SimpleAdapter(Context, List<? extends Map<String, ?>>, int, String[], int[])
      getCount() : int
      getDropDownView(int, View, ViewGroup) : View
      getFilter() : Filter
      getItem(int) : Object
      getItemId(int) : long
      getView(int, View, ViewGroup) : View
      getViewBinder() : ViewBinder
      setDropDownViewResource(int) : void
      setViewBinder(ViewBinder) : void
      setViewImage(ImageView, int) : void
      setViewImage(ImageView, String) : void
      setViewText(TextView, String) : void
```

图 7.3.6 数据适配器类 SimpleAdapter

定义数据适配器 SimpleAdapter 时，共需要 5 个参数，分别是上下文对象、泛型数据、列表布局文件名(不带扩展名)、字段名数组和列表布局项名称数组。定义与使用 SimpleAdapter 的参考代码如下：

```
private Cursor cursor;
ListView listView;
private Map<String,Object> listItem;   //列表项，对应一条记录
private List<Map<String,Object>> listData;   //所有列表数据
//查询数据库并赋值给listData
while(cursor.moveToNext()){
    listItem=new HashMap<String,Object>();   //创建实例
```

```
        //分别获取各字段值给相应变量：id、name和age
        listItem.put("_id",id); //键值对
        listItem.put("name",name);
        listItem.put("age",age);
        listData.add(listItem);   //
}
//创建数据适配器对象并填充数据
SimpleAdapter listAdapter = new SimpleAdapter(this,listData,
        R.layout.list_item,
        new String[]{"_id","name","age"},
        new int[]{R.id.tv_id,R.id.tv_name,R.id.tv_age});
listView.setAdapter(listAdapter);   //绑定数据显示控件
```

注意：

(1) 由于 SimpleAdapter 的一个列表项里包含了多项数据，因此，在使用 ArrayList 定义它的数据项之前，需要使用 HashMap 定义它的列表项；

(2) 创建数据适配器 SimpleAdapter 对象时，其构造方法中共包含 5 个参数；

(3) 通过 ListView 等控件提供的 setAdapter()方法实现数据的绑定。

【例 7.3.1】使用数据适配器 SimpleAdapter 显示数据表。

【程序运行】程序运行后的效果，如图 7.3.7 所示。

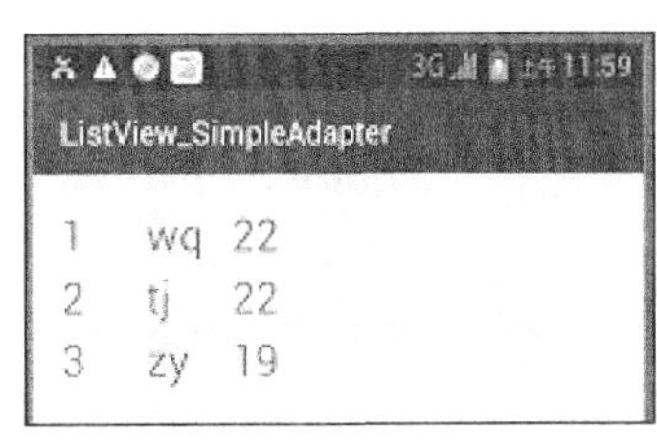

图 7.3.7　工程 ListView_SimpleAdapter 的运行界面

【实现步骤】

(1) 新建名为 ListView_SimpleAdapter 的 Android 应用工程，勾选创建 Activity，工程文件结构(主要部分)如图 7.3.8 所示。

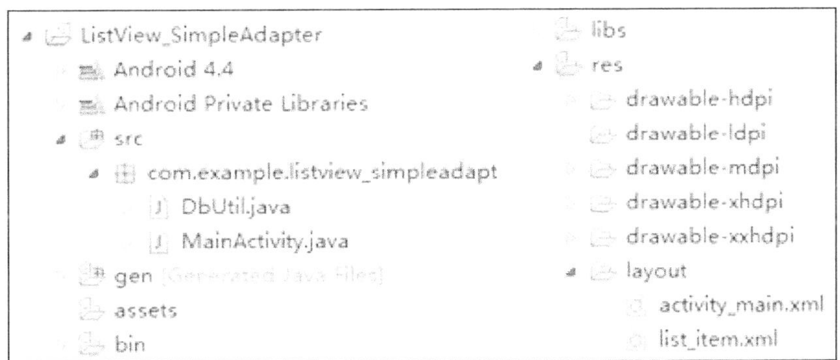

图 7.3.8　ListView_SimpleAdapter 工程文件结构(主要部分)

(2) 建立继承 SQLiteOpenHelper 的数据库工具类 DbUtil.java，其构造方法打开了与数据库 test.db 的连接；在改写的 onCreate()方法里创建数据表 person 并输入 3 条供测试的记录；获取 person 表中所有记录结果集的方法是 allQuery()。文件 DbUtil.java 的代码如下：

```
package com.example.listview_simpleadapter;
import android.database.sqlite.SQLiteOpenHelper;
import android.database.sqlite.SQLiteDatabase;
import android.content.Context;
import android.database.Cursor;
class DbUtil extends SQLiteOpenHelper{   //自定义工具类
        public DbUtil(Context context) {        //构造方法
            super(context, "test.db", null,1);        //建立或打开库
            // TODO Auto-generated constructor stub
        }
        @Override
        public void onCreate(SQLiteDatabase db) {
            // TODO Auto-generated method stub
            String temp="CREATE TABLE person(_id INTEGER PRIMARY KEY
                                AUTOINCREMENT, name VARCHAR(20),age int)";
            db.execSQL(temp);
        /*  ContentValues values = new ContentValues();
            values.put("name", "wq");values.put("age", 22);
            db.insert("person", null, values);
            values.put("name", "tj");values.put("age", 22);
            db.insert("person", null, values);
            values.put("name", "zy");values.put("age", 19);
            db.insert("person", null, values);*/
            db.execSQL("insert into person values(null,'wq',22)");
            db.execSQL("insert into person values(null,'tj',22)");
            db.execSQL("insert into person values(null,'zy',19)");
        }
        @Override
        public void onUpgrade(SQLiteDatabase db, int oldVersion, int newVersion) {
            // TODO Auto-generated method stub
            db.execSQL("DROP TABLE IF EXISTS person");
        }
        public Cursor allQuery(SQLiteDatabase db)   //查询所有记录
        {
            return db.query("person", null,null, null, null, null, "_id ASC");
            //实现记录筛选(过滤)
            //String selectFilter="age<50";
            //return db.query("person", null, selectFilter, null, null, null, "_id ASC");
        }
```

}

(3) 编写主 Activity 对应的布局文件 activity_mail.xml，增加一个 ListView 控件。

(4) 创建与 ListView 列表项对应的布局文件 list_item.xml，采用水平线性布局，包含三个 TextView 控件，分别用来显示数据表的三个字段：_id、name 和 age。

文件 list_item.xml 的代码如下：

```
<?xml version="1.0" encoding="utf-8"?>
<LinearLayout xmlns:android="http://schemas.android.com/apk/res/android"
    android:layout_width="match_parent"
    android:layout_height="match_parent"
    android:orientation="horizontal" >
   <TextView
        android:id="@+id/tv_id"
        android:textSize="20sp"
        android:layout_width="50dp"
        android:layout_height="wrap_content"/>
    <TextView
        android:id="@+id/tvname"
        android:textSize="20sp"
        android:layout_width="50dp"
        android:layout_height="wrap_content"/>
    <TextView
        android:id="@+id/tvage"
        android:textSize="20sp"
        android:layout_width="50dp"
        android:layout_height="wrap_content"/>
</LinearLayout>
```

(5) 在主 Activity 的 MainActivity.java 中，调用工具类 Util.java 的查询方法 allQuery() 返回结果集，建立数据适配器 SimpleAdapter 对象并用结果集数据填充，最后将数据适配器绑定到 ListView 控件上。

```
package com.example.listview_simpleadapter;
import android.app.Activity;
import android.os.Bundle;
import java.util.ArrayList;
import java.util.HashMap;
import java.util.List;
import java.util.Map;
import android.widget.ListView;   //
import android.widget.SimpleAdapter;   //
```

```
import android.database.sqlite.SQLiteDatabase;
import android.database.Cursor;
public class MainActivity extends Activity {
    private Cursor cursor;
    ListView listView;
    private Map<String, Object> listItem;   //列表项，对应一条记录
    private List<Map<String,Object>> listData;  //所有列表数据
    @Override
    public void onCreate(Bundle savedInstanceState) {
        super.onCreate(savedInstanceState);
        setContentView(R.layout.activity_main);
        DbUtil dbUtil = new DbUtil(this);
        SQLiteDatabase db= dbUtil.getWritableDatabase();
        cursor =dbUtil.allQuery(db);      // 查询得到结果集
        listView = (ListView)findViewById(R.id.listView);
        listData = new ArrayList<Map<String,Object>>();   //创建实例
        while (cursor.moveToNext()){
            long id=cursor.getInt(0);   //获取字段值
            String name=cursor.getString(1);
            //int age=cursor.getInt(2);
            int age=cursor.getInt(cursor.getColumnIndex("age"));//另一种方式
            listItem=new HashMap<String,Object>(); //必须在循环体里新建
            listItem.put("_id", id);
            listItem.put("name", name);
            listItem.put("age", age);
            listData.add(listItem);          //添加一条记录
        }
        //创建适配器并填充数据
        SimpleAdapter listAdapter = new SimpleAdapter(this,listData,
                R.layout.list_item,
                new String[]{"_id","name","age"},
                new int[]{R.id.tv_id,R.id.tv_name,R.id.tv_age});
        listView.setAdapter(listAdapter);   //数据绑定显示控件
    }
}
```

下面再介绍一个综合实例：使用 SQLiteDatabase 提供的方法实现对数据库的增加、删除、修改和查询。

【例 7.3.2】数据库的“增/删/改/查”。

【程序运行】程序初次运行后，窗体中出现分别用于输入姓名和年龄的两个 EditText 控件和三个功能按钮，其运行界面如图 7.3.9 所示。

在输入数据后，单击“添加”按钮将数据(记录)写入 SQLite 数据库 mydb.db 的表 friends 中。每添加 1 条记录，该记录都会立即在 ListView 控件里显示。输入 3 条记录后，程序的运行界面如图 7.3.10 所示。

图 7.3.9　MyDbCRUD 工程初次运行界面

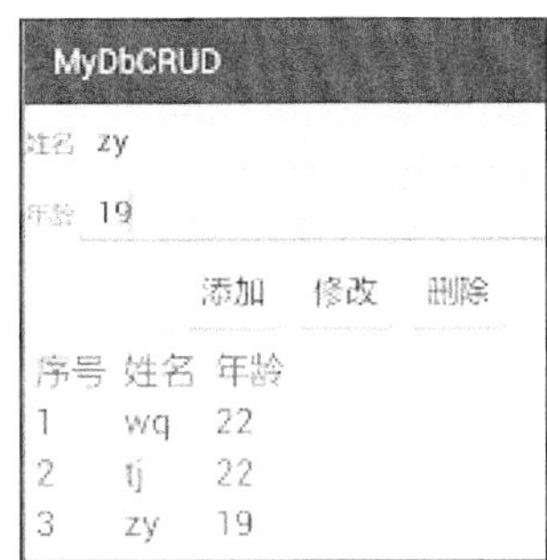

图 7.3.10　输入 3 条记录后的运行界面

【实现步骤】

(1) 新建名为 MyDbCRUD 的 Android 工程，勾选创建 Activity，主 Activity 名称及主布局文件均采用默认值。

(2) 建立与例 7.3.1 中工具类 Util 功能相似的类 DbHelper，相应的类文件 DbHelper.java 的代码如下：

```
package introduction.android.mydbcrud;
/*
 * 本类用于创建或打开数据库mydb.db及表friends
 */
import android.content.Context;
import android.database.sqlite.SQLiteOpenHelper;  //主类1
import android.database.sqlite.SQLiteDatabase;  //主类2
import android.database.sqlite.SQLiteDatabase.CursorFactory;

public class DbHelper extends SQLiteOpenHelper{
    public static final String DATABASE_NAME = "mydb";  //库名
    public static final String TABLE_NAME = "friends";    //表名
    public static final int DATABASE_VERSION = 1;
    public static final int FRIENDS = 1;
    public static final int FRIENDS_ID = 2;
    //对应于表friends的三个字段，加下划线表示该字段不由用户输入
    public static final String ID = "_id";
    public static final String NAME = "name";  //其他字段
    public static final String AGE = "age";
```

```
    public DbHelper(Context context, String name, CursorFactory factory, int version) {
        // TODO Auto-generated constructor stub
        //调用父类的构造方法
        super(context, name, factory, version);
    }
    @Override
    public void onCreate(SQLiteDatabase db) {
        // TODO Auto-generated method stub
        db.execSQL("CREATE TABLE IF NOT EXISTS " +TABLE_NAME + "( _id integer
                    primary key autoincrement," +"name varchar,age integer"+")");
    }
    @Override
    public void onUpgrade(SQLiteDatabase db, int oldVersion, int newVersion) {
        // TODO Auto-generated method stub
        db.execSQL("DROP TABLE IF EXISTS " + TABLE_NAME);    //先删除
        onCreate(db);    //后创建
    }
}
```

(3) 在主布局文件 activity_main.xml 中依次增加 2 组 TextView 和 EditText 控件、3 个 Button 控件和 1 个 ListView 控件，嵌套使用垂直线性布局和水平线性布局，其代码如下：

```
<?xml version="1.0" encoding="utf-8"?>
<LinearLayout xmlns:android="http://schemas.android.com/apk/res/android"
    android:layout_width="fill_parent"
    android:layout_height="fill_parent"
    android:orientation="vertical" >
    <LinearLayout
        android:layout_width="fill_parent"
        android:layout_height="wrap_content"
        android:addStatesFromChildren="true" >
        <TextView
            android:layout_width="wrap_content"
            android:layout_height="wrap_content"
            android:text="姓名"
            android:textColor="?android:attr/textColorSecondary" />
        <EditText
            android:id="@+id/et_name"
            android:layout_width="wrap_content"
            android:layout_height="wrap_content"
            android:layout_weight="1"
            android:singleLine="true" />
```

```
</LinearLayout>
<LinearLayout
    android:layout_width="fill_parent"
    android:layout_height="wrap_content"
    android:addStatesFromChildren="true" >
    <TextView
        android:layout_width="wrap_content"
        android:layout_height="wrap_content"
        android:text="年龄"
        android:textColor="?android:attr/textColorSecondary" />
    <EditText
        android:id="@+id/et_age"
        android:layout_width="wrap_content"
        android:layout_height="wrap_content"
        android:layout_weight="1"
        android:singleLine="true" />
</LinearLayout>
<LinearLayout
    android:layout_width="fill_parent"
    android:layout_height="wrap_content"
    android:addStatesFromChildren="true"
    android:gravity="center" >
    <Button
        android:id="@+id/bt_add"
        android:layout_width="wrap_content"
        android:layout_height="wrap_content"
        android:text="添加"
        android:onClick="addbutton"/>
    <Button
        android:id="@+id/bt_modify"
        android:layout_width="wrap_content"
        android:layout_height="wrap_content"
        android:text="修改"
        android:onClick="updatebutton" />
    <Button
        android:id="@+id/bt_del"
        android:layout_width="wrap_content"
        android:layout_height="wrap_content"
        android:text="删除"
```

```
            android:onClick="updatebutton" />
    </LinearLayout>
    <ListView
        android:gravity="center"
        android:id="@+id/listView"
        android:layout_width="fill_parent"
        android:layout_height="wrap_content"
        android:padding="5dip" >
    </ListView>
```

(4) 建立用于 ListView 控件的布局文件 list_item.xml，其代码与例 7.3.1 相同。此时，工程文件结构(主要部分)如图 7.3.11 所示。

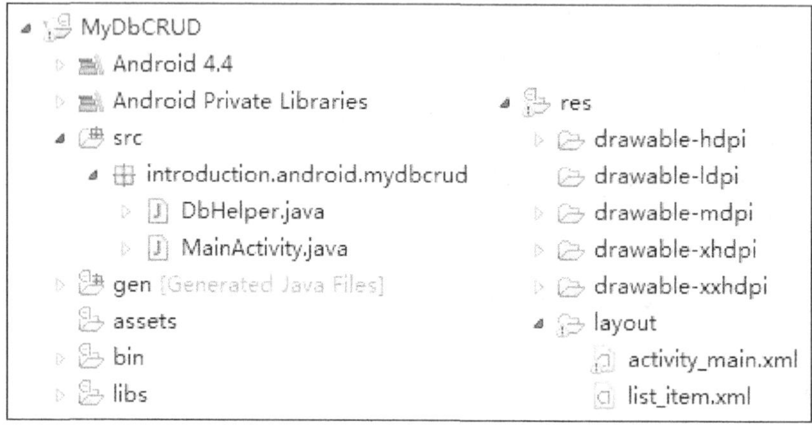

图 7.3.11　MyDbCRUD 工程文件结构(主要部分)

(5) 对数据库进行“增/删/改/查”的代码含于主 Activity 文件 MainActivity.java 里，其代码如下：

```
package introduction.android.mydbcrud;
/*
 * 本例实现了多种方式对数据库的增/删/改/查
 * 显示数据表时使用了SimpleAdapter适配器
*/
import android.app.Activity;    //
import android.os.Bundle;
import java.util.ArrayList;
import java.util.HashMap;
import java.util.Map;
import android.content.ContentValues;    //相关类
import android.database.Cursor;    //
import android.database.sqlite.SQLiteDatabase;    //主类
```

```
import android.view.View;
import android.widget.AdapterView;
import android.widget.Button;
import android.widget.EditText;
import android.widget.Toast;
import android.widget.ListView;   //主类
import android.widget.SimpleAdapter;   //相关类
import android.widget.AdapterView.OnItemClickListener;
public class MainActivity extends Activity {
    private EditText et_name;
    private EditText et_age;
    private Map<String,Object> item;   //
    private ArrayList<Map<String, Object>> data;       //
    private DbHelper dbHelper;
    private SQLiteDatabase db;
    private Cursor cursor;
    private SimpleAdapter listAdapter;     //
    private ListView listview;
    private Button addBtn,updBtn,delBtn;
    private String selId;
    @Override
    public void onCreate(Bundle savedInstanceState) {
        super.onCreate(savedInstanceState);
        setContentView(R.layout.activity_main);
        et_name=(EditText) findViewById(R.id.et_name);
        et_age=(EditText) findViewById(R.id.et_age);
        listview = (ListView) findViewById(R.id.listView);
        addBtn=(Button)findViewById(R.id.bt_add);
        updBtn=(Button)findViewById(R.id.bt_modify);
        delBtn=(Button)findViewById(R.id.bt_del);
        addBtn.setOnClickListener(new Button.OnClickListener(){
            @Override
            public void onClick(View v) {
                // TODO Auto-generated method stub
                dbAdd();  //添加记录
                dbFindAll();
            }
        });
        updBtn.setOnClickListener(new Button.OnClickListener(){
```

```
            @Override
            public void onClick(View v) {
                // TODO Auto-generated method stub
                dbUpdate();   //修改当前记录
                dbFindAll();
            }
        });
        delBtn.setOnClickListener(new Button.OnClickListener(){
            @Override
            public void onClick(View v) {
                // TODO Auto-generated method stub
                dbDel();          //删除当前记录
                dbFindAll();
            }
        });
        //创建工具类的实例
        dbHelper = new DbHelper(this,DbHelper.DATABASE_NAME, null, 1);
        db = dbHelper.getWritableDatabase();        // 以可写入方式打开数据库
        data = new ArrayList<Map<String,Object>>();        //列表类型数据
        dbFindAll();
        //列表项选择监听器
        listview.setOnItemClickListener(new OnItemClickListener() {
            @SuppressWarnings("unchecked")
            @Override
            public void onItemClick(AdapterView<?> parent, View v,
                    int position, long id) {
                // TODO Auto-generated method stub
                Map<String,Object> listItem =(Map<String,Object>) listview
                                            .getItemAtPosition(position);
                et_name.setText((String)listItem.get("name"));
                et_age.setText((String)listItem.get("age"));
                selId=(String)listItem.get("_id");   //
                Toast.makeText(getApplicationContext(), "选择的id是: "+
                                        selId,Toast.LENGTH_SHORT).show();
            }
        });
    }
    protected void dbDel() {
        String where="_id="+selId;
```

```
        int i=db.delete(DbHelper.TABLE_NAME, where, null);
        if(i>0)
            Toast.makeText(getApplicationContext(), "数据删除成功！ ",
                                        Toast.LENGTH_SHORT).show();
        else
            Toast.makeText(getApplicationContext(), "数据删除失败！ ",
                                        Toast.LENGTH_SHORT).show();
    }
    private void showList() {          // 显示记录集
        listAdapter=new SimpleAdapter(this,data,
                R.layout.list_item,
                new String[]{"_id","name","age"},
                new int[]{R.id.tv_id,R.id.tvname,R.id.tvage});
        listview.setAdapter(listAdapter);
    }
    protected void dbUpdate() {
        // TODO Auto-generated method stub
        ContentValues values = new ContentValues();
        values.put("name", et_name.getText().toString().trim());
        values.put("age", et_age.getText().toString().trim());
        String where="_id="+selId;
        int i=db.update(DbHelper.TABLE_NAME, values, where, null);
        if(i>0)
            Toast.makeText(getApplicationContext(), "数据更新成功！ ",
                                        Toast.LENGTH_SHORT).show();
        else
            Toast.makeText(getApplicationContext(), "数据更新失败！ ",
                                        Toast.LENGTH_SHORT).show();
    }
    protected void dbAdd() {      //插入记录
        // TODO Auto-generated method stub
        ContentValues values = new ContentValues();
        values.put("name", et_name.getText().toString().trim());
        values.put("age", et_age.getText().toString().trim());
        //第二参数通常为null
        long rowid=db.insert(DbHelper.TABLE_NAME, null, values);
        if(rowid==-1)
            Toast.makeText(getApplicationContext(), "数据插入失败！ ",
                                        Toast.LENGTH_SHORT).show();
```

```
        else
            Toast.makeText(getApplicationContext(), "数据插入成功！ ",
                                            Toast.LENGTH_SHORT).show();
    }
    protected void dbFindAll() {            //记录查找
        data.clear();   //必须清除
        //生成记录集的两种等价方法
        cursor=db.rawQuery("select * from friends order by _id ASC",null);
        //cursor = db.query(DbHelper.TB_NAME, null, null, null, null, null,"_id ASC");
        item=new HashMap<String,Object>();
        item.put("_id", "序号");item.put("name", "姓名");
        item.put("age", "年龄");data.add(item);
        cursor.moveToFirst();
        while (!cursor.isAfterLast() ) {
            String id=cursor.getString(0);
            String name=cursor.getString(1);
            String age=cursor.getString(2);
            item=new HashMap<String,Object>();
            item.put("_id", id);
            item.put("name", name);
            item.put("age", age);
            data.add(item);
            cursor.moveToNext();
        }
        showList();
    }
}
```

7.3.5 以 DAO 方式编写访问数据库的程序

数据访问对象 DAO，即 database access object。使用 DAO 方式的思想是将数据库操作的业务代码封装到一个类文件的相关方法里。使用 DAO 方式具有如下优点：

- 代码便于重用；
- 代码冗余度低。

注意：如果在 Activity 中直接写访问数据库的相关代码，则不具备 DAO 方式的优点。

【例 7.3.3】使用 DAO 方式创建访问数据库的 Android 工程。

【设计思想】将例 7.3.2 中对数据库操作的业务代码移出并封装到一个类文件的相关方法里，如图 7.3.12 所示。

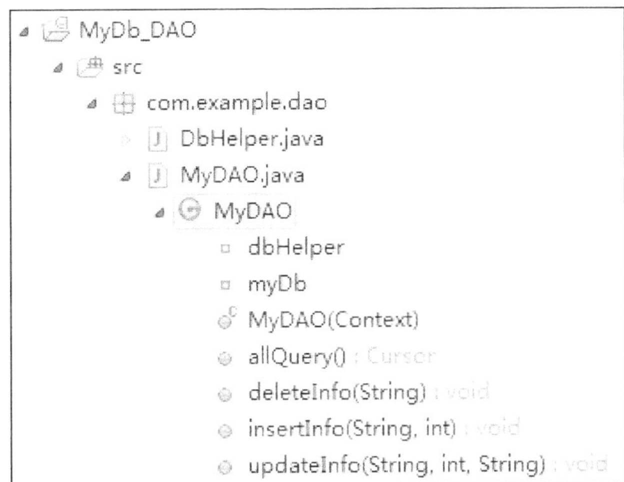

图 7.3.12　访问数据库的 DAO 类的定义

【程序运行】本程序运行时，向数据库表增加 2 条记录并使用 ListView 控件显示，其效果如图 7.3.13 所示。

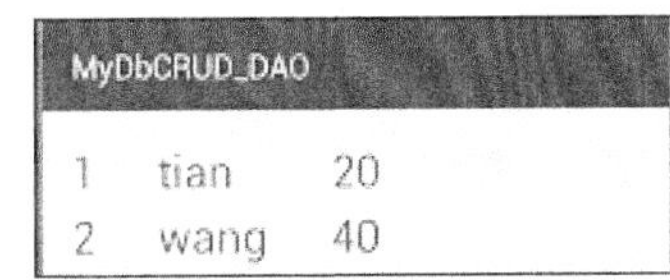

图 7.3.13　例 7.3.3 程序运行效果

【实现步骤】

(1) 新建名为 MyDb_DAO 的 Android 应用工程，勾选创建 Activity，主 Activity 和主布局采用默认的文件名。

(2) 在 src 文件夹里新建一个名为 com.example.dao 的包名，并在其内新建一个名为 MyDAO.java 的标准类文件，其代码如下：

```
package com.example.dao;
/*
 *  本类MyDAO提供了查询数据库表所有记录的方法：allQuery()
 *  插入记录的方法：insertInfo(String name,int age)
 *  删除记录的方法：deleteInfo(String selId)
 *  修改记录的方法：updateInfo(String name,int age,String selId)
 */
import android.content.ContentValues;
import android.content.Context;
import android.database.Cursor;
import android.database.sqlite.SQLiteDatabase;
import android.util.Log;
public class MyDAO {
    private SQLiteDatabase myDb;   //类的成员
    private DbHelper dbHelper;    //类的成员
    public MyDAO(Context context)  {     //构造方法
        dbHelper = new DbHelper(context,"test.db",null,1);
    }
```

```
public Cursor allQuery()       //查询所有记录
{
    myDb = dbHelper.getReadableDatabase();
    return myDb.rawQuery("select * from friends",null);
}

public void insertInfo(String name,int age)    //插入记录
{
    myDb = dbHelper.getWritableDatabase();
    ContentValues values = new ContentValues();
    values.put("name", name);
    values.put("age", age);
    long rowid=myDb.insert(DbHelper.TB_NAME, null, values);
    if(rowid==-1)
        Log.i("myDbDemo", "数据插入失败！ ");
    else
        Log.i("myDbDemo", "数据插入成功！ "+rowid);
}
public void deleteInfo(String selId)    //删除记录
{
    String where="_id="+selId;
    int i=myDb.delete(DbHelper.TB_NAME, where, null);
    if(i>0)
        Log.i("myDbDemo","数据删除成功！ ");
    else
        Log.i("myDbDemo","数据未删除！ ");
}
public void updateInfo(String name,int age,String selId)    //修改记录
{
    ContentValues values = new ContentValues();
    values.put("name", name);
    values.put("age", age);
    String where="_id="+selId;
    int i=myDb.update(DbHelper.TB_NAME, values, where, null);

    //上面几行代码的功能可以用下面的一行代码实现
    //myDb.execSQL("update friends set name = ? ,age = ? where _id = ?",new
                                                    Object[]{name,age,selId});
```

```
        if(i>0)
            Log.i("myDbDemo","数据更新成功！");
        else
            Log.i("myDbDemo","数据未更新！");
    }
}
```

(3) 建立数据适配器对应的布局文件 list_item.xml，其代码与例 7.3.1 相同。此时，工程文件结构(主要部分)如图 7.3.14 所示。

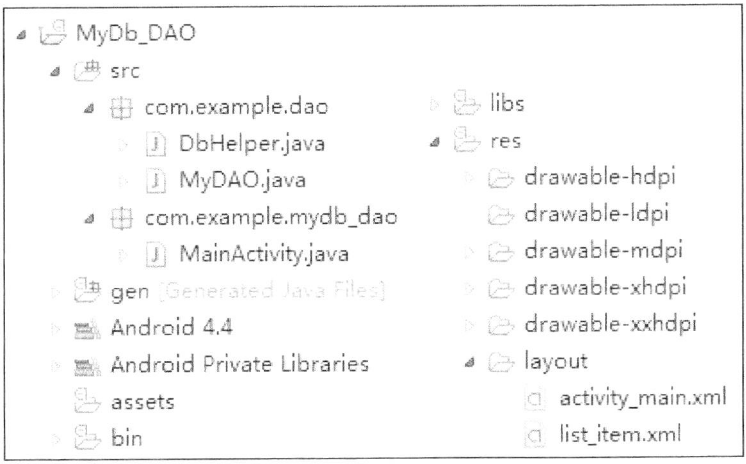

图 7.3.14　使用 DAO 方式开发的工程文件结构(主要部分)

注意：Android 工程的包名(即应用程序的包名，是在清单文件中定义的)，默认情况下与主 Activity 文件的路径一致，与图 7.3.14 中的包名“com.example.dao”(工程的内部包名)是不一样的。

(4) 编写应用程序的主窗口，对应的程序文件 MainActivity.java 的代码如下：

```
package com.example.mydbcrud_dao;
/*
 * 本程序中对数据库的插入和查询操作，使用了MyDAO类的相关方法
 * 首次运行时，增加2条记录并使用ListView控件显示出来
 * 重复部署运行都会增加2条重复的记录，但数据库版本提升后再部署运行时就不会
 */
import com.example.dao.MyDAO;    //
import android.app.Activity;
import android.os.Bundle;
import java.util.ArrayList;
import java.util.HashMap;
import java.util.List;
import java.util.Map;
import android.database.Cursor;
```

```
import android.view.View;
import android.widget.AdapterView;
import android.widget.AdapterView.OnItemClickListener;
import android.widget.ListView;
import android.widget.SimpleAdapter;
import android.widget.TextView;
import android.widget.Toast;
public class MainActivity extends Activity {
    private ListView listView;
    private List<Map<String,Object>> listData;
    private Map<String,Object> listItem;
    @Override
    protected void onCreate(Bundle savedInstanceState) {
        super.onCreate(savedInstanceState);
        setContentView(R.layout.activity_main);
        MyDAO myDAO = new MyDAO(this);    //
        myDAO.insertInfo("tian", 20);     //插入记录
        myDAO.insertInfo("wang", 40); //插入记录
        listView = (ListView)findViewById(R.id.listView);
        listData = new ArrayList<Map<String,Object>>();
        Cursor cursor = myDAO.allQuery();     //
        while (cursor.moveToNext()){
            long id=cursor.getInt(0);   //获取字段值
            String name=cursor.getString(1);
            //int age=cursor.getInt(2);
            int age=cursor.getInt(cursor.getColumnIndex("age"));//另一种方式
            listItem=new HashMap<String,Object>(); //必须在循环体里新建
            listItem.put("_id", id);
            listItem.put("name", name);
            listItem.put("age", age);
            listData.add(listItem);           //添加一条记录
        }
        SimpleAdapter listAdapter = new SimpleAdapter(this,listData,
                R.layout.list_item,
                new String[]{"_id","name","age"},
                new int[]{R.id.tv_id,R.id.tvname,R.id.tvage});
        listView.setAdapter(listAdapter);
        listView.setOnItemClickListener(new OnItemClickListener()
        {
```

```
            @Override
            public void onItemClick(AdapterView<?> arg0, View arg1, int arg2,
                    long arg3) {
                // TODO Auto-generated method stub
                TextView tv = (TextView)arg1.findViewById(R.id.tvname);
                Toast.makeText(MainActivity.this,
                                    tv.getText(),Toast.LENGTH_LONG).show();

            }
        });
    }

}
```

注意：由于在 DAO 类方法中的参数很多是数据表的字段，因此，可以将各个数据表对应地建立一个实体类，这些类通常称为 Model 层。每个实体类除了包含那些字段成员外，只包含 set×××()和 get×××()方法。在 DAO 类的方法中使用实体类对象参数，减少了 DAO 类方法中参数的个数，使程序更加清晰。

7.3.6　使用数据库事务

使用数据库事务的一个典型例子是银行汇款。当一个账户减少金额时，另一个账户增加金额，而不能出现一个账户减少金额后系统抛出异常，导致另一个账户没有增加金额，反之亦然。处理这类同时提交或同时不提交的业务，就是数据库事务。

SQLiteDatabase 提供了处理数据库事务的相关方法。例如，开启一个数据库事务的 beginTransaction()方法、数据库处理完毕的 setTransactionSuccessful()方法、结束数据库事务的 endTransaction()方法等，如图 7.3.15 所示。

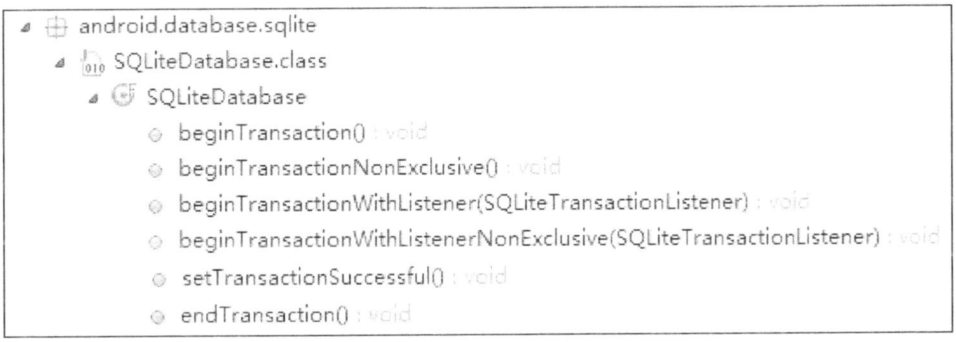

```
android.database.sqlite
    SQLiteDatabase.class
        SQLiteDatabase
            beginTransaction() : void
            beginTransactionNonExclusive() : void
            beginTransactionWithListener(SQLiteTransactionListener) : void
            beginTransactionWithListenerNonExclusive(SQLiteTransactionListener) : void
            setTransactionSuccessful() : void
            endTransaction() : void
```

图 7.3.15　SQLiteDatabase 类提供的数据库事务处理方法

【例 7.3.4】使用数据库事务，实现银行汇款。

【程序运行】使用数据库事务实现银行汇款时，如果在汇出(账户金额减少)和接收(账户金额增加)之间抛出一个运行异常，则汇款业务和收款业务都不会执行，反之都执行。程序运行效果如图 7.3.16 所示。

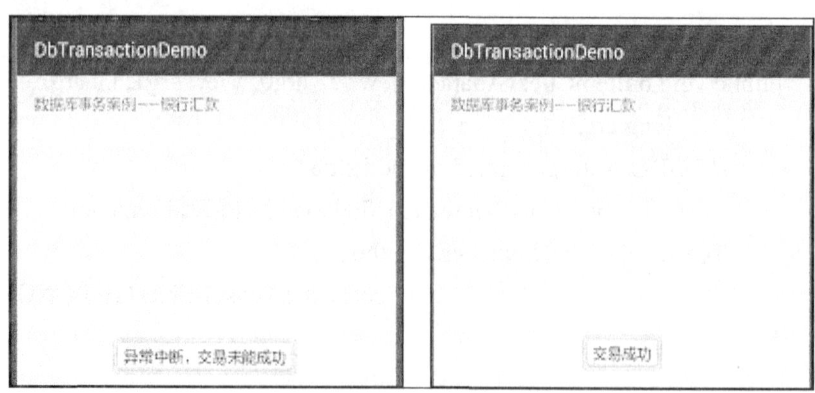

图 7.3.16　银行汇款事务的两个运行界面

初次运行程序，会产生除数为 0 的异常，而屏蔽程序中会产生异常的代码；而再次部署和运行程序，则交易成功。验证方法是查看交易成功前后账户余额的变化，如图 7.3.17 所示。

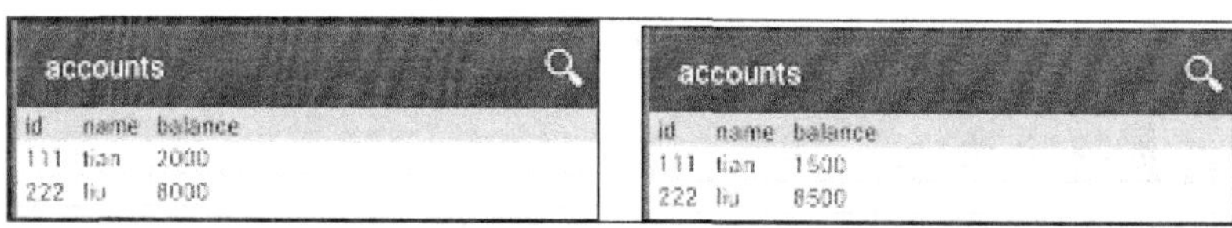

图 7.3.17　交易成功前后账户余额的变化

【实现步骤】

(1) 新建名为 DbTransactionDemo 的 Android 应用工程，勾选创建 Activity 并采用默认的文件名，如图 7.3.18 所示。

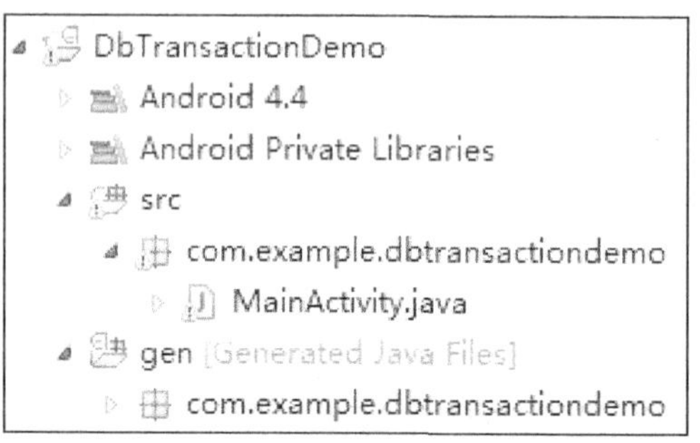

图 7.3.18　一个数据库事务工程的文件结构(部分)

(2) 修改默认的布局文件中 TextView 控件显示的内容为“数据库事务案例——银行汇款”。

(3) MainActivity 中包含了创建、连接及访问 SQLiteDatabase 数据库等操作，文件代码如下：

```
package com.example.dbtransactiondemo;
/*
 * 本程序演示使用数据库事务的方法
```

```
 * 初次运行：在扣款与存款之间会抛出异常，账户余额都不会变化(可查验)
 * 屏蔽产生异常的代码，再次部署运行，数据库事务提交成功，账户余额会同时变化(可查
验)
 * 数据库事务处理代码(开启事务-->设置事务处理成功-->结束事务)要放在
 * try...catch...finally结构中
 */
import android.app.Activity;
import android.os.Bundle;
import android.widget.Toast;
import android.view.Gravity;
import android.database.sqlite.SQLiteOpenHelper; //主类 1
import android.database.sqlite.SQLiteDatabase;   //主类2
import android.content.Context;   //
public class MainActivity extends Activity {
    MyDbOpenHelper helper;
    SQLiteDatabase   db;
    @Override
    public void onCreate(Bundle savedInstanceState) {
        super.onCreate(savedInstanceState);
        setContentView(R.layout.activity_main);
        //创建或连接数据库
        helper = new MyDbOpenHelper(this);
        db = helper.getWritableDatabase();
        setMit();  //执行自定义的事务处理方法
        db.close();
    }
    class MyDbOpenHelper extends SQLiteOpenHelper{   //自定义工具内部类
        public MyDbOpenHelper(Context context) {
            super(context, "test.db", null,1);        //建立或打开库
            // TODO Auto-generated constructor stub
        }
        @Override
        public void onCreate(SQLiteDatabase db) {
            // TODO Auto-generated method stub
            //创建表结构
            db.execSQL("CREATE TABLE accounts(id char(6) ,name VARCHAR(6),
                                                         balance double)");
            //插入2条记录
            db.execSQL("insert into accounts values('111', 'tian',2000)");
```

```
                db.execSQL("insert into accounts values('222', 'liu',8000)");
            }
            @Override
            public void onUpgrade(SQLiteDatabase db, int oldVersion, int newVersion) {
                // TODO Auto-generated method stub
                db.execSQL("DROP TABLE IF EXISTS accounts");
                onCreate(db);   //
            }
        }
        public   void setMit(){   //事务处理方法
            Toast toast;
            db.beginTransaction();   //数据库事务
            try{
                int num = 500;   //汇款金额
                String from = "111";      //汇出账户
                String to = "222";   //接收账户
                db.execSQL("update accounts set balance = balance - ? where id = ?",
                                                            new Object[]{num,from});
                int temp=1/0;   //产生除数为0的异常，屏蔽此行才能完成事务
                db.execSQL("update accounts set balance = balance + ? where id = ?",new
                                                                  Object[]{num,to});
                db.setTransactionSuccessful();   //事务成功标识
                toast=Toast.makeText(MainActivity.this, "交易成功",
                                                            Toast.LENGTH_SHORT);
                toast.setGravity(Gravity.CENTER,0,0);
                toast.show();
            } catch (Exception e){
                e.printStackTrace();
                toast=Toast.makeText(MainActivity.this, "异常中断，
                                              交易未能成功", Toast.LENGTH_SHORT);
                toast.setGravity(Gravity.CENTER,0,0);     //居中
                toast.show();
            } finally{
                db.endTransaction();
            }
        }
    }
```

习　题　7

一、判断题

1. SQLiteOpenHelper 和 Activity 类都提供了 onCreate()方法。
2. 应用程序创建的 SQLite 数据库为该应用私有。
3. 使用 SimpleAdapter 的构造方法创建数据适配器之前，必须先准备好数据。
4. 接口 Cursor 与抽象类 SQLiteOpenHelper 所处的软件包相同。
5. 使用数据库事务能保证一组代码同时执行。
6. 重新部署含有数据库访问的 Android 应用工程,不会删除已经存在的 SQLite 数据库。
7. 在 SD 卡里存储文件采用的是 FAT 文件系统，因此不支持访问模式和权限控制。

二、选择题

1. 使用 SQLiteOpenHelper 的构造方法创建与数据库的连接时，其参数可以设置为 null 的是____。

A. 上下文　　B. 数据库名　　C. 游标工厂　　D. 数据库版本

2. SQLiteDatabase 提供的____方法以完整的 SQL 命令作为参数。

A. insert()　　B. query()　　C. update()　　D. execSQL()

3. 含有 SQLite 数据库访问的 Android 应用中，其数据库存放在应用包名文件夹下的____文件夹里。

A. databases　　B. lib　　C. files　　D. shared_prefs

4. 数据适配器 SimpleAdapter 的构造函数中的参数个数是____。

A. 3　　B. 4　　C. 5　　D. 6

5. 设 Cursor 对象 cursor 的第 3 个字段 age 为 int 类型,则访问该字段的正确方法是____。

A. int age=cursor.getString(2);　　B. int age=cursor.getString(3);

C. int age=cursor.getInt(3);

D. int age=cursor.getInt(cursor.getColumnIndex("age"));

6. 数据库事务的相关方法封装在____类中。

A. SQLiteOpenHelper　　B. SQLiteDatabase

C. ContentValues　　D. Cursor

7. 关于数据存储方式，屏蔽了底层文件操作的是____。

A. SQLite 数据库存储　　B. File 存储

C. SharedPreferences　　D. A 和 C

三、填空题

1. SQLite 数据库文件的扩展名为____。
2. 在 Android 系统中，创建与连接 SQLite 数据库，需要使用抽象类____。
3. 在 Android 的____类中，封装了对 SQLite 数据库及表的操作方法。
4. SQLiteDatabase 提供的 execSQL()方法的返回值类型是____。

5. 当程序提升了数据库版本，则重新部署后会执行____方法。

6. 绑定数据适配器里的数据至数据显示控件(如 ListView)的方法是____。

7. 接口 Cursor 提供的 moveToNext()方法的返回类型是____。

8. 在 Android 的内部存储器里，采用____文件系统，以通过文件访问权限的控制来保证文件的私密性。

实验 7　SQLite 数据库编程

一、实验目的

1. 掌握 SQLite 数据库的建立、连接和更新方法。
2. 掌握对数据库中表结构的增/删/改方法。
3. 掌握对数据库中表内容的增/删/改/查方法。
4. 掌握数据适配器 SimpleAdapter 的使用。
5. 掌握 DAO 方式在数据库中的应用。
6. 掌握数据库事务的使用方法。

二、实验内容及步骤

【预备】访问 http://www.wustwzx.com/android/index.html，下载本次实验内容的源代码(压缩文件)并解压，得到文件夹 ch07，供下面的实验使用。

1. 使用 SQLiteOpenHelper 和 SQLiteDatabase 创建、更新数据库及表(参见例 7.2.1)。

(1) 再次解压 ch07 文件夹里的压缩文件 SQLite_c.zip，得到 Android 工程文件夹 SQLite_c。

(2) 在 Eclipse 中导入 Android 工程 SQLite_c。

(3) 打开源文件 MainActivity.java，查看内部类 MyDbOpenHelper 的定义及功能。

(4) 部署工程到手机后，使用 Root Explore 程序查看手机 data/data/com.example.sqlite_c/databases/test.db 文件的 person 表的结构(只有 2 个字段)。

(5) 修改源程序 MainActivity.java，将数据库的版本从 1 提升到 2，再次部署工程到手机。

(6) 再次使用 Root Explore 程序查看手机 data/data/ com.example.sqlite_c/databases/test.db 文件的结构 person 表的结构(在原有的 2 个字段的基础上多了 1 个字段)。

2. 数据适配器 SimpleAdapter 的使用(参见例 7.3.1)。

(1) 再次解压 ch07 文件夹里的压缩文件 ListView_SimpleAdapter.zip，得到 Android 工程文件夹 ListView_SimpleAdapter。

(2) 在 Eclipse 中导入 Android 工程 ListView_SimpleAdapter。

(3) 打开源文件 DbUtil.java，查看创建 SQLite 数据库及表的相关代码。

(4) 查看 DbUtil.java 上插入记录的两种方法(键值对形式和 SQL 命令)。

(5) 打开源文件 MainActivity.java，查看创建 SimpleAdapter 对象的代码。

(6) 查看 SimpleAdapter 对象绑定 ListView 的代码。

3. 非 DAO 方式实现数据库的“增/删/改/查”(参见例 7.3.2)。

(1) 再次解压 ch07 文件夹里的压缩文件 MyDbCRUD.zip，得到 Android 工程文件夹 MyDbCRUD。

(2) 在 Eclipse 中导入 Android 工程 MyDbCRUD。

(3) 打开源文件 DbHelper.java，查看创建 SQLite 数据库及表的相关代码。

(4) 打开源文件 MainActivity.java，查看添加记录的实现代码。

(5) 部署应用程序，输入 3 条记录。

(6) 查看查询的实现代码，在方法 rawQuery()中使用参数式查询，以获取部分记录。

(7) 再次部署应用程序查验。

(8) 查看删除记录的实现代码。

(9) 查看修改记录的实现代码。

4. 以 DAO 方式实现的数据库访问(参见例 7.3.3)。

(1) 再次解压 ch07 文件夹里的压缩文件 MyDb_DAO.zip，得到 Android 工程文件夹 MyDb_DAO。

(2) 在 Eclipse 中导入 Android 工程 MyDb_DAO。

(3) 查看方法 allQuery()的参数及返回值类型。

(4) 查看方法 updateInfo()的参数、两种实现方式和返回值类型，特别是带参数的 SQL 查询方法的使用。

(5) 将 MyDb_DAO 类的方法与例 7.3.2 的 MainActivity 中的相关方法进行比较。

(6) 部署工程到手机上做运行测试，只显示 2 条记录。

(7) 重新部署工程到手机上做运行测试，会增加 2 条重复的记录。

(8) 打开源程序 DbHelper.java，查看 onUpgrade()方法中的代码(当数据存在表 friends 时，先删除该表，然后再执行 onCreate()方法创建该表)。

(9) 在 MyDAO.java 中，将数据库的版本从 1 修改为 2(即版本提升)，重新部署工程到手机上做运行测试，则只会显示 2 条记录。

5. 数据库事务的使用(参见例 7.3.4)。

(1) 再次解压 ch07 文件夹里的压缩文件 DbTransactionDemo.zip，得到 Android 工程文件夹 DbTransactionDemo。

(2) 在 Eclipse 中导入 Android 工程 DbTransactionDemo。

(3) 打开源文件 MainActivity.java，查看处理数据库事务的 3 个相关方法。

(4) 部署项目到手机运行后，使用 Root Explorer 浏览数据库表中的数据。

(5) 屏蔽 MainActivity.java 中产生异常的那行代码后部署运行项目。

(6) 再次使用 Root Explorer 浏览数据库表，观察其中的数据变化。

三、实验小结及思考

(由学生填写，重点写上机中遇到的问题。)

应用程序间的数据共享

在前面的应用中，一个应用程序创建的数据对于另一个应用程序是不可见的。实际上，也可以将 Android 应用创建的数据暴露到外界，供其他应用程序使用，这样可以达到应用程序间的数据共享，减少数据冗余。之前的文件操作模式(参见第 4.6 节的文件存储和第 4.7 节的 SharedPreferences 存储)，通过指定 Context.MODE_WORLD_READABLE 或 Context.MODE_ WORLD _WRITEABLE 也可以实现对外共享数据。但是，因为数据存储方式的不同，无法使用统一的数据访问方式。

ContentProvider 是实现应用程序间数据共享最标准的方式，也是我们介绍的 Android 四大组件中的最后一个。应用程序通过 ContentResolver 对象访问 ContentProvider 中的数据，该对象提供了持久层数据的 CRUD 方法。本章学习要点如下：

- 掌握使用 ContentProvider 建立内容提供者的方法；
- 掌握通过 ContentResolver 访问 ContentProvider 的方法；
- 掌握 Android 提供的常用公共数据接口；
- 掌握手机联系人数据库的结构信息；
- 通过读取手机联系人的应用，进一步掌握 UI 设计和数据适配器的使用。

8.1 ContentProvider 组件及其相关类

8.1.1 抽象类 ContentProvider(内容提供者)

作为 Android 的四大组件之一，内容提供者 android.content.ContentProvider 是 Android 系统中不同应用程序之间共享数据的接口，用于保存和检索数据。

使用 ContentProvider 访问某个应用程序的数据，可以不必关心某个应用程序的数据存储方式是使用数据库还是使用文件方式，还是通过网络获取数据，这些都不重要，重要的是其他应用程序可以通过一个 ContentProvider 类型的对象来操作某个应用程序的数据。

Android 应用开发者将自己的持久化数据(采用文件存储或 SQLite 数据库存储，通常为后者)公开给其他应用程序，有两种方法：一是定义自己的 ContentProvider 子类，二是将当前应用程序的数据添加到已有的 ContentProvider 中。

新建一个继承于 ContentProvider 的类 TestContentProvider 时，自动创建的源文件 TestContentProvider.java 的代码如下：

```
import android.content.ContentProvider;   //主类
import android.content.ContentValues;     //相关类
import android.database.Cursor;   //相关类
import android.net.Uri;   //相关类
public class TestContentProvider extends ContentProvider {
    @Override
    public boolean onCreate() {
        // TODO Auto-generated method stub
        return false;
    }
    @Override
    public Cursor query(Uri uri, String[] projection, String selection,
            String[] selectionArgs, String sortOrder) {
        // TODO Auto-generated method stub
        return null;
    }
    @Override
    public String getType(Uri uri) {
        // TODO Auto-generated method stub
        return null;
    }
    @Override
    public Uri insert(Uri uri, ContentValues values) {
        // TODO Auto-generated method stub
        return null;
    }
    @Override
    public int delete(Uri uri, String selection, String[] selectionArgs) {
        // TODO Auto-generated method stub
        return 0;
    }
    @Override
    public int update(Uri uri, ContentValues values, String selection,
            String[] selectionArgs) {
        // TODO Auto-generated method stub
        return 0;
    }
}
```

注意：对数据的 CRUD 方法均含有 Uri 类型参数，Uri 将在第 8.1.3 小节中介绍。

Android 定义的抽象类 ContentProvider，如图 8.1.1 所示。

android.content
ContentProvider.class
ContentProvider
onCreate() : boolean
query(Uri, String[], String, String[], String) : Cursor
query(Uri, String[], String, String[], String, CancellationSignal) : Cursor
insert(Uri, ContentValues) : Uri
delete(Uri, String, String[]) : int
update(Uri, ContentValues, String, String[]) : int

图 8.1.1　抽象类 ContentProvider 及其抽象方法

图 8.1.1 也表明，组件 ContentProvider 的四个 CRUD 方法的第一方法参数均为 Uri 类型。

8.1.2　抽象类 ContentResolver(内容解析器)

每个 Activity 都有一个 ContentResolver 对象，并且通过 Activity 的超类 Context(表示上下文)的 getContentResolver()方法获得该对象，如图 8.1.2 所示。

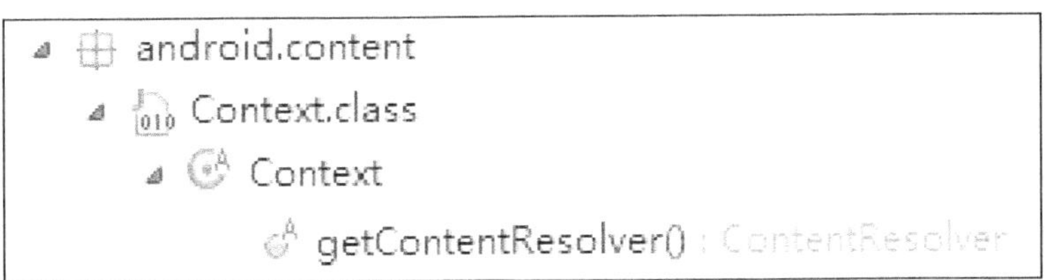

图 8.1.2　ContentResolver 对象的创建方法

android.content.ContentResolver 是与 ContentProvider 相关的一个类，称为内容解析器。通过 ContentResolver 对象访问 ContentProvider 中的数据时，系统要求把 ContentProvider 封装成一个 Uri(通用资源标志符，参见第 5.1.2 小节)。

ContentResolver 也提供了持久层数据的 CRUD 方法，如图 8.1.3 所示。

其中，CRUD 方法中的第一参数也为 Uri 类型，与 ContentProvider 的方法相对应。

注意：

(1) ContentResolver 与 ContenProvider 具有相同的 CRUD 方法。

(2) ContentResolver 的方法 insert(Uri url, ContentValues values)将一条记录的数据插入到 Uri 指定的地方，并返回最新添加那个记录的 Uri。

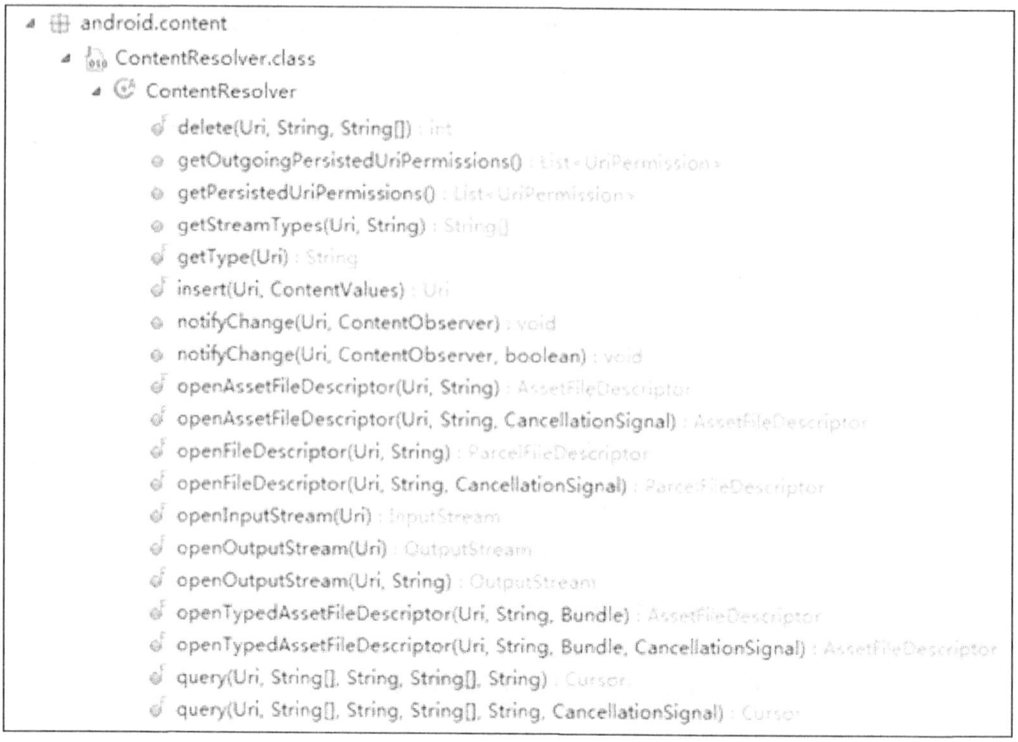

```
android.content
    ContentResolver.class
        ContentResolver
            delete(Uri, String, String[]) : int
            getOutgoingPersistedUriPermissions() : List<UriPermission>
            getPersistedUriPermissions() : List<UriPermission>
            getStreamTypes(Uri, String) : String[]
            getType(Uri) : String
            insert(Uri, ContentValues) : Uri
            notifyChange(Uri, ContentObserver) : void
            notifyChange(Uri, ContentObserver, boolean) : void
            openAssetFileDescriptor(Uri, String) : AssetFileDescriptor
            openAssetFileDescriptor(Uri, String, CancellationSignal) : AssetFileDescriptor
            openFileDescriptor(Uri, String) : ParcelFileDescriptor
            openFileDescriptor(Uri, String, CancellationSignal) : ParcelFileDescriptor
            openInputStream(Uri) : InputStream
            openOutputStream(Uri) : OutputStream
            openOutputStream(Uri, String) : OutputStream
            openTypedAssetFileDescriptor(Uri, String, Bundle) : AssetFileDescriptor
            openTypedAssetFileDescriptor(Uri, String, Bundle, CancellationSignal) : AssetFileDescriptor
            query(Uri, String[], String, String[], String) : Cursor
            query(Uri, String[], String, String[], String, CancellationSignal) : Cursor
```

图 8.1.3 内容解析器类 ContentResolver 的主要方法

8.1.3 内容提供者的 Uri 定义及其相关类(UriMatcher 和 ContentUris)

在第 5.1.1 小节，我们知道，类 android.net.Uri 是用来表示通用资源的。作为内容提供者，还需要提供其他应用访问的路径。为了表示内容提供者这种资源，Android 规定了它的 scheme 为“content”。

Android 系统附带的 ContentProvider 有如下几种。

- Browser：存储如浏览器的信息。
- CallLog：存储通话记录等信息。
- Contacts：存储手机联系人等信息。
- MediaStore：存储媒体文件的信息。
- Settings：存储设备的设置和首选项信息。

Android 2.0 以上版本提供了访问手机联系人信息的多种 URL。例如，访问电话信息的 URI 的常量写法为：

ContactsContract.CommonDataKinds.Phone.CONTENT_URI

为了实现对多张表的灵活访问，通常在外部调用传入的 URI 时，在 ContentProvider 中使用 android.content.UriMatcher 类来解析 URI 地址，即使用工具类 `UriMatcher` 来定义匹配规则。类 UriMatcher 的定义如图 8.1.4 所示。

```
android.content
    UriMatcher.class
        UriMatcher
            NO_MATCH
            UriMatcher(int)
            addURI(String, String, int) : void
            match(Uri) : int
```

图 8.1.4　UriMatcher 类的定义

一个在内容提供者中使用 UriMatcher 来定义匹配规则的示例代码如下：

```
UriMatcher matcher=new UriMatcher(UriMatcher.NO_MATCH);
//下面的com.example.contentprovider.myprovider为内容提供者字符串
//匹配到person时返回1
matcher.addURI("com.example.contentprovider.myprovider","person",1);
//匹配到student时返回2
matcher.addURI("com.example.contentprovider.myprovider","student",2);
DbHelper helper=new DbHelper(this.getContext());
SQLiteDatabase db=helper.getReadableDatabase();
//匹配从使用内容提供者程序中传入的Uri
switch(matcher.match(uri)) {
    case 1:
        return db.query("person",projection,
                              selection,selectionArgs,null,null,sortOrder);
    case 1:
        return db.query("person",projection,
                              selection,selectionArgs,null,null,sortOrder);
    default:
        throw new RuntimeException("Uri不能识别： "+uri);
}
```

定义了上面的匹配规则后，在使用内容提供者的程序中，就可以使用如下的 Uri：

```
Uri uri=Uri.parse("content:// com.example.contentprovider.myprovider/person/10");
```

注意：

(1) 在对 UriMatcher 对象注册时，第一参数为主机名(也叫 Authority)，第二参数为路径名，可以理解为要操作的数据库中表的名字，第三参数为匹配码(匹配不成功时，返回全部数据)；

(2) 在匹配 Uri 时，可以使用“#”代表任意数字，使用“*”代表任意文本。

android.content.ContentUris 也是一个辅助类，其定义如图 8.1.5 所示。

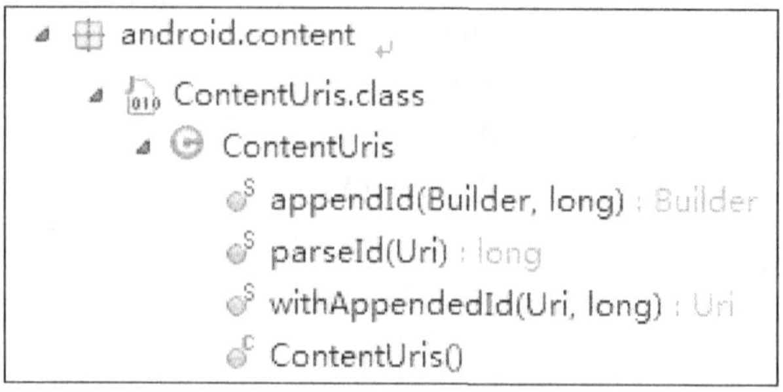

图 8.1.5　ContentUris 的定义

方法 withAppendedId()把 Uri 和 id 连接成一个新的 Uri，这在内容提供者程序的插入记录方法时用到(参见例 8.2.1 的 MyDAO.java 中的增加记录方法)。

ContentUris 类提供的另一个方法 parseId()负责把 Uri 后边的 id 解析出来。

8.2　自定义 ContentProvider 及其使用

8.2.1　在 Android 应用里定义并注册内容提供者

Android 系统支持任意应用程序自己创建的 ContentProvider，以便将本应用程序中的数据提供给其他应用程序，其方法是：在当前应用程序中定义继承 ContentProvider 的子类，并实现 CRUD(参见第 8.1.1 小节)。

如同使用其他的 Android 组件一样，建立自己的 ContentProvider，需要在系统清单文件里注册，其方法是：使用<provider>标签注册，其代码如下：

```
<provider android:name="MyDbProvider"
                    android:authorities="introduction.android.mydbdemo.myfriendsdb
                    android:exported="true" />
```

其中：name 为必填属性，表示 ContentProvider 子类的名称；authorities 也是必填属性，表示其他应用程序访问该 ContentProvider 时的路径，相当于访问 Web 服务器的域名，原则上可以任意定义，但一般还是使用 ContentProvider 子类的存放路径。

注意：

(1) 清单文件中的 <provider>标签还有一个任选属性 permission，用于对本应用提供的数据访问的限定。当省略 permission 属性时，表示任何应用都可以访问本应用提供的数据。

(2) 项目部署后，进入 DDMS 视图，在 LogCat 中 Text 栏里可以查看有如图 8.2.1 所示的信息。

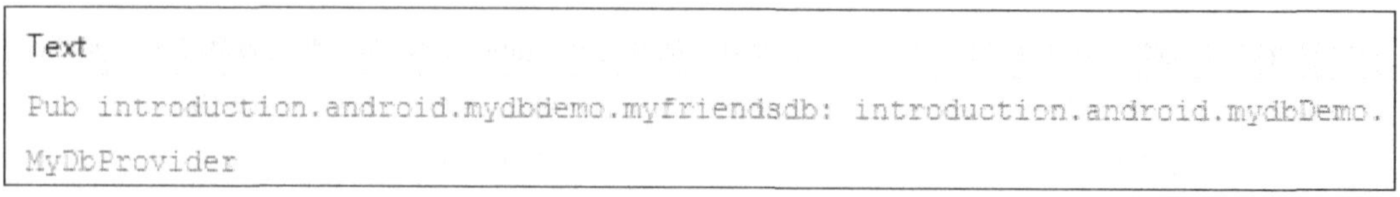

```
Text
Pub introduction.android.mydbdemo.myfriendsdb: introduction.android.mydbDemo.
MyDbProvider
```

图 8.2.1　内容提供者部署成功的提示信息

其中，“Pub”系“Publish”之意，表示发布了一个可供其他应用程序使用的 ContentProvider。

8.2.2　在另一个应用程序里使用内容提供者

在另一个应用程序里使用内容提供者的依据是已经在清单文件中注册的 ContentProvider 提供的访问域名，可以在当前应用程序中使用该 ContentProvider。

【例 8.2.1】内容提供者的定义与使用。

【程序运行】运行内容提供者程序，显示两条记录，退出再运行使用内容提供者程序，增加一条记录，退出后再次运行内容提供者程序，查验刚才增加的一条记录，效果如图 8.2.2 所示。

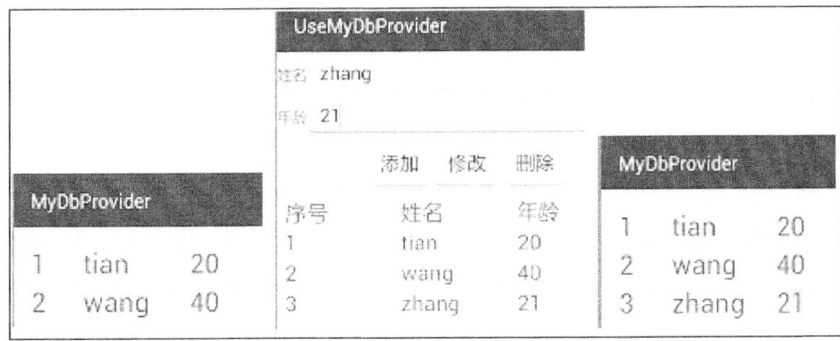

图 8.2.2　内容提供者及其使用

【实现步骤】

(1) 先做好工程“MyDb_DAO”的备份后，在 Eclipse 环境中选中该工程名后按功能键 F2，修改工程名和 values/string.xml 文件中 app_name 值为“MyDbProvider”。

(2) 对工程名“MyDbProvider”应用快捷菜单“Android Tools→Rename Application Package”，修改本应用的包名为“com.example.mydbprovider”。

(3) 在包“com.example.mydbprovider”里新建名为“MyContentProvider”且继承 android.content.ContentProvider 的类文件。此时，工程文件结构如图 8.2.3 所示。

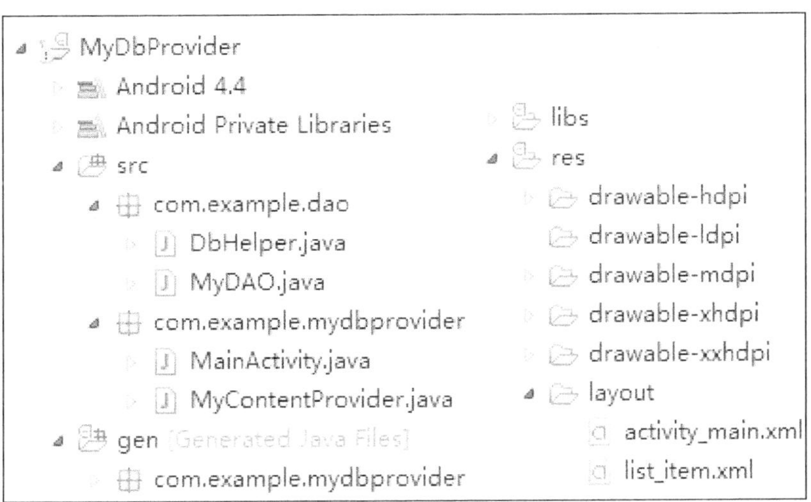

图 8.2.3　内容提供者工程的文件结构

(4) 在清单文件中注册刚才建立的 ContentProvider，其代码如下：

```
<provider    android:name="com.example.mydbprovider.MyContentProvider"
             android:authorities="com.example.mydbprovider.MYPROVIDER"
             android:exported="true" />
```

(5) 编写主 Activity，其文件 MainActivity.java 代码如下：

```
/*
 * 本程序中对数据库的插入操作和查询，使用了MyDAO类的相关方法
 * 程序运行时，增加2条记录并使用ListView控件显示出来
 */
import com.example.dao.MyDAO;    //
import android.app.Activity;
import android.os.Bundle;
import java.util.ArrayList;
import java.util.HashMap;
import java.util.List;
import java.util.Map;
import android.database.Cursor;
import android.view.View;
import android.widget.AdapterView;
import android.widget.AdapterView.OnItemClickListener;
import android.widget.ListView;
import android.widget.SimpleAdapter;
import android.widget.TextView;
import android.widget.Toast;
import com.example.mydbprovider.R;
public class MainActivity extends Activity {
    private ListView listView;
    private List<Map<String,Object>> listData;
    private Map<String,Object> listItem;
    @Override
    protected void onCreate(Bundle savedInstanceState) {
        super.onCreate(savedInstanceState);
        setContentView(R.layout.activity_main);
        MyDAO myDAO = new MyDAO(this);    //上下文作为参数
        Cursor c=myDAO.allQuery();
        if(c.getCount()==0){   //非空时不插入
```

```
            myDAO.insertInfo("tian", 20);    //插入记录
            myDAO.insertInfo("wang", 40); //插入记录
        }
        listView = (ListView)findViewById(R.id.listView);
        listData = new ArrayList<Map<String,Object>>();
        Cursor cursor = myDAO.allQuery();
        while (cursor.moveToNext()){
            long id=cursor.getInt(0);   //获取字段值
            String name=cursor.getString(1);
            //int age=cursor.getInt(2);
            int age=cursor.getInt(cursor.getColumnIndex("age"));//另一种方式
            listItem=new HashMap<String,Object>(); //必须在循环体里新建
            listItem.put("_id", id);
            listItem.put("name", name);
            listItem.put("age", age);
            listData.add(listItem);      //添加一条记录
        }
        SimpleAdapter listAdapter = new SimpleAdapter(this,listData,
                R.layout.list_item,
                new String[]{"_id","name","age"},
                new int[]{R.id.tv_id,R.id.tvname,R.id.tvage});
        listView.setAdapter(listAdapter);
        listView.setOnItemClickListener(new OnItemClickListener(){
            @Override
                public void onItemClick(AdapterView<?> arg0, View arg1, int arg2,
                                                                long arg3) {
                // TODO Auto-generated method stub
                TextView tv = (TextView)arg1.findViewById(R.id.tvname);
                Toast.makeText(MainActivity.this,
                                        tv.getText(),Toast.LENGTH_LONG).show();
            }
        });
    }
}
```

(6) 修改 MyDAO.java 文件，提供被 MyContentProvider 重写方法所调用的若干方法(主要是 CRUD 方法)，其代码如下：

```
package com.example.dao;
/*
 *  本类MyDAO提供了查询数据库表所有记录的方法：allQuery()
```

```
 *  插入记录的方法：insertInfo(String name,int age)
 *  删除记录的方法：deleteInfo(String selId)
 *  修改记录的方法：updateInfo(String name,int age,String selId)
 */
import android.annotation.SuppressLint;
import android.content.ContentUris;
import android.content.ContentValues;
import android.content.Context;
import android.database.Cursor;
import android.database.sqlite.SQLiteDatabase;
import android.net.Uri;
import android.util.Log;
@SuppressLint("SdCardPath")
public class MyDAO {
     private DbHelper dbHelper;   //类的成员
     private SQLiteDatabase myDb;  //类的成员：数据库对象
     //访问内容提供者，与内容提供者定义一致
     private Uri uri=Uri.parse("content://com.example.mydbprovider.MYPROVIDER");   //
     public MyDAO(Context context)  {  //有参构造函数，参数为上下文对象
          dbHelper = new DbHelper(context,"test.db",null,1);
     }
/*   public MyDAO(){     //无参构造函数
          myDb =SQLiteDatabase.openDatabase
                              ("/data/data/com.example.mydbprovider/test.db", null, 0);
     }*/
     public Cursor allQuery()    //查询所有记录
     {
          myDb = dbHelper.getReadableDatabase();
          if(myDb!=null)
               return myDb.rawQuery("select * from friends",null);
          else
               return null;
     }
     //适合ContentProvider里的插入记录方法
     public Uri insertInfo(ContentValues values)
     {
          myDb = dbHelper.getWritableDatabase();
          long rowid=myDb.insert(DbHelper.TB_NAME, null, values);   //
          if(rowid==-1){
```

```
                Log.i("myDbDemo", "数据插入失败！ ");
                return null;}
            else{
                Log.i("myDbDemo", "数据插入成功！ "+rowid);
                Uri insertUri=ContentUris.withAppendedId(uri,rowid);   //
                return insertUri;   //
            }
    }
    public long insertInfo(String name,int age) //普通的插入记录方法
    {
        myDb = dbHelper.getWritableDatabase();
        ContentValues values = new ContentValues();
        values.put("name", name);
        values.put("age", age);
        long rowid=myDb.insert(DbHelper.TB_NAME, null, values);
        if(rowid==-1)
            Log.i("myDbDemo", "数据插入失败！ ");
        else
            Log.i("myDbDemo", "数据插入成功！ "+rowid);
        return rowid;
    }
    //适合ContentProvider里的删除记录方法
    public int deleteInfo(Uri uri, String selection, String[] selectionArgs) {
        return myDb.delete(DbHelper.TB_NAME, selection, selectionArgs);//
    }

    //适合ContentProvider里的更新记录方法
    public int updateInfo(Uri uri, ContentValues values, String selection,
            String[] selectionArgs){
        return myDb.update(DbHelper.TB_NAME, values, selection, selectionArgs); //
    }
}
```

(7) 文件 MyContentProvider.java 提供了对数据库 CRUD 的方法，其代码如下：

```
package com.example.mydbprovider;
/*
 * 自定义ContentProvider
 * 需要调用类DbHelper和MyDAO
 */
```

```
import android.content.ContentProvider;   //
import android.content.ContentValues;   //
import android.content.Context;
import android.database.Cursor;
import android.net.Uri;   //
import com.example.dao.MyDAO;      //
public class MyContentProvider extends ContentProvider {
    private Context context; //
    private   MyDAO mydao;   //
    @Override
    public boolean onCreate() {   //用于初始化
        // TODO: Implement this to initialize your content provider on startup.
        context=this.getContext();   //获取上下文对象
        mydao=new MyDAO(context);   //创建成员(对象)
        return true;
    }
    @Override
    public Cursor query(Uri uri, String[] projection, String selection,
            String[] selectionArgs, String sortOrder) {
        // TODO: Implement this to handle query requests from clients.
        //throw new UnsupportedOperationException("Not yet implemented");
        return mydao.allQuery();   //
    }
    @Override
    public Uri insert(Uri uri, ContentValues values) {
        // TODO: Implement this to handle requests to insert a new row.
        //throw new UnsupportedOperationException("Not yet implemented");
        return mydao.insertInfo(values);   //
    }
    @Override
    public int delete(Uri uri, String selection, String[] selectionArgs) {
        // Implement this to handle requests to delete one or more rows.
        //throw new UnsupportedOperationException("Not yet implemented");
        return mydao.deleteInfo(uri,selection,selectionArgs);   //
    }
    @Override
    public int update(Uri uri, ContentValues values, String selection,
            String[] selectionArgs) {
        // TODO: Implement this to handle requests to update one or more rows.
```

```
        //throw new UnsupportedOperationException("Not yet implemented");
        return mydao.updateInfo(uri, values, selection, null);   //
    }
    @Override
    public String getType(Uri uri) {
        // TODO: Implement this to handle requests for the MIME type of the data
        // at the given URI.
        throw new UnsupportedOperationException("Not yet implemented");
    }
}
```

(8) 部署工程 MyDbProvider 后，按返回键退出主界面。

(9) 新建名为“UseMyDbProvider”的 Android 应用工程，复制上面工程里的两个布局文件到本工程里。此时，工程文件结构如图 8.2.4 所示。

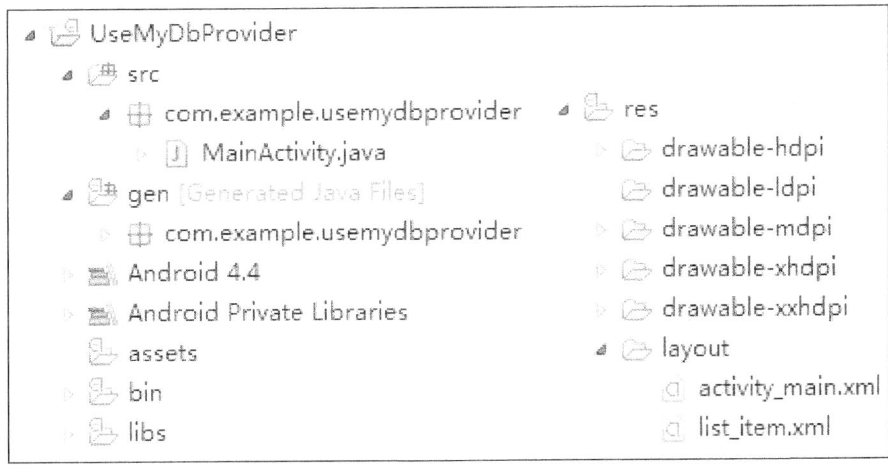

图 8.2.4　使用内容提供者工程的文件结构

(10) 编写 UseMyDbProvider 工程里的 MainActivity.java 文件，利用 MyDbProvider 的 Uri 实现针对 MyDbProvider 工程里的数据库表的 CRUD 操作，其代码如下：

```
package com.example.usemydbprovider;
/*
 * 本例使用Resolver调用数据提供者对数据库进行CRUD
 * 利用工程MyDbProvider中定义的数据提供者com.example.mydbprovider.MYPROVIDER
 * 显示数据库表时使用SimpleAdapter适配器
 */
import java.util.ArrayList;
import java.util.HashMap;
import java.util.Map;
import android.app.Activity;
```

```
import android.content.ContentResolver;//
import android.content.ContentValues;
import android.database.Cursor;
import android.net.Uri;    //
import android.os.Bundle;
import android.view.View;
import android.widget.AdapterView;
import android.widget.AdapterView.OnItemClickListener;
import android.widget.Button;
import android.widget.EditText;
import android.widget.ListView;
import android.widget.SimpleAdapter;
import android.widget.Toast;
public class MainActivity extends Activity {
    private ContentResolver contentResolver;    //
    private Uri uri; //执行CRUD时的参数
    private Cursor cursor=null; //
    private EditText et_name;
    private EditText et_age;
    private Map<String,Object> item;
    private ArrayList<Map<String, Object>> data;
    private SimpleAdapter listAdapter;
    private ListView listview;
    private Button addBtn,updBtn,delBtn;
    private String selId;    //选择项，删除与更新时用
    @Override
    public void onCreate(Bundle savedInstanceState) {
        super.onCreate(savedInstanceState);
        setContentView(R.layout.activity_main);
        et_name=(EditText) findViewById(R.id.et_name);
        et_age=(EditText) findViewById(R.id.et_age);
        listview = (ListView) findViewById(R.id.listView);
        addBtn=(Button)findViewById(R.id.bt_add);
        updBtn=(Button)findViewById(R.id.bt_modify);
        delBtn=(Button)findViewById(R.id.bt_del);
        data = new ArrayList<Map<String,Object>>();
        contentResolver=this.getContentResolver();    //
        uri=Uri.parse("content://com.example.mydbprovider.MYPROVIDER");    //
        dbFindAll();    //显示所有记录
```

```
addBtn.setOnClickListener(new Button.OnClickListener(){  //添加监听器
    @Override
    public void onClick(View v) {
        // TODO Auto-generated method stub
        dbAdd();  //添加记录
        dbFindAll();  //重新显示所有记录
    }
});
delBtn.setOnClickListener(new Button.OnClickListener(){  //删除监听器
    @Override
    public void onClick(View v) {
        // TODO Auto-generated method stub
        dbDel();        //删除当前记录
        dbFindAll();
    }
});
updBtn.setOnClickListener(new Button.OnClickListener(){ //更新监听器
    @Override
    public void onClick(View v) {
        // TODO Auto-generated method stub
        dbUpdate();  //修改当前记录
        dbFindAll();
    }
});
//选择项监听器
listview.setOnItemClickListener(new OnItemClickListener() {
    @SuppressWarnings("unchecked")
    @Override
    public void onItemClick(AdapterView<?> parent, View v,
            int position, long id) {
        // TODO Auto-generated method stub
        Map<String,Object> listItem =(Map<String,Object>)
                                listview.getItemAtPosition(position);
        et_name.setText((String)listItem.get("name"));
        et_age.setText((String)listItem.get("age"));
        selId=(String)listItem.get("_id");
        Toast.makeText(getApplicationContext(), "选择的id是： "
                            +selId,Toast.LENGTH_SHORT).show();
    }
```

```
        });
    }
    protected void dbFindAll() {   //记录查找
        data.clear();   //必须清除
        //利用内容提供者的查询方法得到记录集
        cursor=contentResolver.query(uri,null, null, null, null);   //
        item=new HashMap<String,Object>();
        item.put("_id", "序号"); item.put("name", "姓名");
        item.put("age", "年龄");
        data.add(item);
        cursor.moveToFirst();
        while (!cursor.isAfterLast() ) {
            String id=cursor.getString(0);
            String name=cursor.getString(1);
            String age=cursor.getString(2);
            item=new HashMap<String,Object>();
            item.put("_id", id);item.put("name", name);
            item.put("age", age);
            data.add(item);
            cursor.moveToNext();
        }
        showList();   //显示数据集
    }
    protected void dbAdd() {        //插入方法
        // TODO Auto-generated method stub
        ContentValues values = new ContentValues();   //
        values.put("name", et_name.getText().toString().trim());
        values.put("age", et_age.getText().toString().trim());
        //准备调用内容提供者相应的方法
        Uri insertUri=contentResolver.insert(uri, values);
        //显示表示插入的路径信息insertUri
        Toast.makeText(this, insertUri.toString(),Toast.LENGTH_SHORT).show();
    }
    protected void dbDel() {   //删除方法
        String where="_id="+selId;     //selId在列表项选择监听器给出
        //调用内容提供者里的删除方法
        int i=contentResolver.delete(uri, where,null);
        if(i>0)
                Toast.makeText(this, "数据删除成功！ ",Toast.LENGTH_SHORT).show();
```

```
        else
                Toast.makeText(this, "数据删除失败！ ",Toast.LENGTH_SHORT).show();
    }
    protected void dbUpdate() {    //更新方法
        // TODO Auto-generated method stub
        String where="_id="+selId;     //selId在列表项选择监听器给出
        ContentValues values = new ContentValues();
        values.put("name", et_name.getText().toString().trim());
        values.put("age", et_age.getText().toString().trim());
        int i=contentResolver.update(uri, values, where,null);
        if(i>0)
                Toast.makeText(this, "数据更新成功！ ",Toast.LENGTH_SHORT).show();
        else
                Toast.makeText(getApplicationContext(), "数据更新失败！
                                        ",Toast.LENGTH_SHORT).show();
    }
    private void showList() {          // 显示记录集
        listAdapter=new SimpleAdapter(this,data,
                R.layout.list_item,
                new String[]{"_id","name","age"},
                new int[]{R.id.tv_id,R.id.tvname,R.id.tvage});
        listview.setAdapter(listAdapter);   //绑定控件
    }
}
```

(11) 部署工程“UseMyDbProvider”后做运行测试，并使用 MyDbProvider 工程验证。

8.2.3　使用 Handler 和 AsyncTask 更新 UI 线程

在 Android 组件中，Activity、Service 和 BroadcastReceiver 都工作在主线程上，因此，任何耗时的操作都会降低用户界面的响应速度，甚至导致用户界面失去响应。当用户界面失去响应超过 5 秒后，Android 系统会允许用户强行关闭应用程序。“耗时的处理过程”一般是指复杂的运算过程、大量的文件操作、存在延时的网络通信和数据库操作等。

在 Android 应用中，如果有 5 秒的主线程堵塞，系统就会强退。另外，UI 的更新也只能在主线程中完成。因此，异步处理是不可避免的。Android 提供了类 Handler 和 AsyncTask，用于异步处理，也就是为了不阻塞主线程。

一般地，在主线程中实例化一个 Handler 并实现其内部接口的回调方法（通常用于 UI 更新），在子线程中调用它的发送消息 sendMessage()方法。Handler 及处理的消息类的定义如图 8.2.5 所示。

注意：异步处理本质上也是多线程。

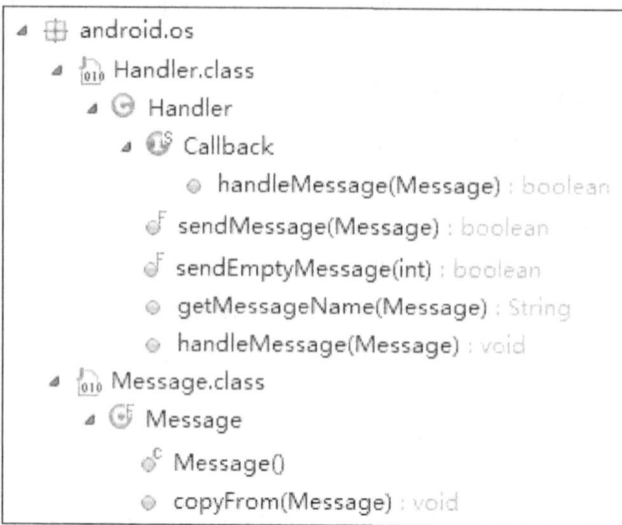

图 8.2.5　Handler 与 Message 类的定义

使用 Handler 的示例，参见例 8.2.2 和例 11.3.1。

定义一个异步任务类 AsyncTask 的实例时，需要重写其抽象方法 doInBackground()和另一个受保护的方法 onPostExecute()，其定义如图 8.2.6 所示。

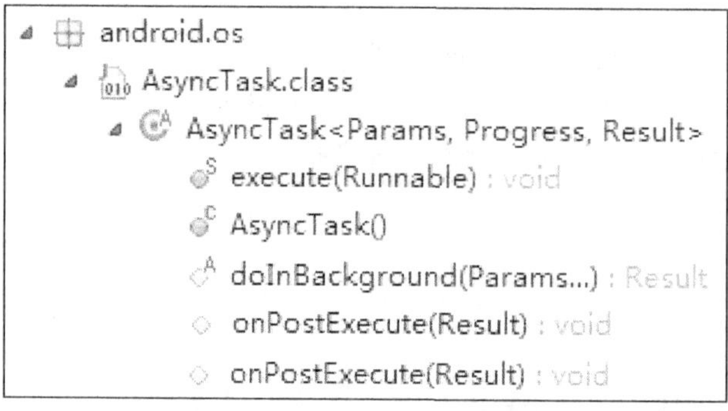

图 8.2.6　异步任务类 AsyncTask 的定义

注意：

(1) Android 的 AsyncTask 比 Handler 更轻量级一些，适用于简单的异步处理。

(2) Android 提供的异步任务类 AsyncTask 与 Java 的线程类 Thread 相对应，使用 AsyncTask 的示例，参见例 8.3.2。

8.2.4　Java 观察者模式与内容观察者 ContentObserver

在 Android 中，很多内容都用到了 Java 提供的观察者模式(observer model)。下面是观察者模式的实现过程：

(1) 编写继承 java.util.Observable 的类，用以创建被观察者；

(2) 编写实现 java.util.Observer 接口的类，用以创建观察者；

(3) 注册观察者，调用 addObserver(Observer observer)；

(4) 在被观察者改变对象内部状态的地方，调用 setChanged()方法，然后调用 notifyObservers(Object)方法，通知被观察者；

(5) 观察者的 update(Object)方法中，对改变做出响应。

在观察者模式中，主要有观察者(Observer)和被观察者(Observable)两类对象。使用 Observable 表示被观察者，这个对象是一个抽象类，只能被继承；使用 Observer 表示观察者，它不是一个类，而是一个接口，观察者可以有多个，实现了该接口类型的对象都是观察者。Java 观察者模式的相关类与接口，如图 8.2.7 所示。

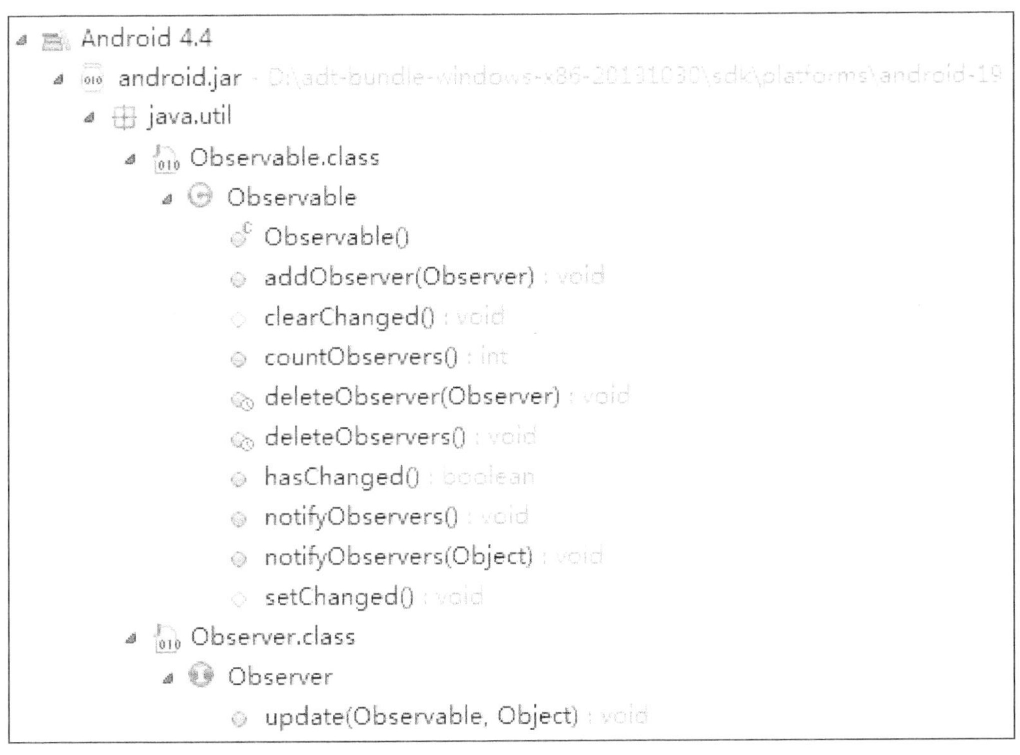

图 8.2.7　Java 观察者模式的相关类与接口

一个 Observable 类型的对象(即被观察者)可以被很多实现了 Observer 接口类型的对象(即观察者)观察，好比被观察者“上电视了”。如果电视机前的你(一个观察者)看到电视里的被观察者做了别的动作，那么电视机前的所有人(所有观察者)将在同一时间看到那个改变了的动作。

注意：可以创建继承 Observable 类同时又实现 Observer 接口类型的对象，这表明观察者同时又是被观察者(或者说被观察者同时也是观察者)。

实现数据库更新时的实时监控的原理是：应用更新时产生广播，内容观察者接收广播后，在其他应用中显示更新数据。Android 提供的内容观察者类，如图 8.2.8 所示。

```
android.database
    ContentObserver.class
        ContentObserver
            ContentObserver(Handler)
            deliverSelfNotifications() : boolean
            dispatchChange(boolean) : void
            dispatchChange(boolean, Uri) : void
            onChange(boolean) : void
            onChange(boolean, Uri) : void
```

图 8.2.8　Android 提供的内容观察者类

【例 8.2.2】Java 观察者模式的使用。

【程序运行】程序运行后，单击“添加观察者”按钮添加一个观察者，此时观察内容为空。在分别输入姓名 kp、年龄 40 和性别男后，单击“改变数据”按钮，此时观察内容显示为刚才输入的内容。再添加一个观察者，修改年龄为 38，再次单击“改变数据”按钮后，所有观察内容都相应地改变，效果如图 8.2.9 所示。

【实现步骤】

(1) 新建一个名为“ObserverModel”的 Android 应用工程，勾选创建 Activity，文件名(含主布局文件名)采用默认值。

(2) 依次建立被观察者类 MyPerson、观察者类 MyObserver 和数据适配器 MyListAdapter，其工程文件结构(主要部分)如图 8.2.10 所示。

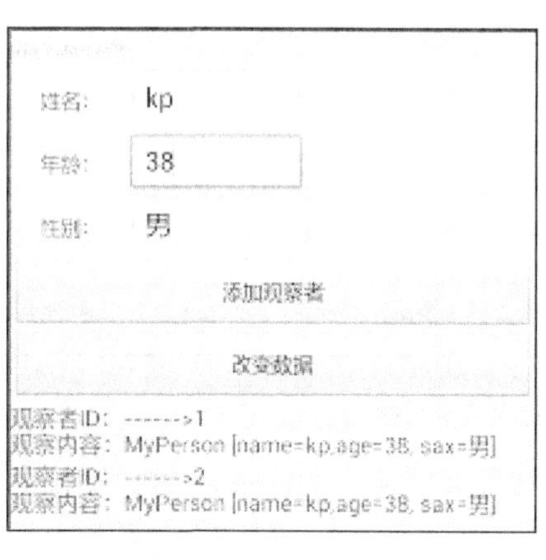

图 8.2.9　添加观察者和改变数据

```
ObserverModel
    Android 4.4
    Android Private Libraries
    src
        com.example.observermodel
            MainActivity.java
            MyListAdapter.java
            MyObserver.java
            MyPerson.java
    gen [Generated Java Files]
        com.example.observermodel
```

图 8.2.10　ObserverModel 工程文件结构 (主要部分)

(3) 编写继承 Observable 类的被观察者类文件 MyPerson.java，其代码如下：

```
package com.example.observermodel;
/*
 * 继承类Observable，创建被观察者类
 * 在所有setXXX()方法里，都使用了父类方法setChanged()和notifyObservers()
```

```
 */
import java.util.Observable;   //被观察者类
public class MyPerson extends Observable {
    private int age;
    private String name;
    private String sax;

    public int getAge() {
        return age;
    }
    public void setAge(int age) {
        this.age = age;
        setChanged();     //改变
        notifyObservers();   //通知观察者
    }
    public String getName() {
        return name;
    }
    public void setName(String name) {
        this.name = name;
        setChanged();
        notifyObservers();
    }
    public String getSax() {
        return sax;
    }
    public void setSax(String sax) {
        this.sax = sax;
        setChanged();
        notifyObservers();
    }
    @Override
    public String toString() {
        return "MyPerson [name=" + name + ",age=" + age + ", sax=" + sax + "]";
    }
}
```

(4) 新建实现接口 Observer 的观察者类文件 MyObserver.java，其代码如下：

```
package com.example. observermodel;
/*
 * 通过实现接口Observer定义观察者类MyObserver
 * 调用了另一个类MyPerson
 */
import java.util.Observer;  //观察者接口
import java.util.Observable;
public class MyObserver implements Observer {
    private int i;
    private MyPerson myPerson;//被观察者对象
    public MyObserver(int i){  //构造方法
        System.out.println("我是观察者---->" + i);
        this.i = i;
    }
    public int getI() {
        return i;
    }
    public void setI(int i) {
        this.i = i;
    }
    //调用其他类
    public MyPerson getMyPerson() {
        return myPerson;
    }
    public void setMyPerson(MyPerson myPerson) {
        this.myPerson = myPerson;
    }
    @Override
    public void update(Observable observable, Object data) {
        System.out.println("观察者---->"+ i +"得到更新！ ");
        this.myPerson = (MyPerson)observable;
        System.out.println(((MyPerson)observable).toString());
    }
}
```

(5) 编写主窗体文件 MainActivity.java，其代码如下：

```
package com.example. observermodel;
package com.example.observermodel;
```

```
/*
 * 程序运行窗体：MainActivity.java
 * 功能是添加观察者、设置被观察者的内容
 */
import java.util.ArrayList;
import java.util.List;
import android.os.Bundle;
import android.os.Handler;   //
import android.os.Message;   //
import android.app.ListActivity;
import android.view.View;
import android.widget.Button;
import android.widget.EditText;
import android.widget.ListView;
import com.example.observermodel.R;

public class MainActivity extends ListActivity {

    private Button add;
    private EditText age,name,sex;
    private MyPerson observable;   //被观察者
    private int i = 1;
    private Button change;
    private ListView lv;
    private List<MyObserver> myObservers;   //观察者、泛型用法
    //使用Handler发送和处理消息(由自己发出去的消息)
    //从其他线程中发送来的消息放入消息队列中，避免线程冲突(常见于更新UI线程)
    private Handler handler = new Handler(new Handler.Callback() {
        @Override
        public boolean handleMessage(Message msg) {
            // 将信息加入list中显示
            MyListAdapter myListAdapter = new MyListAdapter(
                                        MainActivity.this, myObservers);
            lv.setAdapter(myListAdapter);
            return false;
        }
    });

    @Override
```

```
public void onCreate(Bundle savedInstanceState) {
    super.onCreate(savedInstanceState);
    setContentView(R.layout.activity_main);

    name = (EditText)findViewById(R.id.et1);
    age = (EditText)findViewById(R.id.et2);
    sex = (EditText)findViewById(R.id.et3);

    add = (Button) findViewById(R.id.add);
    observable = new MyPerson();
    myObservers = new ArrayList<MyObserver>();
    lv = getListView();
    //添加观察者监听器
    add.setOnClickListener(new View.OnClickListener() {
        @Override
        public void onClick(View v) {
            MyObserver myObserver = new MyObserver(i);    //
            i++;
            observable.addObserver(myObserver);    //
            myObservers.add(myObserver);    //
            handler.sendEmptyMessage(0);
        }
    });
    change = (Button) findViewById(R.id.change);
    //通知数据改变
    change.setOnClickListener(new View.OnClickListener() {
        @Override
        public void onClick(View v) {
            if(!age.getText().toString().equals(null))
                observable.setAge(Integer.valueOf(age.getText().toString()));
            else
                observable.setAge(10 + i);
            if(!name.getText().toString().equals(null))
                observable.setName(name.getText().toString());
            else
                observable.setName("a" + i);
            if(!sex.getText().toString().equals(null))
```

```
                    observable.setSax(sex.getText().toString());
                else
                    observable.setSax("男" + i);
                handler.sendEmptyMessage(0);
            }
        });
    }
}
```

8.3　读取手机联系人信息

8.3.1　手机联系人相关类 ContactsContract

android.provider.ContactsContract 是关于手机联系人信息的类，在 Android 2.0 及以上版本中使用，其内部类 Contacts 提供了访问手机联系人数据库的 URI，如图 8.3.1 所示。

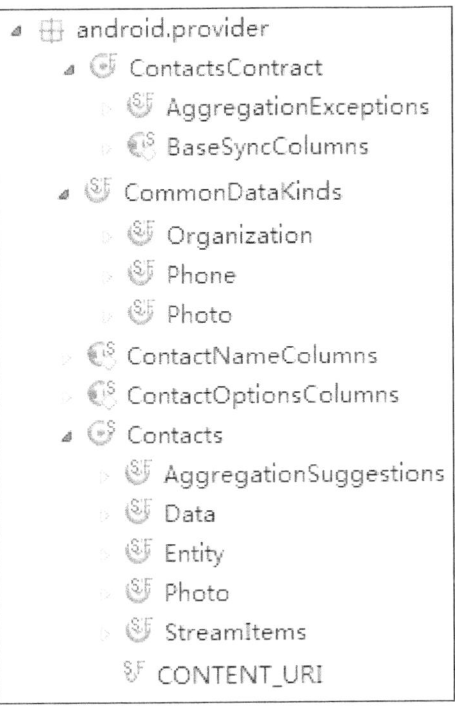

图 8.3.1　手机联系人相关类 ContactsContract 及其内部类

注意：在 android.provider 包内，提供了一个被废弃的类 Contacts(Android 2.0 以下版本使用)，如图 8.3.2 所示。其中，Contacts 前面的符号“ ”表示该类是不推荐使用的。

图 8.3.2　被废弃的手机联系人相关类 Contacts

8.3.2　手机联系人数据库及其相关表

图 8.3.3　手机联系人数据库的存放位置

在手机系统目录 data/data 下的包 com.android.providers.contacts 里，存放手机联系人数据库 contacts2.db，如图 8.3.3 所示。

为了满足手机用户个性化设置的需求，例如动态地创建手机联系人分组、存放多种联系方式(手机、固定电话)等，手机联系人数据库中包含了多张表而不是一张表。手机联系人数据库较好地解决了手机联系人数据的扩展管理问题。

1. 存放手机联系人 id 的表 raw_contacts

存放 2 个手机联系人数据库 contacts2.db 里，表 raw_contacts 的记录信息如图 8.3.4 所示。

_id	contact_id	pinned	display_name
1	1	2147483647	WuZhiXiang
2	2	2147483647	CaoRong

图 8.3.4　一个存放了 2 个手机联系人的 raw_contacts 表

其中，手机联系人 id 存放在字段 contact_id 里。

2. 手机联系人数据表 data

存放 2 个手机联系人数据库 contacts2.db 里，表 data 的记录信息如图 8.3.5 所示。

_id	mimety...	raw_contact_id	data1	data2	data3	data4
	5	1	1 552-764-3858	2		
	8	1	02751012663	1		02751012663
	1	1	707348355@qq.com	1		
	7	1	WuZhixiang	WuZhixiang		
	5	2	1 599-422-0815	2		
	7	2	CaoRong	CaoRong		

图 8.3.5　一个存放了 2 个手机联系人的 data 表

注意：在查询手机联系人信息时，需要先查询表 raw_contacts，取得手机联系人的 id。

3. 手机联系人表 contacts

存放 2 个手机联系人数据库 contacts2.db 里，表 contacts 的记录信息如图 8.3.6 所示。

_id	name_raw_contact_id	photo_id	photo_file_id	custom_ringtone	times_contacted	has_phone_number
1	1				0	1
2	2				0	1

图 8.3.6　一个存放了 2 个手机联系人的 contacts 表

注意：图 8.3.6 所示的表里的字段_id 存放手机联系人的 id，字段 has_phone_number 存放手机联系人的电话个数。此外，还有存放照片的字段 photo_id 和 photo_file_id。

4. 保存数据类型的表 mimetypes

表 mimetypes 存放了 11 种数据类型，如图 8.3.7 所示。

从图 8.3.7 可以看出，E-mail、电话和姓名这三种数据分别用 1、5 和 7 表示。

8.3.3　读取手机联系人程序设计

Android 系统对一系列的公用数据类型提供了对应的 ContentProvider 接口。要访问手机联系人，不必直接对系统的 contacts.db 数据库进行操作，可以通过如下方式获取手机联系人信息 Uri：

```
Uri uri = ContactsContract.Contacts.CONTENT_URI;
```

【例 8.3.1】访问系统提供的公用数据接口，读取并显示手机联系人的姓名及电话。

【程序运行】程序运行时，在主界面的 TextView 控件里显示读到的手机联系人数目，在 ListView 控件里显示所有手机联系人的姓名和电话。程序运行效果，如图 8.3.8 所示。

_id	mimetype
1	vnd.android.cursor.item/email_v2
2	vnd.android.cursor.item/im
3	vnd.android.cursor.item/nickname
4	vnd.android.cursor.item/organization
5	vnd.android.cursor.item/phone_v2
6	vnd.android.cursor.item/sip_address
7	vnd.android.cursor.item/name
8	vnd.android.cursor.item/postal-address_v2
9	vnd.android.cursor.item/identity
10	vnd.android.cursor.item/photo
11	vnd.android.cursor.item/group_membership

图 8.3.7　表 mimetypes 存放的数据类型

图 8.3.8　读取并显示手机联系人程序运行界面

【设计步骤】

(1) 新建名为“ContactsCP”的 Android 应用工程，在出现的对话框中都采用默认选项，勾选创建 Activity，工程文件结构(主要部分)如图 8.3.9 所示。

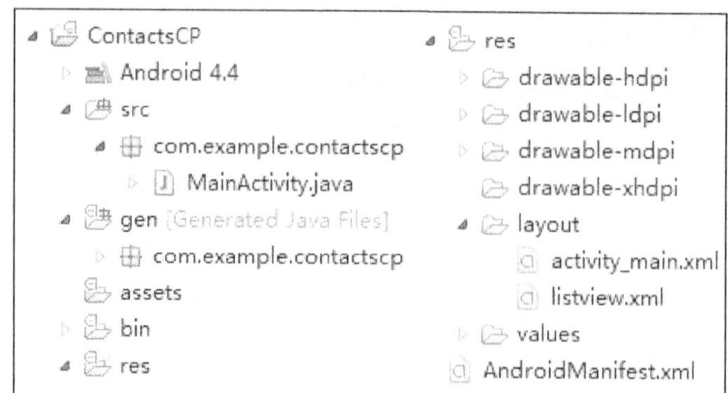

图 8.3.9　ContactsCP 工程文件结构(主要部分)

(2) 主布局文件 activity_main.xml 中，采用垂直线性布局，包含一个 TextView 控件和一个 ListView 控件。ListView 控件使用的布局文件为 listview.xml。

(3) 在清单文件中注册读取手机联系人信息的权限，其代码如下：

```
<uses-permission android:name="android.permission.READ_CONTACTS"/>
```

(4) 编写工程的主 Activity 文件 MainActivity.java，其代码如下：

```
package com.example.contactscp;
/*
 * 本程序读取手机联系人的姓名及电话
 * 要求手机使用Android 2.0以上版本
 */
import android.app.Activity;
import android.os.Bundle;
import android.provider.ContactsContract;  //主类1，手机联系人
import android.net.Uri;  //
import android.content.ContentResolver;  //主类2，内容解析器
import android.widget.ListView;
import android.widget.SimpleAdapter;
import android.widget.TextView;

import com.example.contactscp.R;
import java.util.ArrayList;
import java.util.HashMap;
import java.util.Map;
import android.database.Cursor;

public class MainActivity extends Activity {
    private ListView listView;
    private ContentResolver contentResolver; //;
```

```
private SimpleAdapter listAdapter;
Cursor cursor;
private HashMap<String, String> item;
private ArrayList<Map<String, String>> data;

@Override
public void onCreate(Bundle savedInstanceState) {

    super.onCreate(savedInstanceState);
    setContentView(R.layout.activity_main);
    listView=(ListView)this.findViewById(R.id.listView);
    data = new ArrayList<Map<String,String>>();
    //获取手机联系人的Uri,相当于提供了一个公共的数据库链接
    Uri uri=ContactsContract.Contacts.CONTENT_URI;//内部类
    //android.content.Context提供了抽象方法getContentResolver()方法
    //得到内容(数据)解析器
    contentResolver=this.getContentResolver();
    //抽象类android.content.ContentResolver提供了query()等方法
    cursor=contentResolver.query(uri, null, null, null, null); //得到记录集
    TextView textView=(TextView)this.findViewById(R.id.textView);
    textView.setText("读取到"+String.valueOf(cursor.getCount())+"个联系人"); //
    while(cursor.moveToNext()){
            //先得到字段名_id的索引，然后取得该手机联系人的id值
        Int idFieldIndex=cursor.getColumnIndex("_id");
         //int idFieldIndex=cursor.getColumnIndex(ContactsContract.Contacts._ID);
        int id=cursor.getInt(idFieldIndex);
        //先得到字段名DISPLAY_NAME的索引，然后取得该手机联系人的姓名
        int nameFieldIndex=cursor.getColumnIndex
                                (ContactsContract.Contacts.DISPLAY_NAME);
        String name=cursor.getString(nameFieldIndex);
        //先得到字段名HAS_PHONE_NUMBER的索引，然后取得该手机联系人的电话
        int numCountFieldIndex=cursor.getColumnIndex
                        (ContactsContract.Contacts.HAS_PHONE_NUMBER);
        int numCount=cursor.getInt(numCountFieldIndex);
        String phoneNumber="";
        if(numCount>0){            //手机联系人有至少一个电话号码
            //在类ContactsContract.CommonDataKinds.Phone中根据id查询相应手机
                                                        联系人的所有电话;
            Cursor phonecursor=getContentResolver().query(
```

```
                    ContactsContract.CommonDataKinds.Phone.CONTENT_URI,
                    null, ContactsContract.CommonDataKinds.Phone.CONTACT_ID+"=?",
                    new String[]{Integer.toString(id)}, null);
                if(phonecursor.moveToFirst()){
                //仅读取第一个电话号码
                    int numFieldIndex=phonecursor.getColumnIndex
                            (ContactsContract.CommonDataKinds.Phone.NUMBER);
                    phoneNumber=phonecursor.getString(numFieldIndex);
                }
            }
            item=new HashMap<String,String>(); //必须循环创建
            item.put("name", name);
            item.put("phoneNumber", phoneNumber);
            data.add(item);
        }
        //构造数据适配器
        listAdapter=new SimpleAdapter(this,data,
                android.R.layout.simple_list_item_2,
                new String[]{"name","phoneNumber"},
                new int[]{android.R.id.text1,android.R.id.text2});
        listView.setAdapter(listAdapter);      //显示
    }
}
```

8.3.4 综合应用：群发短信

图 8.3.10 群发消息程序运行的主窗口

目前，一般的手机都提供了群发短信的功能。下面介绍的案例(例 8.3.2)，在选择手机联系人时由于使用复选框设计，因而操作界面更加流畅。

【例 8.3.2】选择手机联系人后群发短信。

【程序运行】在主窗口中，取消了标题栏的显示，在第一行使用水平线性布局，分别放置了用于返回的图像按钮、描述程序功能的文本框和提交选取手机联系人的图像，运行效果如图 8.3.10 所示。

选择手机联系人完毕后，单击右上方的图像按钮“✓”，出现确认对话框。确认后，即可进入系统提供的编辑和发送短信界面。

注意：当手机联系人较多时，出现手机联系人界面需要等待一会儿。

【实现步骤】

(1) 新建名为 GroupSendMessage 的 Android 工程，程序包名默认，勾选创建 Activity。

(2) 建立选择手机联系人时的列表视图 ListItemView.java，其代码如下：

```
package com.example.groupsendmessage;
import android.widget.CheckBox;
import android.widget.ImageView;
import android.widget.TextView;
public class ListItemView {
    public ImageView iv_personGender; //图像
    public TextView tv_consumerName;    //姓名
    public CheckBox selected;    //复选类
}
```

(3) 建立选择手机联系人用到的实体类 Consumer.java，其代码如下：

```
package com.example.groupsendmessage;
//自定义的实体类
class Consumer {
    private String name;
    private String phone;

    public String getName() {
        return name;
    }
    public void setName(String name) {
        this.name = name;
    }
    public String getPhone() {
        return phone;
    }
    public void setPhone(String phone) {
        this.phone = phone;
    }
}
```

(4) 建立继承 SimpleAdapter 的适配器 ChooseAdapter.java，其代码如下：

```
package com.example.groupsendmessage;
/*
 * 使用SimpleAdapter数据适配器
```

```
 */
import java.util.ArrayList;
import java.util.HashMap;
import java.util.Map;
import android.annotation.SuppressLint;
import android.content.Context;
import android.view.LayoutInflater;
import android.view.View;
import android.view.ViewGroup;
import android.widget.CheckBox;
import android.widget.ImageView;
import android.widget.SimpleAdapter;    //
import android.widget.TextView;
import com.example.groupsendmessage.R;
@SuppressLint("UseSparseArrays")
public class ChooseAdapter extends SimpleAdapter {      //
    public Context mContext;
    public LayoutInflater listContainer;
    public ArrayList<Consumer> listData;
    static public String phoneNum;
    public static Map<Integer, Boolean> isSelected;
     // 可以改写显示手机联系人照片
    private static final int[] icons = { R.drawable.male, R.drawable.male };
    HashMap<Integer, View> map = new HashMap<Integer, View>();
    public ChooseAdapter(Context context, ArrayList<Consumer> listItem) {
        super(context, null, 0, null, null);
        mContext = context;
        listContainer = LayoutInflater.from(context);
        this.listData = listItem;
        isSelected = new HashMap<Integer, Boolean>();
        for (int i = 0; i < listItem.size(); i++) {
            isSelected.put(i, false);
        }
    }
    @Override
    public int getCount() {
        // TODO Auto-generated method stub
        return listData.size();
```

```
}
@Override
public Object getItem(int position) {
    // TODO Auto-generated method stub
    return listData.get(position);
}
@Override
public long getItemId(int position) {
    // TODO Auto-generated method stub
    return position;
}
@Override
public View getView(int position, View convertView, ViewGroup parent) {
    ListItemView holder = null;
    // TODO Auto-generated method stub
    if (convertView == null) {
        holder = new ListItemView();
        // 获取list_item布局文件的视图
        convertView = listContainer.inflate(R.layout.list_item,null);
        // 获取控件对象
        holder.selected = (CheckBox) convertView
                .findViewById(R.id.choose_check);
        holder.iv_personGender = (ImageView) convertView
                .findViewById(R.id.choose_personGender);
        holder.tv_consumerName = (TextView) convertView
                .findViewById(R.id.choose_consumerName);
        map.put(position, convertView);// 设置控件集到convertView
        convertView.setTag(holder);
    }
    else {
        holder = (ListItemView) convertView.getTag();
    }
    Consumer consumerInfo = listData.get(position);
    holder.selected.setChecked(isSelected.get(position));
    holder.iv_personGender.setImageResource(icons[0]);
    phoneNum = consumerInfo.getPhone();
    holder.tv_consumerName.setText(consumerInfo.getName()+"
                                        "+consumerInfo.getPhone());
```

```
            return convertView;
        }
        public void refresh(ArrayList<Consumer> listItem) {
            this.listData = listItem;
            this.notifyDataSetChanged();
        }
}
```

(5) 创建与 ListView 列表项对应的布局文件 list_item.xml，其代码如下：

```
<LinearLayout xmlns:android="http://schemas.android.com/apk/res/android"
    xmlns:tools="http://schemas.android.com/tools"
    android:layout_width="fill_parent"
    android:layout_height="wrap_content"
    android:layout_gravity="top"
    android:background="#ffffff"
    android:gravity="center"
    android:orientation="horizontal"
    tools:context=".StartActivity" >
        <ImageView
            android:id="@+id/choose_personGender"
            android:layout_width="72dp"
            android:layout_height="50dp"
            android:contentDescription="@string/hello_world"
            android:src="@drawable/person" />
        <TextView
            android:id="@+id/choose_consumerName"
            android:layout_width="0dp"
            android:layout_height="fill_parent"
            android:layout_weight="1"
            android:gravity="center_vertical"
            android:text="@string/name"
            android:textAppearance="?android:attr/textAppearanceSmall"
            android:textColor="#000000"/>
      <CheckBox
            android:id="@+id/choose_check"
            android:layout_width="wrap_content"
            android:layout_height="wrap_content"
            android:background="@drawable/check_box_style"
```

```
        android:button="@null"
        android:checked="false"
        android:clickable="false"
        android:layout_marginRight="10dp"
        android:focusable="false" />
        </LinearLayout>
```

此时，工程文件结构如图 8.3.11 所示。

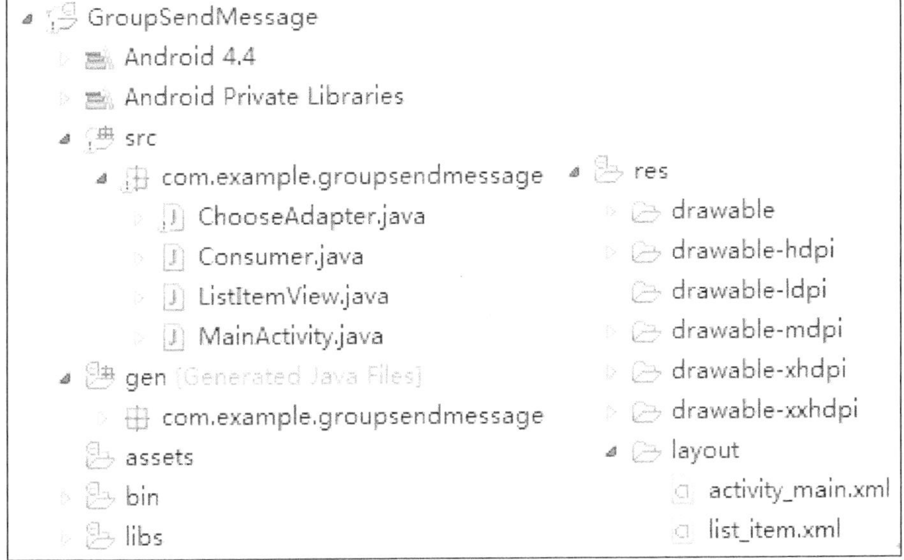

图 8.3.11　群发短信工程文件结构

(6) 对应于主 Activity 的布局文件 activity_main.xml，其代码如下：

```
<LinearLayout xmlns:android="http://schemas.android.com/apk/res/android"
    xmlns:tools="http://schemas.android.com/tools"
    android:layout_width="fill_parent"
    android:layout_height="wrap_content"
    android:background="#ffffff"
    android:clickable="true"
    android:orientation="vertical" >
    <LinearLayout
        android:layout_width="match_parent"
        android:layout_height="45dp"
        android:background="#117CF4" >
        <ImageView
            android:id="@+id/onBack"
            android:layout_width="30dp"
```

```
            android:layout_height="fill_parent"
            android:clickable="true"
            android:src="@drawable/btnclose" />
        <TextView
            android:layout_width="wrap_content"
            android:layout_height="fill_parent"
            android:layout_weight="1"
            android:gravity="center_vertical|center_horizontal"
            android:text="群发消息"
            android:textColor="#ffffff"
            android:textSize="22sp" />
        <ImageButton
            android:id="@+id/imageButton"
            android:layout_width="30dp"
            android:layout_height="fill_parent"
            android:layout_marginRight="10dp"
            android:background="#117CF4"
            android:src="@drawable/ic_action_summit" />
    </LinearLayout>
        <ListView
            android:id="@+id/listView"
            android:layout_width="fill_parent"
            android:layout_height="fill_parent"
            android:divider="#DEDEDE"
            android:dividerHeight="2dp" >
        </ListView>
</LinearLayout>
```

(7) 编写主 Activity 程序文件 MainActivity.java，其代码如下：

```
package com.example.groupsendmessage;
/*
 * 读取手机联系人、群发短信工程：GroupSendMessage
 * 本类包含一个继承AsyncTask的内部类GetContact
 * 当手机联系人较多时，读取有一个等待过程
 * 由于手机运行内存一般较小的特点，所以在本程序中未获取手机联系人的照片信息
 */
import android.app.Activity;
import android.os.Bundle;
import android.content.Intent;
```

```
import android.view.Window;
import android.os.AsyncTask;   //异步处理类
import android.view.View;
import android.view.View.OnClickListener;
import android.widget.Toast;
import android.widget.ImageView;
import android.widget.ImageButton;
import android.widget.ListView;
import android.widget.AdapterView;
import android.widget.AdapterView.OnItemClickListener;
import android.app.AlertDialog;
import android.content.DialogInterface;
import android.provider.ContactsContract;   //
import android.database.Cursor;
import java.util.ArrayList;
import com.example.groupsendmessage.R;
public class MainActivity extends Activity implements OnClickListener {
    private Activity context;
    private ListView listView;
    public static int START = 0;
    public static int LIMIT = 15;
    //存放手机联系人对象数组(姓名+手机号)
    private ArrayList<Consumer> consumers;
    private ArrayList<String> listItemID = new ArrayList<String>();
    private ArrayList<String> nums = new ArrayList<String>();
    private ChooseAdapter myAdapter;   //
    public static String USERID;
    private ImageView backBtn;
    private ImageView imageButton;
    protected void onCreate(Bundle savedInstanceState) {
        // TODO Auto-generated method stub
        super.onCreate(savedInstanceState);
        //取消对应用程序标题栏的显示
        this.requestWindowFeature(Window.FEATURE_NO_TITLE);
        setContentView(R.layout.activity_main);

        context = this;
        backBtn = (ImageView) findViewById(R.id.onBack);
        backBtn.setOnClickListener(this);
        imageButton = (ImageButton) findViewById(R.id.imageButton);
```

```
        imageButton.setOnClickListener(this);

        consumers = new ArrayList<Consumer>();   //使用了实体类
        new GetContact().execute();
    }
    private void initView() {    //
        listView = (ListView) findViewById(R.id.listView);
        myAdapter = new ChooseAdapter(context, consumers);
        listView.setAdapter(myAdapter);   //
        listView.setItemsCanFocus(false);
        listView.setChoiceMode(ListView.CHOICE_MODE_MULTIPLE);
        //列表项选择监听器
        listView.setOnItemClickListener(new OnItemClickListener() {
            @Override
            public void onItemClick(AdapterView<?> parent, View view,
                    int position, long id) {
                    Toast.makeText(MainActivity.this,"点击了id为" + position + "的
                                    联系人！ ", Toast.LENGTH_SHORT).show();
                // TODO Auto-generated method stub
                ListItemView mlistItem = (ListItemView) view.getTag();
                // mlistItem.selected.setChecked(true);
                boolean check = mlistItem.selected.isChecked();
                if (check) {
                    mlistItem.selected.setChecked(false);
                    check = false;
                }
                else {
                    mlistItem.selected.setChecked(true);
                    check = true;
                }
                ChooseAdapter.isSelected.put(position, check);
            }
        });
        this.refreshListView(consumers);   //自定义的刷新视图方法
        if (consumers.size() == 0) {
            Toast.makeText(MainActivity.this, "未找到相关记录",
                                            Toast.LENGTH_SHORT).show();
        }
    }
    //考虑到目前手机联系人数量日前增多，使用异步任务类避免阻塞主线程
```

```
class GetContact extends AsyncTask<String, Integer, String> {
    @Override
    protected void onPreExecute() {
        // TODO Auto-generated method stub
        super.onPreExecute();
        Toast.makeText(MainActivity.this, "联系人信息加载中，稍候...",
                Toast.LENGTH_SHORT).show();
    }
    @Override
    //可变参数(JDK 5以后的新特性)，可以为离散变量或数组
    protected String doInBackground(String... arg0) {    //
        // TODO Auto-generated method stub
        getContacts();//获取所有的手机联系人并放在自定义的Consumers类中
        return null;
    }
    @Override
    protected void onPostExecute(String result) {
        // TODO Auto-generated method stub
        super.onPostExecute(result);
        initView();    //
    }
}

private void refreshListView(ArrayList<Consumer> datas) {    //刷新
    // TODO Auto-generated method stub
    listView.setClickable(true);
    myAdapter.refresh(datas);
    if (datas.size() == 0) {
        listView.setClickable(false);
    }
}

public void set() {        //在统计选择的基础上确认
    listItemID.clear();
    for (int i = 0; i < ChooseAdapter.isSelected.size(); i++) {
        if (ChooseAdapter.isSelected.get(i)) {
            listItemID.add(consumers.get(i).getName());
            nums.add(consumers.get(i).getPhone());
        }
```

```
        }
        if (listItemID.size() == 0) {       //未勾选手机联系人
            new AlertDialog.Builder(this).setTitle("没有选中任何记录")
                    .setPositiveButton("确定", null).show();
        }
        else {
            StringBuilder sb = new StringBuilder();
            if (listItemID.size() == 0) {
                return;
            }
            sb.append("收件人:" + listItemID.get(0));
            String num = nums.get(0);
            for (int i = 1; i < listItemID.size(); i++) {
                sb.append("," + listItemID.get(i));
                num += ";" + nums.get(i);
            }
            final String nums = num;
            System.out.println("nums--->" + nums);
            new AlertDialog.Builder(this)
                    .setTitle("是否确定？")
                    .setMessage(sb.toString())
                    .setPositiveButton("确定",
                            new DialogInterface.OnClickListener() {
                                @Override
                                public void onClick(DialogInterface arg0,
                                        int arg1) {
                                    // TODO Auto-generated method stub
                                    Intent intent = new Intent(
                                                    Intent.ACTION_VIEW);
                                    intent.putExtra("address", nums);
                                    //准备使用系统的短信程序
                                    intent.setType("vnd.android-dir/mms-sms");
                                    intent.putExtra("sms_body", "");
                                    startActivity(intent);   //调用
                                }
                            }).setNegativeButton("取消", null).show();
        }
    }
```

```
@Override      //参见本类定义时 implement OnClickListener
public void onClick(View arg0) {
    switch (arg0.getId()) {
    case R.id.onBack:
        this.finish();    //返回
        break;
    case R.id.imageButton:
        set();        //选择手机联系人并发送短信
        break;
    }
}

private void getContacts() {   // 获得所有的手机联系人
    // TODO Auto-generated method stub
    Consumer tmpConsumer;   //声明自定义实体类对象
    consumers.clear();
    Cursor cur =this.getContentResolver().query(
            ContactsContract.Contacts.CONTENT_URI,null,null,null,
            ContactsContract.Contacts.DISPLAY_NAME
                                    + " COLLATE LOCALIZED ASC");
    // 循环遍历
    if (cur.moveToFirst()) {
        int idColumn = cur.getColumnIndex(ContactsContract.Contacts._ID);
        int displayNameColumn = cur
                .getColumnIndex(ContactsContract.Contacts.DISPLAY_NAME);
        do {
            tmpConsumer = new Consumer();
            // 获得手机联系人的ID号
            String contactId = cur.getString(idColumn);
            // 获得手机联系人姓名
            String disPlayName = cur.getString(displayNameColumn);
            // 查看该手机联系人有多少个电话号码。如果没有，返回值为0
            int phoneCount =
            cur.getInt(cur.getColumnIndex(ContactsContract.
                                    Contacts.HAS_PHONE_NUMBER));
            tmpConsumer.setName(disPlayName);
            if (phoneCount > 0) {
                // 获得手机联系人的电话号码
                Cursor phones = getContentResolver().query(
```

```
                    ContactsContract.CommonDataKinds.Phone.
                                        CONTENT_URI,null,
                    ContactsContract.CommonDataKinds.Phone.CONTACT_ID
                                    + " = " + contactId, null, null);
            if (phones.moveToFirst()) {
                String tmpPhone = phones.getString(phones.getColumnIndex
                    (ContactsContract.CommonDataKinds.Phone.NUMBER));
                if(tmpPhone.length() == 11&&tmpPhone.substring
                                                (0, 1).equals("1"))
                    tmpConsumer.setPhone(tmpPhone);
            }
        }
        if(tmpConsumer.getPhone() != null)
            consumers.add(tmpConsumer);
    } while (cur.moveToNext());
  }
 }
}
```

(8) 在工程清单文件里，注册读取手机联系人的权限，其代码如下：

```
<uses-permission android:name="android.permission.READ_CONTACTS"/>
```

(9) 部署工程到手机后做运行测试。

习　题　8

一、判断题

1. 不能限定对已经注册的 ContentProvider 组件的访问。
2. 手机联系人数据包含在一张表内。
3. 类 UriMatcher 和 ContentUris 应出现在使用内容提供者的程序里。
4. 任何使用内容提供者的程序，都要使用类 UriMatcher。
5. 编写读取手机联系人应用程序，必须知道手机联系人数据库名字。

二、选择题

1. 在 Eclipse 中新建一个继承于 ContentProvider 的子类时，不会被自动导入的类是____。

   A. android.content.ContentValues　　B. android.database.Cursor

   C. android.app.Activity　　D. android.net.Uri

2. 在工程清单文件中对内容提供者配置的<provider>标签内，不必使用的属性是____。

   A. android:name　　B. android:authorities

   C. android:exported　　D. android:label

3. 检索手机联系人是否有电话，应检索手机联系人数据库中的____表。

   A. raw_contacts　　B. data　　C. contacts　　D. mimetypes

4. Handler 位于软件包 andoid. ____内。

   A. content　　B. app　　C. os　　D. provider

5. 创建 Handler 对象时，需要实现其内部接口____。

   A. Callback　　B. Feedback　　C. Return　　D. Loop

6. 下列不属于异步任务类 AsyncTask 提供的方法的是____。

A. onPreExecute()　B. doInBackground()　　C. start()　　D. onPostExecute()

三、填空题

1. 在清单文件中注册自定义的 ContentProvider 组件所使用的标签是____。
2. 定义内容提供者时的 Uri 表示访问资源的____位置。
3. 表示内容提供者时的 Uri 的 scheme 为____。
4. 与 Android 2.0 以上版本中的手机联系人数据库名相对应的类名为____。
5. 创建 Handler 对象时，需要重写内部接口方法____。

实验 8 使用内容提供者实现应用程序间的数据共享

一、实验目的

1. 掌握使用 ContentProvider 建立内容提供者的方法。
2. 掌握通过 ContentResolver 访问 ContentProvider 的方法。
3. 掌握手机联系人数据库的主要表及其关系。
4. 掌握对手机联系人数据库的编程。

二、实验内容及步骤

【预备】访问 http://www.wustwzx.com/android/index.html，单击“实验 8”超链接，下载本次实验内容的源代码并解压(得到文件夹 ch08)，供研读和调试使用。

1. 掌握内容提供者的建立与使用方法。

(1) 再次解压 ch08 内的压缩文件 ContentProvider.zip，得到两个 Android 应用工程，导入工程 MyDbProvider 和 UseMyDbProvider。

(2) 查看工程 MyDbProvider 的清单文件对内容提供者的配置标签<provider>中的三个属性及属性值。

(3) 查看在 MyDbProvider.java 的 onCreate()方法中创建 MyDAO 对象的代码。

(4) 在 MyDbProvider 工程中，分别查看 MyDAO.java 中的两个插入方法及 insertInfo()在 MainActivity.java 和 MyDbProvider.java 中的使用(方法重载：方法名称相同但参数类型和返回值不同)。

(5) 分别部署两个工程到手机上，初次运行 MyDbProvider 时，只显示 2 条记录，然后运行工程 UseMyDbProvider，插入一条记录，观察 Toast 显示的该记录的 Uri，然后再运行工程 MyDbProvider 进行查验。

(6) 运行工程 UseMyDbProvider，分别做记录修改和删除操作后，再运行工程 MyDbProvider 进行查验。

(7) 修改 MyDbProvider 工程 ContentProvider 的配置，设置 android:exported="false"后重新部署，验证运行工程 UseContentProvider 时将自动挂断。

2. 了解手机联系人数据库的相关表，访问系统提供的公用数据接口，读取并显示手机联系人的姓名及电话(参见例 8.3.1)。

(1) 从作者教学网站下载手机联系人数据库 contact2.db，然后使用 SQLiteSpy 软件打开。

(2) 查看表 raw_contacts 的结构，验证它含有表示手机联系人姓名的字段 display_name。

(3) 分别查看手机联系人数据表 data 和手机联系人表 contacts 的信息，比较其结构差异。

(4) 再次解压 ch08 内的压缩文件 ContactsCP.zip，导入工程 ContactsCP。

(5) 屏蔽工程清单文件中读取手机联系人的权限，部署工程到手机时异常中止，查看 LogCat 中的错误信息(没有权限)，然后取消屏蔽。

(6) 打开源文件 MainActivity，查看访问手机联系人信息的 Uri。

(7) 查看获取手机联系人 id、姓名和电话的代码。

(8) 查看使用适配器及 ListView 控件显示手机联系人信息的相关代码。

3. 选择手机联系人后群发短信(参见例 8.3.2)。

(1) 再次解压 ch08 文件夹的压缩文件 GroupSendMessage.zip，导入工程 GroupSendMessage。

(2) 打开源文件 MainActivity.java，查看取消 Android 应用标题栏显示的代码。

(3) 查看获取手机联系人信息方法 getContacts()的实现代码。

(4) 查看对选中的手机联系人发送短信方法 set()的实现代码，并与第 5.2 节介绍的短信方法相比较。

(5) 部署工程到手机并做运行测试。

4. 了解观察者模式，监视数据的变化。

(1) 再次解压 ch08 内的压缩文件 ObserverModel.zip，导入工程 ObserverModel。

(2) 查看继承 Observable 的被观察者类 MyPerson 的源代码，并与实体类文件比较其差别(每个 set×××()方法里多了 2 个方法 setChanged()和 notifyObservers())。

(3) 查看通过实现接口 Observer 定义的观察者类 MyObserver 的源代码。

(4) 查看 MainActivity.java 程序中使用 Handler 处理消息的代码。

(5) 部署工程到手机后做运行测试。

三、实验小结及思考

(由学生填写，重点写上机中遇到的问题。)

第9章 Android 近距离通信技术及其应用

电话和短信，是手机的远距离通信功能。Android 手机还有近距离的通信功能，通过手机的 WiFi 网卡、蓝牙设备和 NFC 设备等，可以实现 Android 设备之间的近距离通信。本章学习要点如下：

- 掌握手机 WiFi 的使用及 Android 的 WiFi 编程；
- 掌握手机蓝牙的使用及 Android 蓝牙编程；
- 了解 Android 手机 NFC 功能的应用。

9.1 WiFi 通信

9.1.1 WiFi 简介

使用 WiFi 上网是目前很常见的无线宽带上网方式。WiFi 是一种局域网协议，是为改善基于 IEEE 802.11 标准的无线网络产品之间的互通性而出现的。WiFi 的工作频段为 2.4 GHz，传输速率为 2 Mbit/s。

WiFi MAC 是终端的物理地址，在局域网中辨别终端的标识，采用十六进制的表达形式(6 字节 48 位)。

WiFi 地址，是无线路由器发出信号的地址。

出于安全考虑，每个无线路由器里都可以设置其使用 WiFi 的密码，它是用于登录无线网络的依据。

注意：在手机检测到的 WiFi 信号中，带锁表示设置了使用密码。

9.1.2 Android 对 WiFi 的支持

在包 android.net 内，提供了三个 WiFi 类：WiFi 管理器类 WifiManager、扫描结果类 ScanResult 和 WiFi 信息类 WifiInfo。

WifiManager 主要提供了获取 WiFi 状态、打开或关闭 WiFi、扫描 WiFi 并获取结果和获得当前 WiFi 连接等方法，如图 9.1.1 所示。

android.net.wifi
WifiManager.class
WifiManager
getWifiState() : int
isWifiEnabled() : boolean
setWifiEnabled(boolean) : boolean
startScan() : boolean
getScanResults() : List<ScanResult>
getConnectionInfo() : WifiInfo

图 9.1.1　WifiManager 的常用方法

在调用 WifiManager 类的 getScanResults()方法前，必须先使用 WifiManager 类的 startScan()方法，扫描结果是一个列表类型 List<ScanResult>。

ScanResult 类包含了 WiFi 名称(对应于 SSID 字段)、WiFi 地址(对应于 BSSID 字段，表示无线路由器的 MAC 地址)、WiFi 频率(对应于 frequency 字段)和 WiFi 信号强度(对应于 level 字段)等，如图 9.1.2 所示。

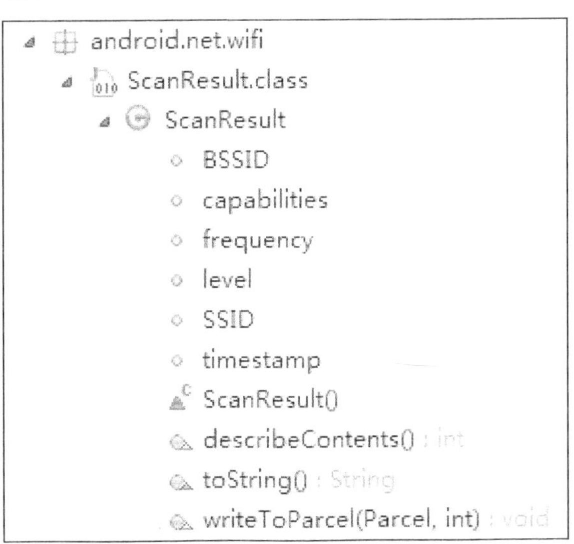

图 9.1.2　ScanResult 类的定义

通过 WifiManager 类的 getConnectionInfo()方法实时获取当前连接的 WiFi 的有关信息，其结果是一个 WifiInfo 类型。

WifiInfo 类提供了获取当前 WiFi 名称方法 getSSID()、WiFi 地址方法 getBSSID()、WiFi 信号强度方法 getRssi()、手机 WiFi 网卡的 MAC 地址方法 getMacAddress()等，如图 9.1.3 所示。

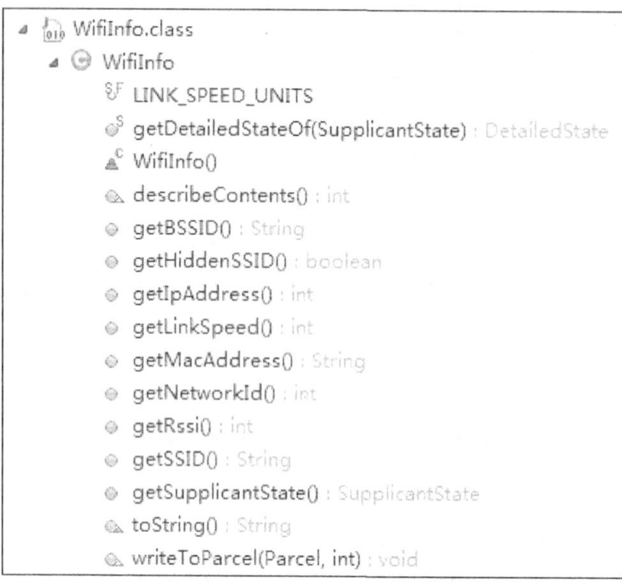

图 9.1.3　WifiInfo 类的常用方法

注意：经比较可以看出，WifiInfo 类中的某些方法与 ScanResult 类的某些属性相对应。

9.1.3　一个 WiFi 应用实例

下面介绍一个例子，以掌握对 WiFi 相关类的使用。

【例 9.1.1】WiFi 的基本操作。

【程序运行】程序运行时，主界面上出现四个功能按钮。单击“检查 WiFi”按钮，会出现一个显示 WiFi 状态的 Toast 消息。如果 WiFi 没有开启，单击“打开 WiFi”按钮，然后单击“扫描 WiFi”按钮，其扫描结果在一个 TextView 控件内显示，程序运行效果如图 9.1.4 所示。

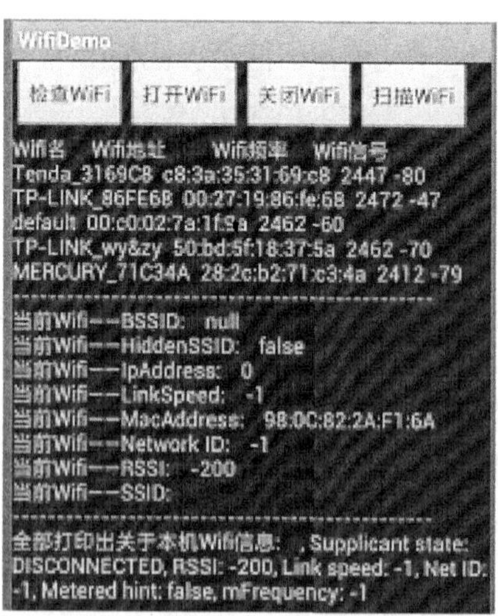

图 9.1.4　WiFi 基本操作程序运行界面

【设计步骤】

(1) 新建名为"WifiDemo"的 Android 应用工程，在出现的对话框中都采用默认选项，勾选创建 Activity，工程文件结构(主要部分)如图 9.1.5 所示。

图 9.1.5　工程文件结构(主要部分)

(2) 在布局文件中，采用 ScrollView 卷轴视图实现垂直线性布局(当文本框拥有很多内容、一屏显示不完时，可通过滚动条来显示全部内容)，其中的四个命令按钮为一个水平线性布局，文本框控件位于命令按钮下方，布局文件 activity_main.xml 代码如下：

```
<?xml version="1.0" encoding="utf-8"?>
<ScrollView xmlns:android="http://schemas.android.com/apk/res/android"
    android:id="@+id/mScrollView" android:layout_width="fill_parent"
    android:layout_height="wrap_content" android:scrollbars="vertical">
<LinearLayout
    android:layout_width="fill_parent"
    android:layout_height="fill_parent"
    android:orientation="vertical" >
    <LinearLayout
        android:layout_width="wrap_content"
        android:layout_height="wrap_content"
        android:orientation="horizontal" >
        <Button
            android:id="@+id/check_bt"
            android:layout_width="wrap_content"
            android:layout_height="wrap_content"
            android:text="检查WiFi" />
        <Button
            android:id="@+id/open_bt"
            android:layout_width="wrap_content"
            android:layout_height="wrap_content"
            android:text="打开WiFi" />
        <Button
            android:id="@+id/close_bt"
            android:layout_width="wrap_content"
            android:layout_height="wrap_content"
            android:text="关闭WiFi" />
         <Button
            android:id="@+id/search_bt"
```

```
            android:layout_width="wrap_content"
            android:layout_height="wrap_content"
            android:text="扫描WiFi" />
    </LinearLayout>
    <TextView
        android:id="@+id/text"
        android:layout_width="wrap_content"
        android:layout_height="wrap_content"
        android:text="null"/>
</LinearLayout>
</ScrollView>
```

(3) 在清单文件中注册使用 WiFi 的两个权限，其代码如下：

```
<uses-permission android:name="android.permission.CHANGE_WIFI_STATE" />
    <uses-permission android:name="android.permission.ACCESS_WIFI_STATE" />
```

(4) 编写工程的主 Activity 文件 MainActivity.java，其代码如下：

```
package com.example.wifi;
/*
*本程序演示了WiFi操作
*/
import java.util.List;
import android.app.Activity;
import android.content.Context;
import android.net.wifi.ScanResult;   //
import android.net.wifi.WifiInfo;   //
import android.net.wifi.WifiManager;   //
import android.os.Bundle;
import android.view.View;
import android.view.View.OnClickListener;
import android.widget.Button;
import android.widget.ScrollView;   //
import android.widget.TextView;
import android.widget.Toast;
import com.example.wifidemo.R;
public class MainActivity extends Activity {
    private Button open_bt, close_bt, check_bt, search_bt;
    private TextView textView;
    private WifiManager wifiManager;     //主类
```

```
private WifiInfo wifiInfo;
private ScanResult scanResult;
private List<ScanResult> WifiList;
private StringBuffer stringBuffer = new StringBuffer();
public void onCreate(Bundle savedInstanceState) {
    super.onCreate(savedInstanceState);
    setContentView(R.layout.activity_main);
    check_bt = (Button) findViewById(R.id.check_bt);
    check_bt.setOnClickListener(new check_btListener());
    open_bt = (Button) findViewById(R.id.open_bt);
    open_bt.setOnClickListener(new open_btListener());
    search_bt = (Button) findViewById(R.id.search_bt);
    search_bt.setOnClickListener(new search_btListener());
    close_bt = (Button) findViewById(R.id.close_bt);
    close_bt.setOnClickListener(new close_btListener());

    //scrollView = (ScrollView) findViewById(R.id.mScrollView);
    textView = (TextView) findViewById(R.id.text);
}

class check_btListener implements OnClickListener {   //检查
    public void onClick(View v) {
        // TODO Auto-generated method stub
        wifiManager = (WifiManager) WiFiDemoActivity.this
                                .getSystemService(Context.WIFI_SERVICE);     //
    Toast.makeText(WiFiDemoActivity.this,"当前网卡状态为： "+
                                getWiFiState(), Toast.LENGTH_SHORT).show();

    }
}

class open_btListener implements OnClickListener {   //打开
    public void onClick(View v) {
        // TODO Auto-generated method stub
        wifiManager = (WifiManager) WiFiDemoActivity.this
                .getSystemService(Context.WIFI_SERVICE);
        wifiManager.setWifiEnabled(true);
        Toast.makeText(WiFiDemoActivity.this,"当前网卡状态为： "
                                + getWiFiState(), Toast.LENGTH_SHORT).show();

    }
}
```

```
class close_btListener implements OnClickListener {    //关闭
    public void onClick(View v) {
        // TODO Auto-generated method stub
        wifiManager = (WifiManager) WiFiDemoActivity.this
                            .getSystemService(Context.WIFI_SERVICE);
        wifiManager.setWifiEnabled(false);    //
        Toast.makeText(WiFiDemoActivity.this,"当前网卡状态为: " +
                        getWiFiState(), Toast.LENGTH_SHORT).show();
    }
}
class search_btListener implements OnClickListener {    扫描
    public void onClick(View v) {
        // TODO Auto-generated method stub
        wifiManager.startScan();    //扫描
        WifiList = wifiManager.getScanResults();    //所有WiFi列表
        wifiInfo = wifiManager.getConnectionInfo();    //当前WiFi
        if (stringBuffer != null) {
            stringBuffer = new StringBuffer();
        }
        stringBuffer=stringBuffer.append("Wifi名").append("        ")
                        .append("Wifi地址").append("            ")
                        .append("Wifi频率").append("        ")
                        .append("Wifi信号").append("\n");
        if (WifiList != null) {
            for (int i = 0; i < WifiList.size(); i++) {
                scanResult = WifiList.get(i);
                stringBuffer = stringBuffer
                        .append(scanResult.SSID).append("    ")
                        .append(scanResult.BSSID).append("    ")
                        .append(scanResult.frequency).append(" ")
                        .append(scanResult.level).append("\n");
                textView.setText(stringBuffer.toString());
            }
            stringBuffer = stringBuffer.append("
            -----------------------------------------").append("\n");
            textView.setText(stringBuffer.toString());
            stringBuffer = stringBuffer
                    .append("当前Wifi—SSID").append(":        ")
                    .append(wifiInfo.getSSID()).append("\n")
```

```
                    .append("当前Wifi—BSSID").append(":      ")
                    .append(wifiInfo.getBSSID()).append("\n")
                    .append("当前Wifi—MacAddress").append(":      ")
                    .append(wifiInfo.getMacAddress()).append("\n")
                    .append("当前Wifi—HiddenSSID").append(":      ")
                    .append(wifiInfo.getHiddenSSID()).append("\n")
                    .append("当前Wifi—IpAddress").append(":      ")
                    .append(wifiInfo.getIpAddress()).append("\n")
                    .append("当前Wifi—LinkSpeed").append(":      ")
                    .append(wifiInfo.getLinkSpeed()).append("\n")
                    .append("当前Wifi—Network ID").append(":      ")
                    .append(wifiInfo.getNetworkId()).append("\n")
                    .append("当前Wifi—RSSI").append(":      ")
                    .append(wifiInfo.getRssi()).append("\n")
                    .append("-----------------------------------------")
                    .append("\n").append("全部打印出关于本机Wifi信息")
                    .append(":      ").append(wifiInfo.toString());
            textView.setText(stringBuffer.toString());
        }
    }
}
public String getWiFiState(){
    String temp=null;
    switch (wifiManager.getWifiState()) {     // WiFi网卡共有四种状态
    case 0:
            temp="Wifi正在关闭......";
            break;
    case 1:
            temp="Wifi已经关闭！ ";
            break;
    case 2:
            temp="Wifi正在打开......";
            break;
    case 3:
            temp="Wifi已经打开";
            break;
    default:
            break;
    }
```

```
        return temp;
    }
}
```

注意：当扫描到的WiFi信源较多使得一屏显示不完时，本例使用的方法是在布局文件中使用卷轴视图ScrollView。其实，也可以设置文本框的手势动作，参见第4.3.1小节和例11.1.1。

9.2　蓝牙通信 Bluetooth

9.2.1　Bluetooth 简介

蓝牙(Bluetooth)是一种支持设备短距离(一般10 m内) 通信的无线电技术，能在包括移动电话、PDA、无线耳机、笔记本电脑、相关外设等众多设备之间进行无线信息交换。

蓝牙采用分散式网络结构以及快跳频和短包技术，支持点对点及一点对多点通信，工作在全球通用的2.4GHz ISM(即工业、科学、医学)频段，其数据速率为1 Mbps。蓝牙采用时分双工传输方案实现全双工传输。

注意：蓝牙通信可以是双向的，这不同于WiFi通信。另外，蓝牙通信与WiFi使用的频段也不同。

9.2.2　Android 对 Bluetooth 的支持

蓝牙适配器就是各种数码产品能适用蓝牙设备的接口转换器，它采用了全球通用的短距离无线连接技术。利用蓝牙技术，能够有效地简化移动通信终端设备之间的通信，也能够成功地简化设备与因特网之间的通信，从而数据传输变得更加迅速高效，为无线通信拓宽道路。

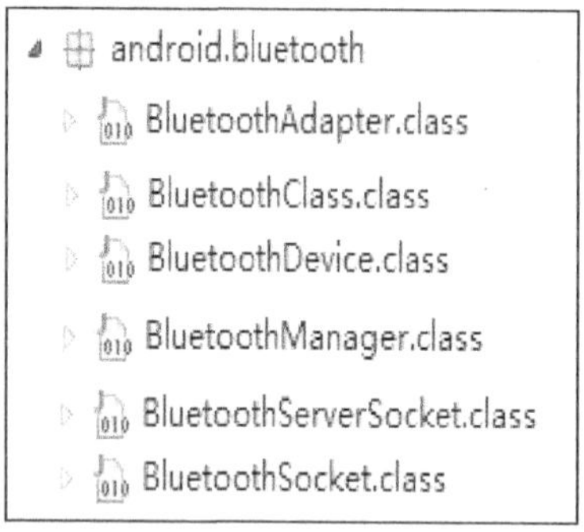

- android.bluetooth
  - BluetoothAdapter.class
  - BluetoothClass.class
  - BluetoothDevice.class
  - BluetoothManager.class
  - BluetoothServerSocket.class
  - BluetoothSocket.class

图9.2.1　android.bluetooth包里的主要类

包android.bluetooth提供的主要类包括蓝牙适配器类BluetoothAdapter、蓝牙装置类BluetoothDevice、蓝牙管理器类BluetoothManager、蓝牙服务器端类BluetoothServerSocket、蓝牙客户端类BluetoothSocket等，如图9.2.1所示。

蓝牙适配器类BluetoothAdapter代表本地的蓝牙适配器设备，让用户执行基本的蓝牙任务。例如： 初始化设备的搜索，查询可匹配的设备集，使用一个已知的MAC地址来初始化一个BluetoothDevice类，创建一个 BluetoothServerSocket类以监听其他设备对本机的连接请求等。

为了得到这个代表本地蓝牙适配器的 BluetoothAdapter 对象，需要调用getDefaultAdapter()这一静态方法。在拥有本地适配器以后， 用户可以获得一系列的

BluetoothDevice 对象，这些对象代表所有拥有 getBondedDevices()方法且已经匹配的设备；用 startDiscovery()方法来开始设备的搜寻；或者创建一个 BluetoothServerSocket 类，通过 listenUsingRfcommWithServiceRecord(String, UUID) 方法来监听新来的连接请求。BluetoothAdapter 类的定义，如图 9.2.2 所示。

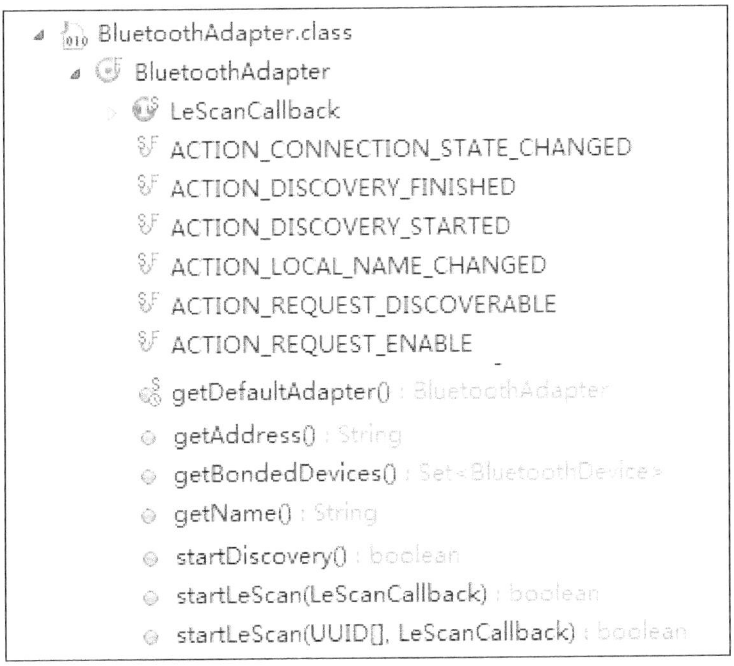

```
BluetoothAdapter.class
  BluetoothAdapter
    LeScanCallback
    ACTION_CONNECTION_STATE_CHANGED
    ACTION_DISCOVERY_FINISHED
    ACTION_DISCOVERY_STARTED
    ACTION_LOCAL_NAME_CHANGED
    ACTION_REQUEST_DISCOVERABLE
    ACTION_REQUEST_ENABLE
    getDefaultAdapter() : BluetoothAdapter
    getAddress() : String
    getBondedDevices() : Set<BluetoothDevice>
    getName() : String
    startDiscovery() : boolean
    startLeScan(LeScanCallback) : boolean
    startLeScan(UUID[], LeScanCallback) : boolean
```

图 9.2.2　蓝牙适配器类的定义

注意：智能手机一般都配有蓝牙适配器，但大部分台式计算机(包括笔记本)没有蓝牙适配器。要想实现计算机与手机的蓝牙通信，需要在计算机上添加蓝牙适配器。目前，USB 型的蓝牙适配器很方便地为计算机增加了蓝牙功能。

蓝牙管理器类用来管理远程蓝牙设备，主要包括显示所有蓝牙配对设备的列表、配对新设备的扫描、设置本机的蓝牙设备等。BluetoothManager 的定义如图 9.2.3 所示。

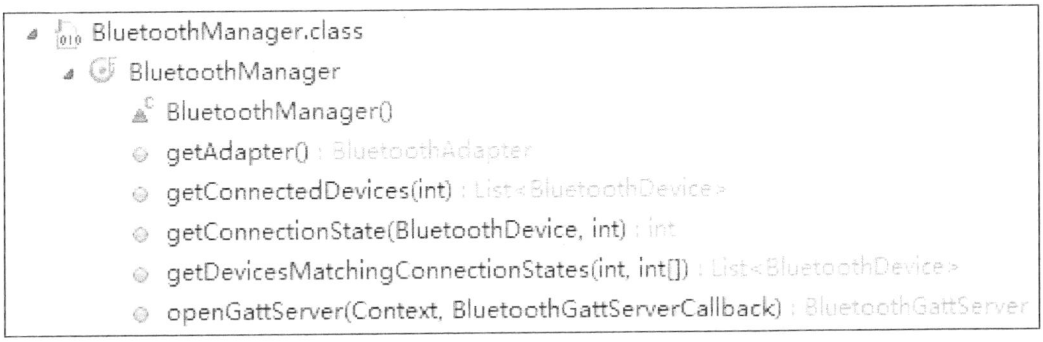

```
BluetoothManager.class
  BluetoothManager
    BluetoothManager()
    getAdapter() : BluetoothAdapter
    getConnectedDevices(int) : List<BluetoothDevice>
    getConnectionState(BluetoothDevice, int) : int
    getDevicesMatchingConnectionStates(int, int[]) : List<BluetoothDevice>
    openGattServer(Context, BluetoothGattServerCallback) : BluetoothGattServer
```

图 9.2.3　蓝牙管理器类的定义

蓝牙装置类 BluetoothDevice 的定义如图 9.2.4 所示。

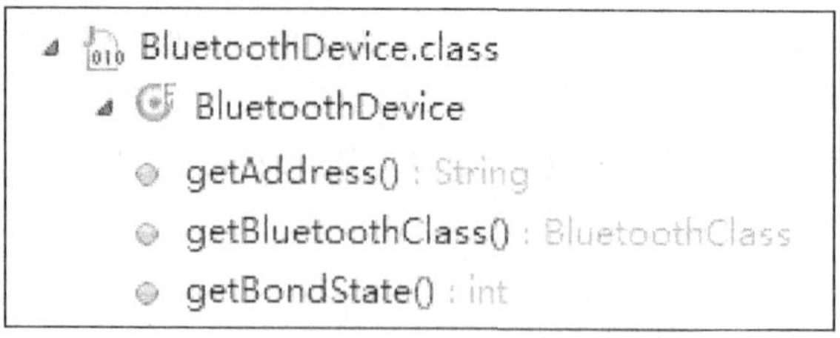

BluetoothDevice.class
BluetoothDevice
getAddress() : String
getBluetoothClass() : BluetoothClass
getBondState() : int

图 9.2.4　蓝牙装置类的定义

用于蓝牙 Socket 通信的两个类分别为 BluetoothSocket(客户端，即发出请求的那一端)和 BluetoothServerSocket(服务器端，即被请求连接的那一端)，其定义分别如图 9.2.5 和图 9.2.6 所示。

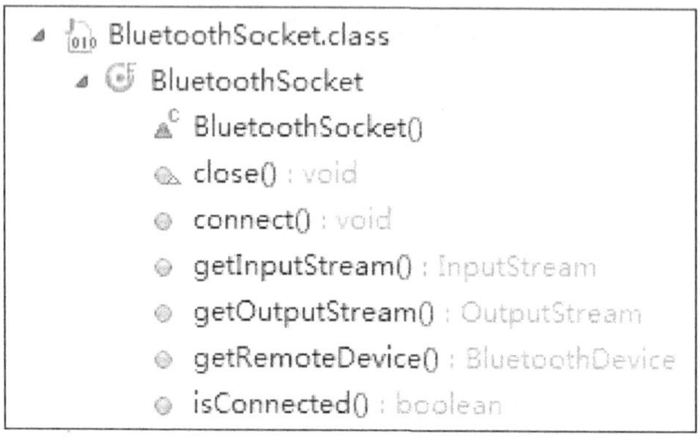

BluetoothSocket.class
BluetoothSocket
BluetoothSocket()
close() : void
connect() : void
getInputStream() : InputStream
getOutputStream() : OutputStream
getRemoteDevice() : BluetoothDevice
isConnected() : boolean

图 9.2.5　BluetoothSocket 类的定义

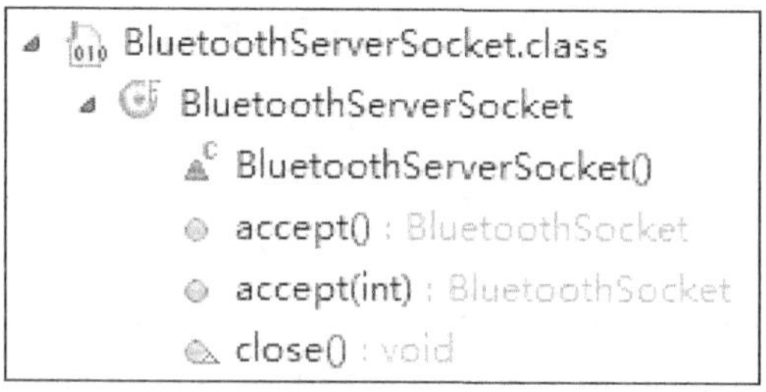

BluetoothServerSocket.class
BluetoothServerSocket
BluetoothServerSocket()
accept() : BluetoothSocket
accept(int) : BluetoothSocket
close() : void

图 9.2.6　BluetoothServerSocket 类的定义

蓝牙 Socket 通信是比较底层的网络编程，是跨平台的编程方式。

注意：手机蓝牙会话的服务程序，使用了 BluetoothSocket 建立蓝牙连接，这与 Java 的 Socket 通信是类似的，参见第 11.4 节。

在应用程序中，请求或建立蓝牙连接并传递数据，需要使用 BLUETOOTH 权限；初始化设备发现功能或更改蓝牙设置，则需要 BLUETOOTH_ADMIN 权限。在清单文件中，需要配置与蓝牙相关权限的代码：

```
<uses-permissionandroid:name="android.permission.BLUETOOTH_ADMIN" />
<uses-permissionandroid:name="android.permission.BLUETOOTH" />
```

当蓝牙开关未打开时，使用请求的代码如下：

```
if (!bluetoothAdapter.isEnabled()) {
        // 若当前设备蓝牙功能未开启，则开启蓝牙功能
        Intent intent = new Intent(BluetoothAdapter.ACTION_REQUEST_ENABLE);
        startActivityForResult(intent, REQUEST_ENABLE_BT);   //
 }
```

执行上述代码的效果，如图 9.2.7 所示。

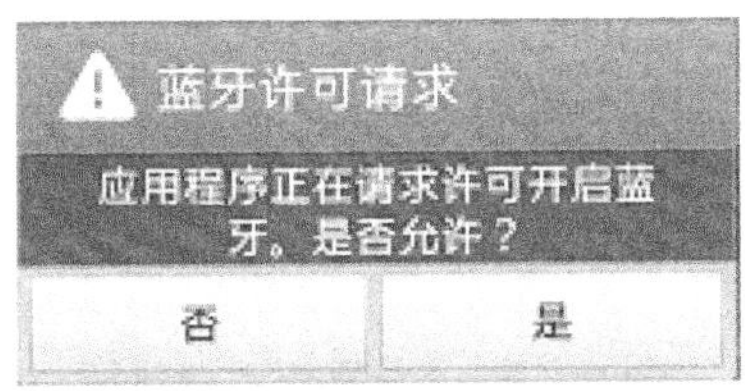

图 9.2.7　请求打开蓝牙功能

9.2.3　蓝牙聊天实例

利用手机的蓝牙功能，我们实现了在两个蓝牙手机之间的聊天功能。

注意：手机蓝牙通信并不需要移动数据网络或 WiFi 网络的支持。

【例 9.2.1】使用手机蓝牙实现聊天功能。

【程序运行】将程序部署到手机并运行后，窗口标题栏显示“无连接”，单击手机菜单键后，再选择“我的好友”菜单项，进入另一个选择已经配对好友的界面，效果如图 9.2.8 所示。

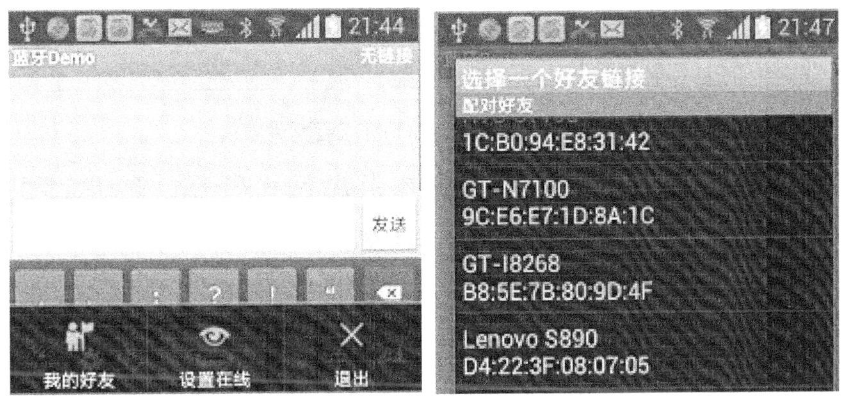

图 9.2.8　蓝牙聊天程序开始运行时效果

选择一个好友后，将返回至主 Activity，并在标题栏右边显示连接的手机。此时，即可与连接的手机使用蓝牙聊天，效果如图 9.2.9 所示。

【设计步骤】

(1) 新建名为 BluetoothChat 的 Android 应用工程，创建名为 BluetoothChat 的主 Activity，其他选项采用默认值，工程文件结构如图 9.2.10 所示。

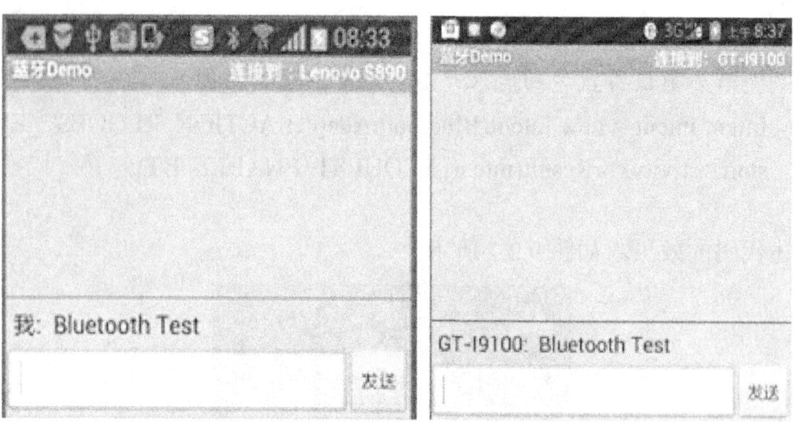

图 9.2.9　蓝牙聊天运行效果

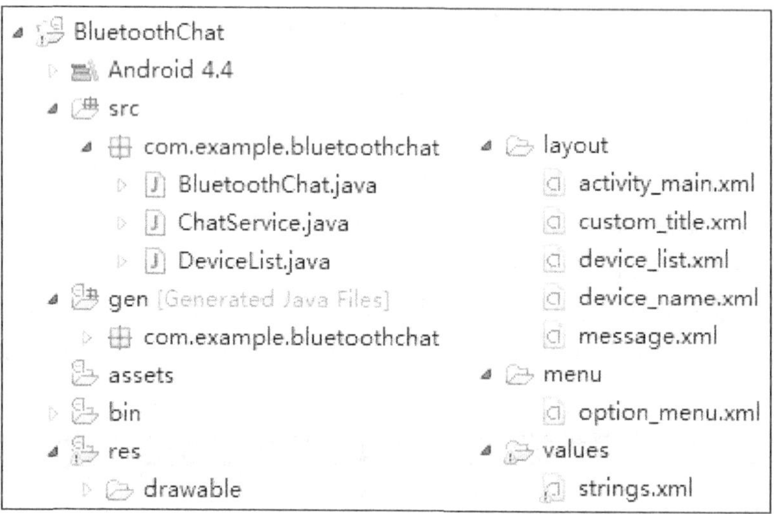

图 9.2.10　BluetoothChat 工程文件结构

(2) 程序运行过程中的状态描述文本及配色代码等，含于工程资源文件 res/values/strings.xml 内，其代码如下：

```
<?xml version="1.0" encoding="utf-8"?>
<resources>
    <string name="app_name">蓝牙Demo</string>
    <string name="send">发送</string>
    <string name="not_connected">你没有链接一个设备</string>
    <string name="bt_not_enabled_leaving">蓝牙不可用，离开聊天室</string>
    <string name="title_connecting">链接中...</string>
    <string name="title_connected_to">连接到：</string>
    <string name="title_not_connected">无链接</string>
    <string name="scanning">搜索好友中...</string>
    <string name="select_device">选择一个好友链接</string>
```

```
    <string name="none_paired">没有配对好友</string>
    <string name="none_found">附近没有发现好友</string>
    <string name="title_paired_devices">配对好友</string>
    <string name="title_other_devices">其它可连接好友</string>
    <string name="button_scan">搜索好友</string>
    <string name="connect">我的好友</string>
    <string name="discoverable">设置在线</string>
    <string name="back">退出</string>
    <string name="startVideo">开始聊天</string>
    <string name="stopVideo">结束聊天</string>
    <drawable name="bg">#ADD8E6</drawable>
    <drawable name="bg01">#87CEFA</drawable>
</resources>
```

(3) 修改默认的主布局文件 activity_main.xml，它包含了一个 TextView 控件、一个 EditText 控件和一个 Button 控件，使用垂直线性布局和水平线性布局的嵌套。

(4) 分别建立主 Activity 的标题布局文件 custom_title.xml 及菜单设置文件 option_menu.xml、选择好友(即已经配对过的蓝牙设备)界面的布局文件 device_list.xml。

(5) 编写 Activity 程序 DeviceList.java，选取与之会话的蓝牙设备，其文件代码如下：

```
package com.example.bluetoothchat;
/*
 *  通过本Activity选取与之会话的蓝牙设备
 */
import java.util.Set;
import android.annotation.SuppressLint;
import android.annotation.TargetApi;
import android.app.Activity;
import android.bluetooth.BluetoothAdapter;   //
import android.bluetooth.BluetoothDevice;   //
import android.content.BroadcastReceiver;
import android.content.Context;
import android.content.Intent;
import android.content.IntentFilter;     //
import android.os.Build;
import android.os.Bundle;
import android.view.View;
import android.view.Window;
import android.view.View.OnClickListener;
import android.widget.AdapterView;    //
import android.widget.AdapterView.OnItemClickListener;
```

```
import android.widget.ArrayAdapter;
import android.widget.Button;
import android.widget.ListView;
import android.widget.TextView;
import com.example.bluetoothchat.R;
@TargetApi(Build.VERSION_CODES.ECLAIR)
@SuppressLint({ "NewApi", "InlinedApi" })
public class DeviceList extends Activity {
    public static String EXTRA_DEVICE_ADDRESS = "device_address";
    private BluetoothAdapter mBtAdapter;
    private ArrayAdapter<String> mPairedDevicesArrayAdapter;
    private ArrayAdapter<String> mNewDevicesArrayAdapter;
    @TargetApi(Build.VERSION_CODES.ECLAIR)
    @SuppressLint({ "NewApi", "InlinedApi" })
    @Override
    protected void onCreate(Bundle savedInstanceState) {
        super.onCreate(savedInstanceState);
        requestWindowFeature(Window.FEATURE_INDETERMINATE_PROGRESS);
        setContentView(R.layout.device_list);    //
        setResult(Activity.RESULT_CANCELED);
        Button scanButton = (Button) findViewById(R.id.button_scan);
        scanButton.setOnClickListener(new OnClickListener() {
            public void onClick(View v) {
                doDiscovery();
                v.setVisibility(View.GONE);
            }
        });
        mPairedDevicesArrayAdapter = new ArrayAdapter<String>(this,
                                                R.layout.device_name);
        mNewDevicesArrayAdapter = new ArrayAdapter<String>(this,
                                                R.layout.device_name);
        ListView pairedListView = (ListView) findViewById(R.id.paired_devices);
        pairedListView.setAdapter(mPairedDevicesArrayAdapter);
        pairedListView.setOnItemClickListener(mDeviceClickListener);
        ListView newDevicesListView = (ListView)findViewById(R.id.new_devices);
        newDevicesListView.setAdapter(mNewDevicesArrayAdapter);
        newDevicesListView.setOnItemClickListener(mDeviceClickListener);
        IntentFilter filter = new IntentFilter(BluetoothDevice.ACTION_FOUND);
        this.registerReceiver(mReceiver, filter);
```

```
        filter = new IntentFilter(BluetoothAdapter.ACTION_DISCOVERY_FINISHED);
        this.registerReceiver(mReceiver, filter);
        mBtAdapter = BluetoothAdapter.getDefaultAdapter();    //
        Set<BluetoothDevice> pairedDevices = mBtAdapter.getBondedDevices();//
        if (pairedDevices.size() > 0) {
            findViewById(R.id.title_paired_devices).setVisibility(View.VISIBLE);
            for (BluetoothDevice device : pairedDevices) {
                mPairedDevicesArrayAdapter.add(device.getName() + "\n"
                        + device.getAddress());
            }
        }
            else {
            String noDevices = getResources().getText(R.string.none_paired)
                    .toString();
            mPairedDevicesArrayAdapter.add(noDevices);
        }
    }
    @SuppressLint("NewApi")
    @Override
    protected void onDestroy() {
        super.onDestroy();
        if (mBtAdapter != null)
            mBtAdapter.cancelDiscovery();
        this.unregisterReceiver(mReceiver);
    }
    private void doDiscovery() {
        setProgressBarIndeterminateVisibility(true);
        setTitle(R.string.scanning);
        findViewById(R.id.title_new_devices).setVisibility(View.VISIBLE);
        if (mBtAdapter.isDiscovering())
            mBtAdapter.cancelDiscovery();
        mBtAdapter.startDiscovery();
    }
    private OnItemClickListener mDeviceClickListener = new OnItemClickListener() {
        public void onItemClick(AdapterView<?> av, View v, int arg2, long arg3) {
            mBtAdapter.cancelDiscovery();
            String info = ((TextView) v).getText().toString();
            String address = info.substring(info.length() - 17);
            Intent intent = new Intent();
```

```
                intent.putExtra(EXTRA_DEVICE_ADDRESS, address);
                setResult(Activity.RESULT_OK, intent);
                finish();
            }
        };
        @SuppressLint("NewApi")
        private final BroadcastReceiver mReceiver = new BroadcastReceiver() {
            @SuppressLint("NewApi")
            @Override
            public void onReceive(Context context, Intent intent) {
                String action = intent.getAction();
                if (BluetoothDevice.ACTION_FOUND.equals(action)) {
                    BluetoothDevice device = intent
                            .getParcelableExtra(BluetoothDevice.EXTRA_DEVICE);
                    if (device.getBondState() != BluetoothDevice.BOND_BONDED)
                        mNewDevicesArrayAdapter.add(device.getName() + "\n"
                                                + device.getAddress());
                } else if (BluetoothAdapter.ACTION_DISCOVERY_FINISHED.equals(action)) {
                    setProgressBarIndeterminateVisibility(false);
                    setTitle(R.string.select_device);
                    if (mNewDevicesArrayAdapter.getCount() == 0) {
                        String noDevices = getResources().getText(
                                            R.string.none_found).toString();
                        mNewDevicesArrayAdapter.add(noDevices);
                    }
                }
            }
        };
}
```

(6) 编写蓝牙会话的服务程序 ChatService.java，其文件代码如下：

```
package com.example.bluetoothchat;
/*
 * 本程序ChatService是蓝牙会话的服务程序
 * UUID(universally unique identifier)，即通用唯一标识符
 * 使用Handler处理消息
 * 使用多种线程
 * 使用Java提供的同步方法(使用修饰符：synchronized)
 */
```

```
import java.io.IOException;
import java.io.InputStream;
import java.io.OutputStream;
import java.util.UUID;  //
import android.annotation.SuppressLint;
import android.bluetooth.BluetoothAdapter;  //
import android.bluetooth.BluetoothDevice;   //
import android.bluetooth.BluetoothServerSocket;  //
import android.bluetooth.BluetoothSocket;  //
import android.content.Context;
import android.os.Bundle;
import android.os.Handler;  //
import android.os.Message;
@SuppressLint("NewApi")
public class ChatService {  //
    private static final String NAME = "BluetoothChat";
    // UUID是一个128位长的数字，一般用十六进制表示
    //算法的核心思想是结合机器的网卡、当地时间、一个随机数来生成
    private static final UUID MY_UUID = UUID.fromString(
                                        "fa87c0d0-afac-11de-8a39-0800200c9a66");
    private final BluetoothAdapter mAdapter; //
    private final Handler mHandler;  //
    private AcceptThread mAcceptThread;  //
    private ConnectThread mConnectThread;  //
    private ConnectedThread mConnectedThread;  //
    private int mState;
    public static final int STATE_NONE = 0;
    public static final int STATE_LISTEN = 1;
    public static final int STATE_CONNECTING = 2;
    public static final int STATE_CONNECTED = 3;

    public ChatService(Context context, Handler handler) {    //构造方法
        mAdapter = BluetoothAdapter.getDefaultAdapter();  //
        mState = STATE_NONE;
        mHandler = handler;
    }
    private synchronized void setState(int state) {
        mState = state;
        mHandler.obtainMessage(BluetoothChat.MESSAGE_STATE_CHANGE, state, -1)
                .sendToTarget();
```

```
    }
    public synchronized int getState() {
        return mState;
    }
    public synchronized void start() {
        if (mConnectThread != null) {
            mConnectThread.cancel();
            mConnectThread = null;
        }
        if (mConnectedThread != null) {
            mConnectedThread.cancel();
            mConnectedThread = null;
        }
        if (mAcceptThread == null) {
            mAcceptThread = new AcceptThread();
            mAcceptThread.start();
        }
        setState(STATE_LISTEN);
    }
    // 取消 CONNECTING 和 CONNECTED 状态下的相关线程
    // 运行新的 mConnectThread 线程
    public synchronized void connect(BluetoothDevice device) {
        if (mState == STATE_CONNECTING) {
            if (mConnectThread != null) {
                mConnectThread.cancel();
                mConnectThread = null;
            }
        }
        if (mConnectedThread != null) {
            mConnectedThread.cancel();
            mConnectedThread = null;
        }
        mConnectThread = new ConnectThread(device);
        mConnectThread.start();
        setState(STATE_CONNECTING);
    }
    // 开启一个 ConnectedThread 来管理对应的当前连接
    @SuppressLint("NewApi")
    public synchronized void connected(BluetoothSocket socket,
```

```
            BluetoothDevice device) {
        if (mConnectThread != null) {
            mConnectThread.cancel();
            mConnectThread = null;
        }
        if (mConnectedThread != null) {
            mConnectedThread.cancel();
            mConnectedThread = null;
        }
        if (mAcceptThread != null) {
            mAcceptThread.cancel();
            mAcceptThread = null;
        }
        mConnectedThread = new ConnectedThread(socket);
        mConnectedThread.start();
        Message msg = mHandler.obtainMessage
                                    (BluetoothChat.MESSAGE_DEVICE_NAME);
        Bundle bundle = new Bundle();
        bundle.putString(BluetoothChat.DEVICE_NAME, device.getName());
        msg.setData(bundle);
        mHandler.sendMessage(msg);
        setState(STATE_CONNECTED);
    }
    // 停止所有相关线程
    public synchronized void stop() {
        if (mConnectThread != null) {
            mConnectThread.cancel();
            mConnectThread = null;}
        if (mConnectedThread != null) {
            mConnectedThread.cancel();
            mConnectedThread = null;}
        if (mAcceptThread != null) {
            mAcceptThread.cancel();
            mAcceptThread = null;}
        setState(STATE_NONE);
    }
    // 在 STATE_CONNECTED 状态下，调用 mConnectedThread 里的 write() 方法
    public void write(byte[] out) {
        ConnectedThread r;
```

```
        synchronized (this) {
            if (mState != STATE_CONNECTED)
                return;
            r = mConnectedThread;
        }
        r.write(out);
    }
    // 连接失败的时候处理，通知UI，并设为 STATE_LISTEN 状态
    private void connectionFailed() {
        setState(STATE_LISTEN);
        Message msg = mHandler.obtainMessage(BluetoothChat.MESSAGE_TOAST);
        Bundle bundle = new Bundle();
        bundle.putString(BluetoothChat.TOAST, "链接不到设备");
        msg.setData(bundle);
        mHandler.sendMessage(msg);
    }
    // 当连接失败的时候，设为 STATE_LISTEN 状态并通知 UI
    private void connectionLost() {
        setState(STATE_LISTEN);
        Message msg = mHandler.obtainMessage(BluetoothChat.MESSAGE_TOAST);
        Bundle bundle = new Bundle();
        bundle.putString(BluetoothChat.TOAST, "设备链接中断");
        msg.setData(bundle);
        mHandler.sendMessage(msg);
    }
    // 创建监听线程，准备接收新连接
    private class AcceptThread extends Thread {
        private final BluetoothServerSocket mmServerSocket;
        public AcceptThread() {
            BluetoothServerSocket tmp = null;
            try {
                tmp = mAdapter
                        .listenUsingRfcommWithServiceRecord(NAME, MY_UUID);//
            }
            catch (IOException e) {
                //
            }
            mmServerSocket = tmp;
        }
```

```
public void run() {
    setName("AcceptThread");
    BluetoothSocket socket = null;
    while (mState != STATE_CONNECTED) {
        try {
            socket = mmServerSocket.accept(); // 使用阻塞方式
        }
        catch (IOException e) {
            break;
        }
        if (socket != null) {
            synchronized (ChatService.this) {
                switch (mState) {
                case STATE_LISTEN:
                case STATE_CONNECTING:
                    connected(socket, socket.getRemoteDevice());
                    break;
                case STATE_NONE:
                case STATE_CONNECTED:
                    try {
                        socket.close();
                    }
                    catch (IOException e) {
                        //
                    }
                    break;
                }
            }
        }
    }
}
public void cancel() {
    try {
        mmServerSocket.close();
    }
    catch (IOException e) {
        //
    }
}
```

```
}
// 连接线程
private class ConnectThread extends Thread {
    private final BluetoothSocket mmSocket;
    private final BluetoothDevice mmDevice;
    //构造方法
    public ConnectThread(BluetoothDevice device) {
        mmDevice = device;
        BluetoothSocket tmp = null;
        try {
            //android与其他蓝牙模块连接时，需要输入UUID
            tmp = device.createRfcommSocketToServiceRecord(MY_UUID);   //
        }
        catch (IOException e) {
            //
        }
        mmSocket = tmp;
    }
    public void run() {
        setName("ConnectThread");
        mAdapter.cancelDiscovery();
        try {
            mmSocket.connect();
        }
        catch (IOException e) {
            connectionFailed();
            try {
                mmSocket.close();
            }
            catch (IOException e2) {
                //
            }
            ChatService.this.start();
            return;
        }
        synchronized (ChatService.this) {
            mConnectThread = null;
        }
        connected(mmSocket, mmDevice);
```

```
        }
        public void cancel() {
            try {
                mmSocket.close();
            }
            catch (IOException e) {
                //
            }
        }
    }
    // 双方蓝牙连接后一直运行的线程
    private class ConnectedThread extends Thread {
        private final BluetoothSocket mmSocket;
        private final InputStream mmInStream;
        private final OutputStream mmOutStream;
        //构造方法
        public ConnectedThread(BluetoothSocket socket) {
            mmSocket = socket;
            InputStream tmpIn = null;
            OutputStream tmpOut = null;
            try {
                tmpIn = socket.getInputStream();
                tmpOut = socket.getOutputStream();
            }
            catch (IOException e) {
                //
            }
            mmInStream = tmpIn;
            mmOutStream = tmpOut;
        }
        // 使用阻塞模式的 InputStream.read() 循环读取输入流
        public void run() {   //
            byte[] buffer = new byte[1024];
            int bytes;
            while (true) {
                try {
                    bytes = mmInStream.read(buffer);
                    mHandler.obtainMessage(BluetoothChat.MESSAGE_READ,
                                                   bytes,-1, buffer).sendToTarget();
```

```
                }
                catch (IOException e) {
                    connectionLost();
                    break;
                }
            }
        }
        //将聊天消息写入输出流传输至对方
        public void write(byte[] buffer) {
            try {
                mmOutStream.write(buffer);
                mHandler.obtainMessage(BluetoothChat.MESSAGE_WRITE,
                                            -1, -1,buffer).sendToTarget();
            }
            catch (IOException e) {
                //
            }
        }
        //关闭连接的 Socket
        public void cancel() {
            try {
                mmSocket.close();
            }
            catch (IOException e) {
                //
            }
        }
    }
}
```

(7) 编写蓝牙会话的主 Activity 程序 BluetoothChat.java，其文件代码如下：

```
package com.example.bluetoothchat;
/*
 * 本程序演示了使用手机蓝牙实现会话功能
 * 通过定义菜单来调用另一个Activity: DeviceList
 * 调用了使用BluetoothSocket开发的蓝牙服务程序(包含了Java多线程): ChatService
 */
import android.app.Activity;
import android.bluetooth.BluetoothAdapter;
```

```
import android.bluetooth.BluetoothDevice;
import android.content.Intent;
import android.os.Bundle;
import android.os.Handler;
import android.os.Message;
import android.view.KeyEvent;
import android.view.Menu;
import android.view.MenuInflater;
import android.view.MenuItem;
import android.view.View;
import android.view.View.OnClickListener;
import android.view.Window;
import android.view.inputmethod.EditorInfo;
import android.widget.ArrayAdapter;
import android.widget.Button;
import android.widget.EditText;
import android.widget.ListView;
import android.widget.TextView;
import android.widget.Toast;
import com.example.bluetoothchat.R;
public class BluetoothChat extends Activity {
    public static final int MESSAGE_STATE_CHANGE = 1;
    public static final int MESSAGE_READ = 2;
    public static final int MESSAGE_WRITE = 3;
    public static final int MESSAGE_DEVICE_NAME = 4;
    public static final int MESSAGE_TOAST = 5;
    public static final String DEVICE_NAME = "device_name";   //
    public static final String TOAST = "toast";
    private static final int REQUEST_CONNECT_DEVICE = 1;
    private static final int REQUEST_ENABLE_BT = 2;
    private TextView mTitle;
    private ListView mConversationView;
    private EditText mOutEditText;
    private Button mSendButton;
    private String mConnectedDeviceName = null;
    private ArrayAdapter<String> mConversationArrayAdapter;
    private StringBuffer mOutStringBuffer;
    private BluetoothAdapter mBluetoothAdapter = null;
    private ChatService mChatService = null;
```

```
@Override
public void onCreate(Bundle savedInstanceState) {
    super.onCreate(savedInstanceState);
    // 设置窗口布局为自定义标题
    requestWindowFeature(Window.FEATURE_CUSTOM_TITLE);
    setContentView(R.layout.activity_main);
    // 设置窗口标题布局文件
    getWindow().setFeatureInt(Window.FEATURE_CUSTOM_TITLE,
            R.layout.custom_title);   //标题布局栏

    mTitle = (TextView) findViewById(R.id.title_left_text);
    mTitle.setText(R.string.app_name);
    mTitle = (TextView) findViewById(R.id.title_right_text);
    // 得到本地蓝牙适配器
    mBluetoothAdapter = BluetoothAdapter.getDefaultAdapter();
    // 若当前设备不支持蓝牙功能
    if (mBluetoothAdapter == null) {
        Toast.makeText(this, "蓝牙不可用", Toast.LENGTH_LONG).show();
        finish();
        return;
    }
}
@Override
public void onStart() {
    super.onStart();
    if (!mBluetoothAdapter.isEnabled()) {
        // 若当前设备蓝牙功能未开启，则开启蓝牙功能
        Intent enableIntent = new Intent(
                BluetoothAdapter.ACTION_REQUEST_ENABLE);
        startActivityForResult(enableIntent, REQUEST_ENABLE_BT);
    }
    else {
        if (mChatService == null)
            setupChat();
    }
}
@Override
public synchronized void onResume() {
    super.onResume();
```

```
        if (mChatService != null) {
            if (mChatService.getState() == ChatService.STATE_NONE) {
                mChatService.start();
            }
        }
    }
    private void setupChat() {    //设置会话
        mConversationArrayAdapter = new ArrayAdapter<String>(this,
                                                R.layout.message);
        mConversationView = (ListView) findViewById(R.id.in);
        mConversationView.setAdapter(mConversationArrayAdapter);
        mOutEditText = (EditText) findViewById(R.id.edit_text_out);
        mOutEditText.setOnEditorActionListener(mWriteListener);
        mSendButton = (Button) findViewById(R.id.button_send);
        mSendButton.setOnClickListener(new OnClickListener() {
            public void onClick(View v) {
                TextView view = (TextView) findViewById(R.id.edit_text_out);
                String message = view.getText().toString();
                sendMessage(message);
            }
        });
        mChatService = new ChatService(this, mHandler);    //创建服务对象
        mOutStringBuffer = new StringBuffer("");
    }
    @Override
    public synchronized void onPause() {
        super.onPause();
    }
    @Override
    public void onStop() {
        super.onStop();
    }
    @Override
    public void onDestroy() {
        super.onDestroy();
        if (mChatService != null)
            mChatService.stop();
    }
    private void ensureDiscoverable() {
```

```
        if (mBluetoothAdapter.getScanMode() != BluetoothAdapter.
                                SCAN_MODE_CONNECTABLE_DISCOVERABLE) {
            Intent discoverableIntent = new Intent(
                    BluetoothAdapter.ACTION_REQUEST_DISCOVERABLE);
            discoverableIntent.putExtra(
                    BluetoothAdapter.EXTRA_DISCOVERABLE_DURATION, 300);
            startActivity(discoverableIntent);
        }
    }
    private void sendMessage(String message) {
        if (mChatService.getState() != ChatService.STATE_CONNECTED) {
            Toast.makeText(this, R.string.not_connected,
                                            Toast.LENGTH_SHORT).show();
            return;
        }
        if (message.length() > 0) {
            byte[] send = message.getBytes();
            mChatService.write(send);
            mOutStringBuffer.setLength(0);
            mOutEditText.setText(mOutStringBuffer);
        }
    }
    private TextView.OnEditorActionListener mWriteListener =
                                    new TextView.OnEditorActionListener() {
        public boolean onEditorAction(TextView view, int actionId,
                                                    KeyEvent event) {
            if (actionId == EditorInfo.IME_NULL
                    && event.getAction() == KeyEvent.ACTION_UP) {
                String mcssage = view.getText().toString();
                sendMessage(message);
            }
            return true;
        }
    };
    private final Handler mHandler = new Handler() {    //避免线程冲突
        @Override
        public void handleMessage(Message msg) {        //消息处理
            switch (msg.what) {
            case MESSAGE_STATE_CHANGE:
```

```
        switch (msg.arg1) {
        case ChatService.STATE_CONNECTED:
        //显示连接的蓝牙设备
            mTitle.setText(R.string.title_connected_to);
            mTitle.append(mConnectedDeviceName);
            mConversationArrayAdapter.clear();
            break;
        case ChatService.STATE_CONNECTING:
            mTitle.setText(R.string.title_connecting);
            break;
        case ChatService.STATE_LISTEN:
        case ChatService.STATE_NONE:
            mTitle.setText(R.string.title_not_connected);
            break;
        }
        break;
    case MESSAGE_WRITE:
        byte[] writeBuf = (byte[]) msg.obj;
        String writeMessage = new String(writeBuf);
        mConversationArrayAdapter.add("我：  " + writeMessage);
        break;
    case MESSAGE_READ:
        byte[] readBuf = (byte[]) msg.obj;
        String readMessage = new String(readBuf, 0, msg.arg1);
        mConversationArrayAdapter.add(mConnectedDeviceName + ":  "
                                                        + readMessage);
        break;
    case MESSAGE_DEVICE_NAME:
        //连接到的蓝牙设备
        mConnectedDeviceName = msg.getData().getString(DEVICE_NAME);
        Toast.makeText(getApplicationContext(),"链接到 " +
                    mConnectedDeviceName, Toast.LENGTH_SHORT).show();
        break;
    case MESSAGE_TOAST:
        Toast.makeText(getApplicationContext(),
                msg.getData().getString(TOAST), Toast.LENGTH_SHORT)
                .show();
        break;
    }
```

```
        }
    };
    public void onActivityResult(int requestCode, int resultCode, Intent data) {
        switch (requestCode) {
        case REQUEST_CONNECT_DEVICE:
            if (resultCode == Activity.RESULT_OK) {
                String address = data.getExtras().getString(
                        DeviceList.EXTRA_DEVICE_ADDRESS);
                BluetoothDevice device = mBluetoothAdapter
                        .getRemoteDevice(address);
                mChatService.connect(device);
            }
            break;
        case REQUEST_ENABLE_BT:
            if (resultCode == Activity.RESULT_OK) {
                setupChat();
            } else {
                Toast.makeText(this, R.string.bt_not_enabled_leaving,
                        Toast.LENGTH_SHORT).show();
                finish();
            }
        }
    }
    @Override
    public boolean onCreateOptionsMenu(Menu menu) {
        MenuInflater inflater = getMenuInflater();
        inflater.inflate(R.menu.option_menu, menu);   //创建菜单
        return true;
    }
    @Override
    public boolean onOptionsItemSelected(MenuItem item) {   //选项后处理
        switch (item.getItemId()) {
        case R.id.scan:
            //Activity的显式调用
            Intent serverIntent = new Intent(this, DeviceList.class);   //
            startActivityForResult(serverIntent, REQUEST_CONNECT_DEVICE);
            return true;
        case R.id.discoverable:
            ensureDiscoverable();
```

```
                return true;
            case R.id.back:
                finish();
                System.exit(0);
                return true;
            }
            return false;
        }
}
```

9.3　近场通信 NFC

9.3.1　NFC 简介

NFC 是英文 near field communication 的缩写，是一种近距离、非接触式识别的无线通信技术，由飞利浦公司和索尼公司共同开发，通常有效通信距离在 4 厘米以内，通信速率为 106～848 Kbit/s，工作频率为 13.65 MHz。

通过 NFC 技术，可以实现 Android 设备与 NFC Tag(target 的简写)或其他 Android 设备之间小批量数据的传输。NFC 可以在移动设备、消费类电子产品、PC 和智能控件工具间进行近距离无线通信。

NFC 通信总是由一个发起者(initiator)和一个被动式接受者(passive target)组成。通常发起者主动发送电磁场可以为被动式接受者提供电源。因此,被动式接受者(简称 Tag) 可以有非常简单的形式,比如标签、芯片卡和纽扣等形式。

具有 NFC 功能的手机与 Tag 如图 9.3.1 所示。

图 9.3.1　具有 NFC 功能的手机与 Tag

NFC 芯片作为组成 RFID 模块的一部分装在手机上，手机背后有一块 NFC 的区域，如图 9.3.2 所示。

图 9.3.2　手机背后的 NFC 区域

在 Android NFC 应用中，Android 手机通常是通信中的发起者和 NFC 的读写器。Android 手机也可以模拟 NFC 通信的接收者，从而实现点到点的通信(P2P)。

和其他无线通信方式(如 Bluetooth)相比，NFC 支持的通信带宽和距离要小得多，但是它成本低，如价格标签可能只有几分钱，也不需要搜寻设备、配对等，通信在靠近的瞬间完成。

注意：

(1) 并不是所有手机都具有 NFC 功能，只有高端手机(如三星 S4)才支持，如图 9.3.3 所示。

图 9.3.3　三星 S4 手机的 NFC 开关设置

(2) 金融 IC 卡，就是我们常说的芯片卡，由于卡与读写器双向认证，因此芯片卡复制的难度极高，具备很强的抗攻击能力。此外，金融 IC 卡的稳定性也比磁条卡更好，不会出现消磁的情况。

9.3.2 Android 对 NFC 的支持

Android 对 NFC 的支持主要在 android.nfc 和 android.nfc.tech 两个包中，如图 9.3.4 所示。

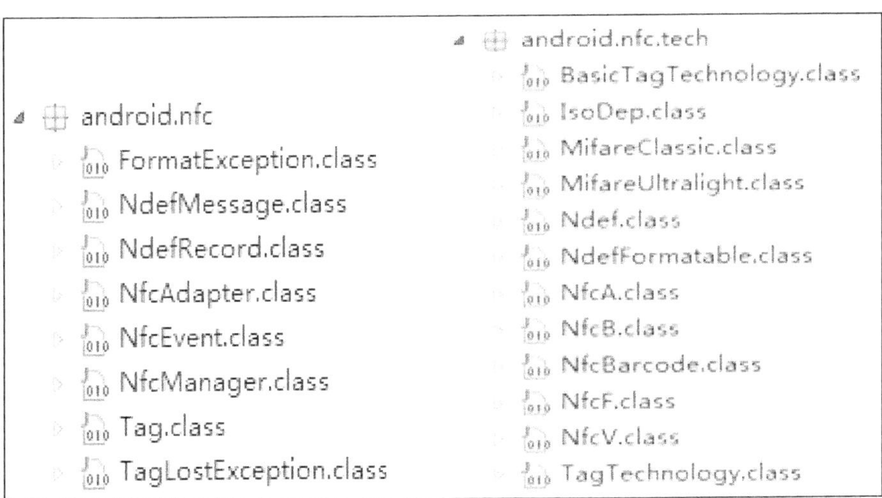

图 9.3.4　Android 提供的 NFC 软件包

在手机的 NFC 应用中，需要在清单文件中添加 NFC 权限及设备特征，其代码如下：

```
<uses-permission android:name="android.permission.NFC" />
<uses-feature    android:name="android.hardware.nfc" android:required="true" />
```

注意： Android Market 会根据 uses-feature 过滤所有手机设备不支持的应用。通过使用<uses-feature>元素，一个应用可以指定它所支持的硬件型号。

NfcManager 类可以用来管理 Android 设备中指出的所有 NFC Adapter，但由于大部分 Android 设备只支持一个 NFC Adapter，可使用该类提供的 getDefaultAdapter() 来获取系统支持的 NFC Adapter。

注意： NfcAdapter 类也提供了静态方法 getDefaultAdapter()来获取 NfcAdapter 对象。

当 Android 设备检测到一个 Tag 时，会创建一个 Tag 对象，将其放在 Intent 对象，然后发送到相应的 Activity。

NfcAdapter 类可以用来定义一个 Intent，在系统检测到 NFC Tag 时通知事先定义的 Activity,以实现对 Tag 设备的读写操作。

类 NdefMessage 和 NdefRecord 是 NFC forum 定义的数据格式。

在包 android.nfc.tech 中，定义了可以对 Tag 设备进行读写操作的类。

9.3.3 一个 NFC 应用实例：读写 Tag 标签

下面介绍一个读写 Tag 标签的实例。

【例 9.3.1】使用手机 NFC 功能，读写 Tag 标签。

一个对 NFC Tag 设备进行读写的工程 TestNFC，其工程文件结构(部分)如图 9.3.5 所示。

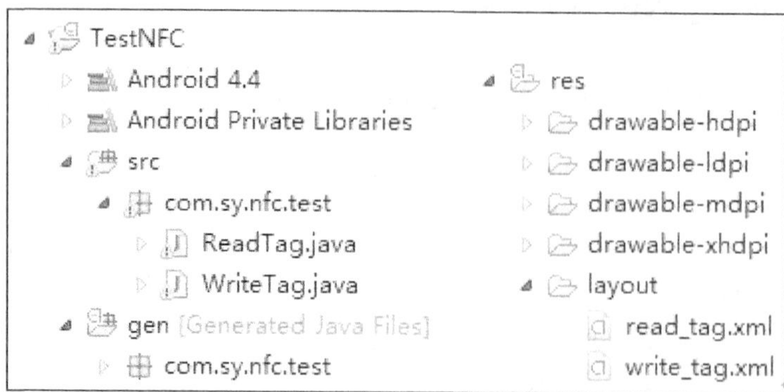

图 9.3.5　TestNFC 工程文件结构(部分)

图 9.3.6　主 Activity 运行效果

其中，ReadTag.java 是主 Activity，实现读取 Tag 设备的信息；WriteTag.java 是一个供主 Activity 调用的 Activity，实现对 Tag 设备的写入。

程序运行后，主 Activity 的界面效果如图 9.3.6 所示。

将 NFC Tag(如芯片卡)贴近手机的 NFC 区域，会在主窗口中显示读取的芯片信息。单击主 Activity 里的“写入标签”按钮后，进入写入 Activity 界面，提示输入要写入 Tag 标签的内容，单击“写入标签”按钮后，提示需要将 Tag 标签靠近手机的 NFC 区域(非接触式)，操作界面效果如图 9.3.7 所示。

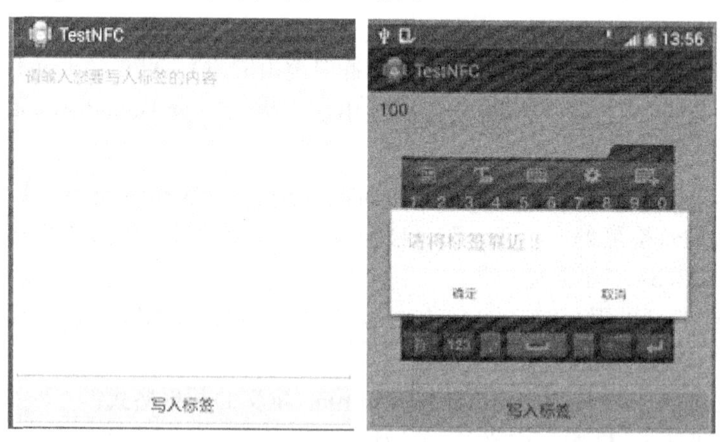

图 9.3.7　写 Tag 的 Activity

ReadTag.java 的文件代码如下：

```
package com.sy.nfc.test;
/*
 * 本程序在手机上运行时，需要手机具有NFC功能
```

```
*/
import android.app.Activity;
import android.os.Bundle;
import android.nfc.NfcAdapter;      //主类
import android.nfc.NdefMessage;
import android.nfc.NdefRecord;
import android.nfc.tech.NfcA;
import android.nfc.tech.MifareClassic;
import android.content.Intent;
import android.app.PendingIntent;
import android.content.IntentFilter;
import android.view.View;
import android.view.View.OnClickListener;
import android.widget.Button;
import android.widget.TextView;
import android.widget.Toast;
import android.os.Parcelable;
import android.annotation.SuppressLint;
@SuppressLint("NewApi")
  public class ReadTag extends Activity {
    private NfcAdapter nfcAdapter;
    private TextView resultText;
    private PendingIntent pendingIntent;
    private IntentFilter[] mFilters;
    private String[][] mTechLists;
    private Button mJumpTagBtn;
    private boolean isFirst = true;
    @SuppressLint("NewApi")
    @Override
    protected void onCreate(Bundle savedInstanceState) {
        super.onCreate(savedInstanceState);
        // 获取NFC适配器，判断设备是否支持NFC功能
        nfcAdapter = NfcAdapter.getDefaultAdapter(this);
        if (nfcAdapter == null) {
            Toast.makeText(this, getResources().getString(R.string.no_nfc),
                                                    Toast.LENGTH_SHORT).show();
            finish();
            return;
        }
```

```
        else if (!nfcAdapter.isEnabled()) {
            Toast.makeText(this, getResources().getString(R.string.open_nfc),
                                                Toast.LENGTH_SHORT).show();
            finish();
            return;
        }

        setContentView(R.layout.read_tag);
        resultText = (TextView) findViewById(R.id.resultText); // 显示结果
        mJumpTagBtn = (Button) findViewById(R.id.jump); // 写入标签按钮
        mJumpTagBtn.setOnClickListener(new WriteBtnOnClick());

        //创建一个PendingIntent对象，以便Android系统能够在扫描到NFC标签时
        //用它来封装NFC标签的详细信息
        pendingIntent = PendingIntent.getActivity(this, 0, new Intent(this,
                getClass()).addFlags(Intent.FLAG_ACTIVITY_SINGLE_TOP), 0);
        // 做一个IntentFilter，过滤你想要的action
        IntentFilter ndef = new IntentFilter(NfcAdapter.ACTION_TECH_DISCOVERED);
        ndef.addCategory("*/*");
        mFilters = new IntentFilter[] { ndef };      // 过滤器
        //如果Android设备支持MIFARE，提供对MIFARE Classic目标的属性和I/O操作
        //允许扫描的标签类型
        mTechLists = new String[][] {
                new String[] { MifareClassic.class.getName() },
                new String[] { NfcA.class.getName() } };
    }

    @SuppressLint("NewApi")
    @Override
    protected void onResume() {
        // TODO Auto-generated method stub
        super.onResume();
        //设定intentfilter和techlist。
        //如果两个都为null就代表优先接收任何形式的TAG action
        //也就是说，系统会主动发TAG intent
        nfcAdapter.enableForegroundDispatch(this, pendingIntent, mFilters,mTechLists);
        if (isFirst) {
            if (NfcAdapter.ACTION_TECH_DISCOVERED.equals(getIntent().getAction())) {
                String result = processIntent(getIntent());
                resultText.setText(result);
```

```
            }
            isFirst = false;
        }
    }
    //实现onNewIntent回调方法来处理扫描到的NFC标签的数据
    @Override
    protected void onNewIntent(Intent intent) {
        // TODO Auto-generated method stub
        super.onNewIntent(intent);
        if (NfcAdapter.ACTION_TECH_DISCOVERED.equals(intent.getAction())) {
            String result = processIntent(intent);
            resultText.setText(result);
        }
    }
    //获取标签中的内容
    @SuppressLint("NewApi")
    private String processIntent(Intent intent) {
        Parcelable[] rawmsgs = intent.getParcelableArrayExtra(
                                        NfcAdapter.EXTRA_NDEF_MESSAGES);
        NdefMessage msg = (NdefMessage) rawmsgs[0];
        NdefRecord[] records = msg.getRecords();
        String resultStr = new String(records[0].getPayload());
        return resultStr;
    }
    //写入按钮单击事件
    class WriteBtnOnClick implements OnClickListener {
        @Override
        public void onClick(View v) {
            // TODO Auto-generated method stub
            switch (v.getId()) {
            case R.id.jump:
                Intent intent = new Intent(ReadTag.this, WriteTag.class);
                startActivity(intent);
            default:
                break;
            }
        }
    }
}
```

WriteTag.java 的文件代码如下：

```
package com.sy.nfc.test;
import android.app.Activity;
import android.os.Bundle;
import android.nfc.NfcAdapter;   //主类
import android.nfc.NdefMessage;
import android.nfc.NdefRecord;
import android.nfc.Tag;   //
import android.nfc.tech.MifareClassic;
import android.nfc.tech.Ndef;
import android.nfc.tech.NdefFormatable;
import android.nfc.tech.NfcA;
import android.app.AlertDialog;
import android.app.PendingIntent;   //
import android.content.DialogInterface;
import android.content.Intent;
import android.content.IntentFilter;
import android.view.View;
import android.view.View.OnClickListener;
import android.widget.Button;
import android.widget.EditText;
import android.widget.Toast;
import java.io.IOException;
import android.annotation.SuppressLint;
//写入标签
@SuppressLint("NewApi")
public class WriteTag extends Activity {
    private IntentFilter[] mWriteTagFilters;
    private NfcAdapter nfcAdapter;
    PendingIntent pendingIntent;
    String[][] mTechLists;
    Button writeBtn;
    boolean isWrite = false;
    EditText mContentEditText;
    @Override
    protected void onCreate(Bundle savedInstanceState) {
        // TODO Auto-generated method stub
        super.onCreate(savedInstanceState);
```

```
        setContentView(R.layout.write_tag);
        writeBtn = (Button) findViewById(R.id.writeBtn);
        writeBtn.setOnClickListener(new WriteOnClick());
        mContentEditText = (EditText) findViewById(R.id.content_edit);
        // 获取NFC适配器，判断设备是否支持NFC功能
        nfcAdapter = NfcAdapter.getDefaultAdapter(this);
        if (nfcAdapter == null) {
            Toast.makeText(this, getResources().getString(R.string.no_nfc),
                    Toast.LENGTH_SHORT).show();
            finish();
            return;
        }
        else if (!nfcAdapter.isEnabled()) {
            Toast.makeText(this, getResources().getString(R.string.open_nfc),
                                                        Toast.LENGTH_SHORT).show();
            finish();
            return;
        }
        pendingIntent = PendingIntent.getActivity(this, 0, new Intent(this,
                getClass()).addFlags(Intent.FLAG_ACTIVITY_SINGLE_TOP), 0);
        // 写入标签权限
        IntentFilter writeFilter = new IntentFilter(
                                    NfcAdapter.ACTION_TECH_DISCOVERED);
        mWriteTagFilters = new IntentFilter[] { writeFilter };
        mTechLists = new String[][] {
                new String[] { MifareClassic.class.getName() },
                new String[] { NfcA.class.getName() } };// 允许扫描的标签类型
    }
    class WriteOnClick implements OnClickListener {
        @Override
        public void onClick(View v) {
            // TODO Auto-generated method stub
            isWrite = true;
            AlertDialog.Builder builder = new AlertDialog.Builder(
                                        WriteTag.this).setTitle("请将标签靠近！");
            builder.setNegativeButton("确定",
                    new DialogInterface.OnClickListener() {
                        @Override
                        public void onClick(DialogInterface dialog, int which) {
```

```
                        // TODO Auto-generated method stub
                        dialog.dismiss();
                        mContentEditText.setText("");
                        isWrite = false;
                        WriteTag.this.finish();
                    }
                });
        builder.setPositiveButton("取消",
                new DialogInterface.OnClickListener() {
                    @Override
                    public void onClick(DialogInterface dialog, int which) {
                        // TODO Auto-generated method stub
                        dialog.dismiss();
                        isWrite = false;
                    }
                });
        builder.create();
        builder.show();
    }
}

@Override
protected void onResume() {
    // TODO Auto-generated method stub
    super.onResume();
    nfcAdapter.enableForegroundDispatch(this, pendingIntent,
                                        mWriteTagFilters, mTechLists);
}

// 写入模式时，才执行写入操作
@Override
protected void onNewIntent(Intent intent) {
    // TODO Auto-generated method stub
    super.onNewIntent(intent);
    if (isWrite == true     && NfcAdapter.ACTION_TECH_DISCOVERED
                                        .equals(intent.getAction())) {
        Tag tag = intent.getParcelableExtra(NfcAdapter.EXTRA_TAG);
        NdefMessage ndefMessage = getNoteAsNdef();
        if (ndefMessage != null) {
```

```
                writeTag(getNoteAsNdef(), tag);
            } else {
                showToast("请输入您要写入标签的内容");
            }
        }
    }
    // 根据文本生成一个NdefRecord
    private NdefMessage getNoteAsNdef() {
        String text = mContentEditText.getText().toString();
        if (text.equals("")) {
            return null;
        } else {
            byte[] textBytes = text.getBytes();
            NdefRecord textRecord = new NdefRecord(NdefRecord.TNF_MIME_MEDIA,
                    "image/jpeg".getBytes(), new byte[] {}, textBytes);
            return new NdefMessage(new NdefRecord[] { textRecord });
        }
    }
    // 写入Tag标签
    boolean writeTag(NdefMessage message, Tag tag) {
        int size = message.toByteArray().length;
        try {
            Ndef ndef = Ndef.get(tag);
            if (ndef != null) {
                ndef.connect();
                if (!ndef.isWritable()) {
                    showToast("Tag不允许写入");
                    return false;
                }
                if (ndef.getMaxSize() < size) {
                    showToast("文件大小超出容量");
                    return false;
                }
                ndef.writeNdefMessage(message);
                showToast("写入数据成功.");
                return true;
            } else {
                NdefFormatable format = NdefFormatable.get(tag);
                if (format != null) {
```

```
                    try {
                        format.connect();
                        format.format(message);
                        showToast("格式化tag并且写入message");
                        return true;
                    } catch (IOException e) {
                        showToast("格式化tag失败.");
                        return false;
                    }
                } else {
                    showToast("Tag不支持NDEF");
                    return false;
                }
            }
        } catch (Exception e) {
            showToast("写入数据失败");
        }
        return false;
    }
    private void showToast(String text) {
        Toast.makeText(this, text, Toast.LENGTH_SHORT).show();
    }
}
```

习 题 9

一、判断题

1. WiFi 和 Bluetooth 是手机的基本配置。
2. 如果手机配置中有 NFC，则在“设置”程序中可以找到。
3. 笔记本计算机共享手机的移动数据连接，只能使用 WiFi 连接方式。
4. 蓝牙通信使用 HTTP 协议。
5. Android 不支持蓝牙功能。
6. 目前的 Android 手机都具有 NFC 功能。
7. 蓝牙聊天程序设计必须使用多线程。

二、选择题

1. 为了获取手机连接 WiFi 使用的物理地址，应使用 WifiInfo 类的____方法。

   A. getMacAddress()　　B. getSSID()

   C. getBSSID()　　D. getIpAddress()

2. 获取已经配对过的蓝牙设备列表，需要使用____类提供的方法 getBondedDevice()。

   A. Bluetooth　　B. BluetoothManager

   C. BluetoothAdapter　　D. BluetoothDevice

3. 下列选项中，涉及 Internet 网络的是____。

   A. WiFi　　B. GPS　　C. Bluetooth　　D. NFC

4. 下列不属于 android.bluetooth 包的选项是____。

   A. BluetoothAdapter　　B. BluetoothSocket

   C. BluetoothServerSocket　　D.UUID

三、填空题

1. WiFi 状态共有____种。
2. 在得到所有 WiFi 信息列表前，必须先使用 WifiManager 的____方法。
3. 为了得到接收到的 WiFi 信息列表，需要先使用 WifiManager 的 startScan()方法，然后再使用____方法。
4. 获取手机当前连接的 WiFi 信源的相关信息，需要使用 WifiManager 的____方法。
5. 为了得到本地蓝牙适配器，需要调用 BluetoothAdapter 类的静态方法____。
6. 手机蓝牙会话时，数据传递的形式是____。

实验 9　Android 近距离通信技术及其应用

一、实验目的

1. 掌握手机 WiFi 的使用和 Android 提供的 WiFi 支持。
2. 掌握手机蓝牙的使用和 Android 提供的蓝牙支持。
3. 了解 Android 对手机 NFC 的支持。

二、实验内容及步骤

【预备】访问 http://www.wustwzx.com/android/index.html，单击“实验 9”超链接，下载本次实验内容的源代码并解压(得到文件夹 ch09)，供研读和调试使用。

1. 掌握手机 WiFi 的使用和 Android 提供的 WiFi 支持。

(1) 运行手机自带的设置程序，熟悉 WiFi 的打开、关闭、扫描、选择等操作。

(2) 打开手机的移动数据连接，运行手机自带的“网络分享和便携式热点”程序(三星手机在设置程序的“更多”选项里，联想手机是名为“个人热点”的应用程序)，选择 WiFi 连接方式，共享手机的移动数据连接。

(3) 再次解压 ch09 文件夹的压缩文件 WifiDemo.zip，导入工程 WifiDemo。

(4) 打开主 Activity 源文件，查看获取 WifiManager 对象的代码。

(5) 查看扫描 WiFi 信源类方法及保存信源类方法。

(6) 部署工程到手机上运行测试。

2. 掌握手机蓝牙的使用和 Android 提供的蓝牙支持。

(1) 运行手机自带的设置程序，熟悉蓝牙的打开、关闭、扫描、配对、传送文件等操作。

(2) 再次解压 ch09 内的压缩文件 BluetoothChat.zip，导入工程 BluetoothChat。

(3) 查看清单文件中注册使用 Buletooth 的权限。

(4) 查看主窗体文件 BuletoothChat.java 中检查蓝牙是否可用及打开蓝牙的方法。

(5) 查看菜单事件里显式调用 Activity 程序 DeviceList，用以选取与之会话的蓝牙设备。

(6) 查看蓝牙会话服务程序 ChatService.java 的实现代码。

(7) 分别部署工程到两部手机上运行，对其中一部手机使用菜单键选择与之会话的另一部手机，建立好连接后做会话测试。

3. 了解 Android 对手机 NFC 的支持。

(1) 再次解压 ch09 文件夹的压缩文件 TestNFC.zip，导入工程 TestNFC。

(2) 打开主 Activity 源文件 ReadTag.java，查看检测设备是否支持 NFC 的代码。

(3) 查看写入 Tag 的 Activity 文件 WriteTag.java 中延期意图的使用。

(4) 部署工程到手机上运行测试(如果手机有 NFC 功能的话)。

三、实验小结及思考

(由学生填写，重点写上机中遇到的问题。)

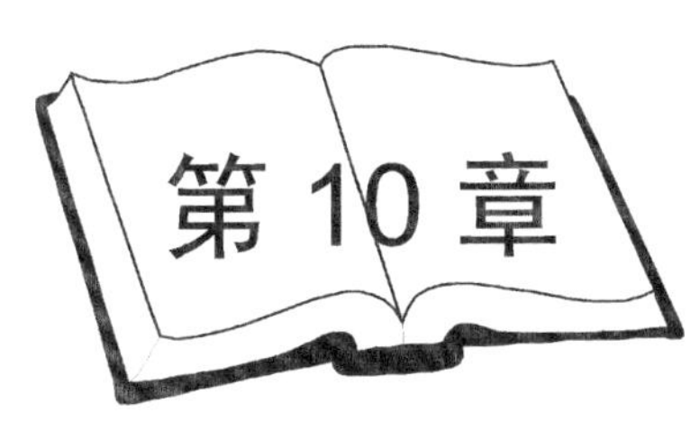

第10章 位置服务与地图应用开发

位置服务(location based services，简称 LBS)又称定位服务，是指通过 GPS 卫星或者网络(GPRS 或 WiFi)获取各种终端的地理坐标(经度和纬度)，在电子地图平台的支撑下，为用户提供基于位置导航、查询的一种信息服务。位置服务是 Android 设备的一个重要功能，本章学习要点如下：

- Android 提供的 GPS 定位的系统服务及相关类；
- Android 提供的 WiFi 定位的系统服务及相关类；
- 百度提供定位服务的相关类及其方法；
- 百度地图应用。

10.1 位置服务概述

10.1.1 基于位置的服务 LBS

位置服务是移动设备的一个重要功能，Android 提供了基于位置的服务 LBS，共有如下三种定位方式。

1. GPS 定位

GPS 是全球定位系统(global positioning system)的英文缩写，是 20 世纪 70 年代由美国陆海空三军联合研制的新一代空间卫星导航系统。基于 GPS 的定位方式是利用手机上的 GPS 定位模块将自己的位置信号发送到定位后台来实现手机定位的。

手机 GPS 定位是基于手机 GPS 定位模块(芯片)与空中的 GPS 卫星之间的通信实现的。由于卫星的位置精确，在 GPS 观测中，我们可以得到卫星到接收机之间的距离，然后利用三维坐标中的距离公式和 3 颗卫星就可以组成 3 个方程式，解出观测点的位置坐标(x, y, z)。考虑到卫星的时钟与接收机的时钟之间的误差，因此，至少需要 4 颗卫星信号才能进行 GPS 定位。

注意：

(1) 相对于下面介绍的两种网络定位方式(WiFi 定位和基站定位)，GPS 定位具有较高的精度。

(2) GPS 定位方式仅在户外有效，因为在室内或在地下通道内，手机 GPS 信号不正常。

2. WiFi 定位

WiFi 是 wireless fidelity 的英文缩写，表示无线相容认证。

WiFi 能够对用户进行定位，因为在 Android、iOS 和 Windows Phone 这些手机操作系统中内置了位置服务。由于每一个 WiFi 热点都有一个独一无二的 Mac 地址(手机 WiFi 网卡的 Mac 地址)，智能手机开启 WiFi 后就会自动扫描附近热点并上传其位置信息，这样就建立了一个庞大的热点位置数据库，这个数据库是对用户进行定位的关键。

注意：

(1) WiFi 定位和 GPRS 定位，统称为网络定位。

(2) 网络定位响应速度快、费电少，但精度没有 GPS 定位高。

3. 基站定位

基站定位是通过移动数据中的 GPRS (general packet radio service，通用分组无线服务技术)网络，利用基站对手机的距离的测算来确定手机位置的。因此，基站定位也称 GPRS 定位。

GPRS 是一种基于 GSM 系统的无线分组交换技术，提供了端到端的、广域的无线 IP 连接。基站定位服务一般应用于手机用户，手机基站定位服务是通过网络运营商(主要有中国电信、中国移动和中国联通)的移动网络获取移动终端用户的位置信息(经纬度坐标)，在电子地图平台的支持下，为用户提供相应服务的一种增值业务。

注意：使用 GPRS 网络定位的 Android 应用程序所需要的权限与 WiFi 定位所需要的权限相同。

10.1.2 Android API 提供的位置包

手机定位是指能得到手机持有人所在地理位置的经纬度。Android 系统提供了用于计算地理数据的相关软件包(类)。在 Android API 的位置包 android.location 里，主要有位置类 Location、位置管理器类 LocationManager、位置提供者类 LocationProvider 和位置监听器接口 LocationListener 等，如图 10.1.1 所示。

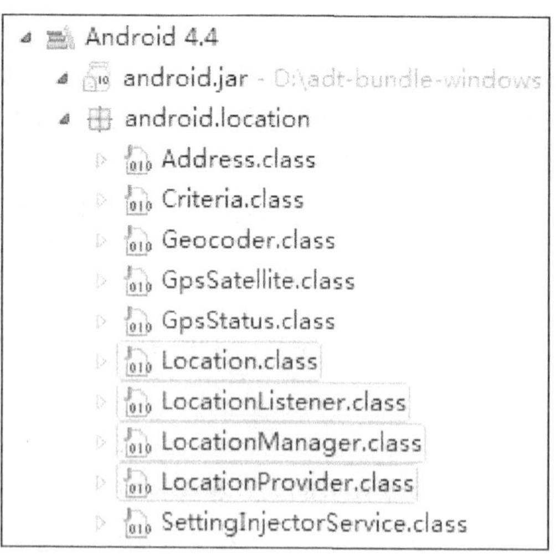

图 10.1.1　Android 标准 API 中的 location 包

为了实现定位信息的实时显示，需要使用位置管理器类 LocationManager 和位置监听器接口 LocationListener。

要让 Android 应用程序跟踪手机位置的变化，需要调用 LocationManager 实例对象的 requestLocationUpdates()方法，这个方法需要四个参数：

- 位置提供者的类型；
- 多久更新一次位置，以毫秒为单位；
- 位置变化至少多少才更新一次位置，以米为单位；
- 用于处理位置变化的监听器对象。

其中，位置监听器接口 LocationListener 需要实现的几个方法，如图 10.1.2 所示。

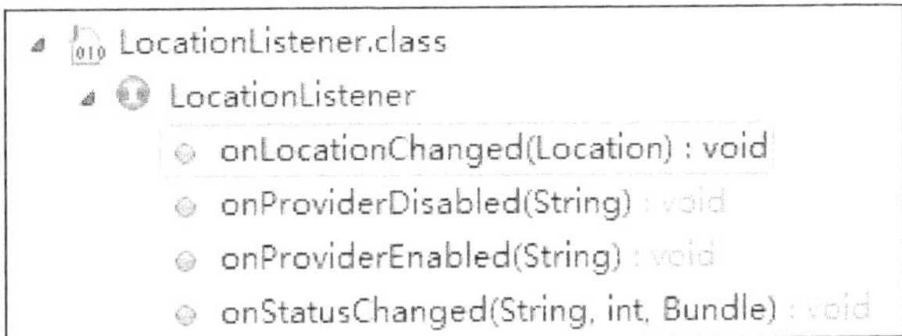

图 10.1.2　Android 位置监听接口

在 Android 程序中，需要实现接口 LocationListener 的相关方法如下：

```
@Override
public void onLocationChanged(Location arg0) {
    // TODO Auto-generated method stub
}
@Override
public void onProviderDisabled(String arg0) {
    // TODO Auto-generated method stub
}
@Override
public void onProviderEnabled(String arg0) {
    // TODO Auto-generated method stub
}
@Override
public void onStatusChanged(String arg0, int arg1, Bundle arg2) {
    // TODO Auto-generated method stub
}
```

10.1.3　Google Map APIs 与 Baidu Map API

1. Google Map APIs

Google Map APIs 是指在标准 Android API 的基础上增加了用于开发地图应用的 Map

API。Google Map APIs 定义了一系列用于在 Google Map 上显示、控制和层叠信息的功能类。

Google Map APIs 不属于标准 Android SDK 的标准库组件，需要单独下载，下载方法是使用 Android SDK Manager，如图 10.1.3 所示。

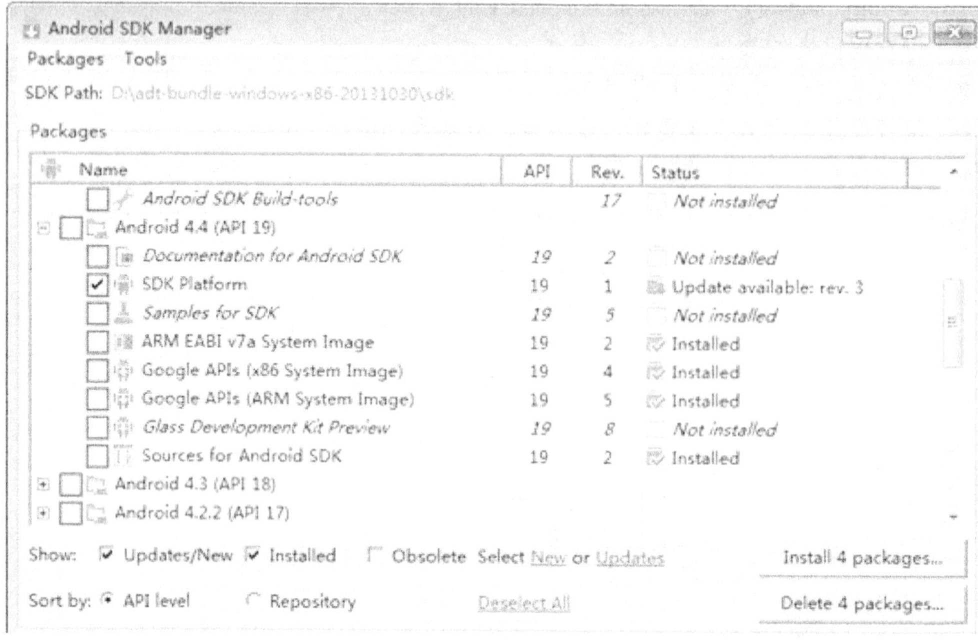

图 10.1.3 Android SDK Manager

新建地图应用项目时，Google Map APIs 与标准 Android 项目操作的不同在于编译 API 的选取上，如图 10.1.4 所示。

New Android Application

New Android Application

The prefix 'com.example.' is meant as a placeholder and should not be used

Application Name: MapTest

Project Name: MapTest

Package Name: com.example.maptest

Minimum Required SDK: API 8: Android 2.2 (Froyo)

Target SDK: API 18

Compile With: Google APIs x86 (Google Inc.) (API 19)

Theme: API 19: Android 4.4 (KitKat)

Google APIs (Google Inc.) (API 19)

Google APIs x86 (Google Inc.) (API 19)

图 10.1.4 新建地图项目时使用 Google APIs

生成的项目中，在原来的 Android 4.4 前加上了 Google APIs，主要是增加了用于地图开发的 maps.jar 包，如图 10.1.5 所示。

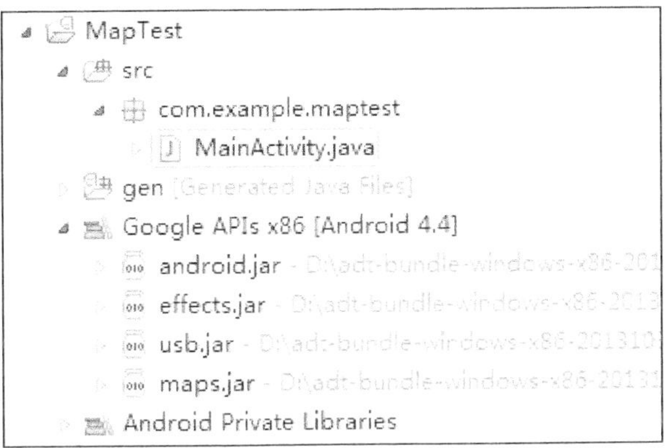

图 10.1.5　地图项目可使用的 jar 包

注意：maps.jar 包，主要提供了地图应用的相关类，不包含定位的相关类(因为标准的 Android 包已经提供了定位的相关类与接口)。

展开 maps.jar，可见它只含有一个软件包，该包所含的类(或接口)如图 10.1.6 所示。

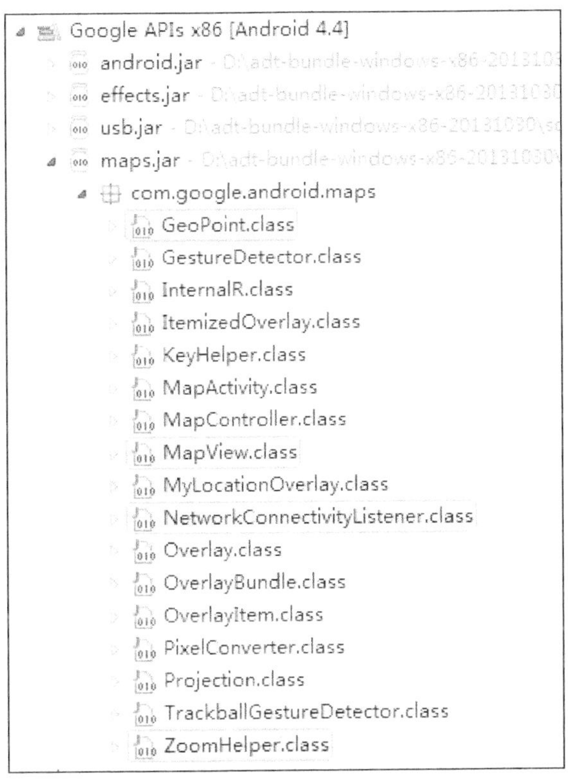

图 10.1.6　Google APIs 提供的地图 jar 包及类

2. Baidu Map API

目前，广大 Android 应用开发人员纷纷使用百度公司推出的定位服务 SDK 和地图服务的 API，详见第 10.3 节。

10.2 常用的定位方式与网络管理器类

10.2.1 Android GPS 定位及实例

使用 GPS 定位，首先，需要在清单文件中注册如下获取位置精确数据的权限：

```
<uses-permission android:name="android.permission.ACCESS_FINE_LOCATION" />
```

使用 Android 提供的标准 GPS 定位的主要步骤如下：

- 使用当前 Activity 拥有的方法 getSystemService(Context.LOCATION_SERVICE)得到一个位置管理器 LocationManager 对象；
- 使用 LocationManager 的 getLastKnownLocation(LocationManager.GPS_PROVIDER) 方法得到一个 Location 对象，该方法参数 LocationManager.GPS_PROVIDER 表明是使用 GPS 服务的；
- 使用 Location 对象具有的方法 getLongitude()和 getLatitude()获得经纬度数据。

注意：

(1) 通过 LocationManager 提供的方法 isProviderEnabled(LocationManager.GPS_PROVIDER)可以检查 GPS 是否开启。若未开启，进入手机的设置程序，其代码如下：

```
LocationManager lm = (LocationManager)
currentActivity.getSystemService(Context.LOCATION_SERVICE);
//若GPS未开启
if(!lm.isProviderEnabled(LocationManager.GPS_PROVIDER)){
//Toast.makeText(currentActivity, "请开启GPS！ ", Toast.LENGTH_SHORT).show();
//开启设置GPS的界面
Intent intent = new Intent(Settings.ACTION_LOCATION_SOURCE_SETTINGS);
startActivity(intent);
//设置完成后需要按返回键才能回到应用程序界面
}
```

(2) 如果希望能动态获取经纬度数据，则需要对 LocationManager 对象应用如下方法：

```
LocationManager lm=this. .getSystemService(Context.LOCATION_SERVICE);
lm.requestLocationUpdates(LocationManager.GPS_PROVIDER,
                                        0, 0,new MyLocationListener());
```

在位置监听器内通过调用 Location 的相关类输出定位信息(经纬度、海拔高度等)，其中，第四参数为 LocationListener 的接口类型。

【例 10.2.1】使用 GPS 定位功能，获取当前位置的经纬度。

【程序运行】程序运行时，先检测手机是否开启 GPS，若没有开启 GPS 功能，则先运行“设置”程序。退出“设置”程序后，需要等待手机的 GPS 模块启动(一般需要到室外)。

当 GPS 信号正常后，单击“获取 GPS 信息”按钮就会显示手机当前位置的经纬度(联想手机的 GPS 模块在单击“获取 GPS 信息”按钮后才启动)，如图 10.2.1 所示。

图 10.2.1　GPS 定位界面

【设计步骤】

(1) 新建名为 GPSLocation 的 Android 应用工程，勾选创建 Activity，工程文件结构(主要部分)如图 10.2.2 所示。

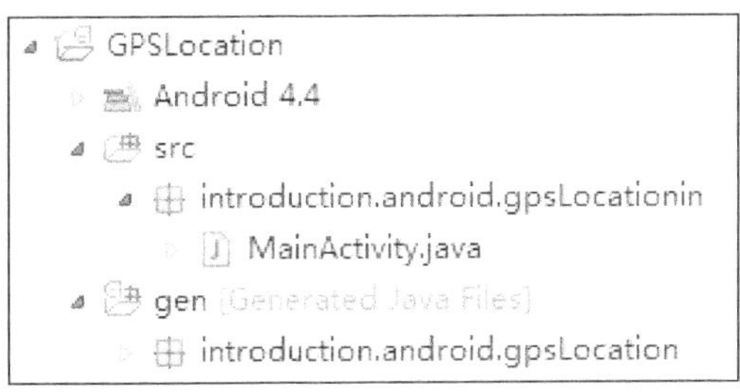

图 10.2.2　GPSLocation 工程文件结构(主要部分)

(2) 在清单文件中，添加使用 GPS 的权限：

```
<uses-permission android:name="android.permission.ACCESS_FINE_LOCATION"/>
```

(3) 按照图 10.2.1 所示的样式，修改布局文件 activity_main.xml，设计相应的布局。

(4) 编写源程序，文件 MainActivity.java 的代码如下：

```
package introduction.android.gpsLocationin;
/*
 * 本工程GPSLocation的功能是使用GPS定位，实时显示手机的经纬度
*/
import android.app.Activity;
import android.content.Context;
import android.content.Intent;
import android.location.Location;
import android.location.LocationListener;
import android.location.LocationManager;
import android.os.Bundle;
import android.provider.Settings;
import android.util.Log;
```

```
import android.view.View;
import android.view.View.OnClickListener;
import android.widget.Button;
import android.widget.TextView;
import android.widget.Toast;
import introduction.android.gpslocation.R;

public class MainActivity extends Activity {
    private Button btn_listen;
    private TextView tv_01,tv_02;
    LocationManager lm;
    @Override
    public void onCreate(Bundle savedInstanceState) {
        super.onCreate(savedInstanceState);
        setContentView(R.layout.activity_main);
        lm =(LocationManager) this.getSystemService(Context.LOCATION_SERVICE);
            if(!lm.isProviderEnabled(Lm.GPS_PROVIDER)){
            Toast.makeText(MainActivity.this,"请开启GPS服务
                                                    ",Toast.LENGTH_LONG).show();
            Intent myintent = new Intent
                                    (Settings.ACTION_LOCATION_SOURCE_SETTINGS);
            startActivity(myintent);
        }
        btn_listen=(Button) findViewById(R.id.btn_listen);
        tv_01=(TextView) findViewById(R.id.tv_01);
        tv_02=(TextView) findViewById(R.id.tv_02);

        btn_listen.setOnClickListener(new OnClickListener() {
            @Override
            public void onClick(View v) {
                lm.requestLocationUpdates(LocationManager.GPS_PROVIDER, 0, 0,
                                    new MyLocationListener());    //应用位置监听器
                }
            }
        });
    }
    class MyLocationListener implements LocationListener{ //位置监听器
        @Override
        public void onLocationChanged(Location location) {
            // TODO Auto-generated method stub
```

```
            tv_01.setText("您当前位置的经度为: "+location.getLongitude());
            tv_02.setText("您当前位置的纬度为: "+location.getLatitude());
        }
        @Override
        public void onProviderDisabled(String provider) {
            //在provider被用户关闭时调用
            Log.i("GpsLocation","provider被关闭! ");
        }
        @Override
        public void onProviderEnabled(String provider) {
            //在provider被用户开启后调用
            Log.i("GpsLocation","provider被开启! ");
        }
        @Override
        public void onStatusChanged(String provider, int status, Bundle extras) {
            //当provider的状态在OUT_OF_SERVICE、TEMPORARILY_UNAVAILABLE
和AVAILABLE之间发生变化时调用
            Log.i("GpsLocation","provider状态发生改变! ");
        }
    }
}
```

(5) 部署工程到手机，做运行测试。

10.2.2　网络连接及状态相关类

前面介绍的 GPS 定位方式不需要使用网络连接，但是 WiFi 定位和 GPRS 定位这两种定位方式需要使用网络连接，因此，这里先介绍与网络连接相关的类。

Android 提供了与网络连接相关的类，用于判定是否连接、连接状态，它们位于软件包 android.net 里。使用 WiFi 定位或 GPRS 定位时，需要引入的几个主要软件包(类)如下：

```
import android.net.ConnectivityManager;
import android.net.NetworkInfo;
import android.net.NetworkInfo.State;
```

使用上述类的代码如下：

```
    //是否有网络连接(GPRS或WiFi)
    ConnectivityManager cm;
    cm =(ConnectivityManager)getSystemService(Context.CONNECTIVITY_SERVICE);
    NetworkInfo networkInfo = cm.getActiveNetworkInfo();
    if (networkInfo==null)
```

```
        Toast.makeText(this,"没有网络连接",Toast.LENGTH_LONG).show();
    else
        Toast.makeText(this,"有网络连接",Toast.LENGTH_LONG).show();
```

检测到程序没有联网的提示界面如图 10.2.3 所示。

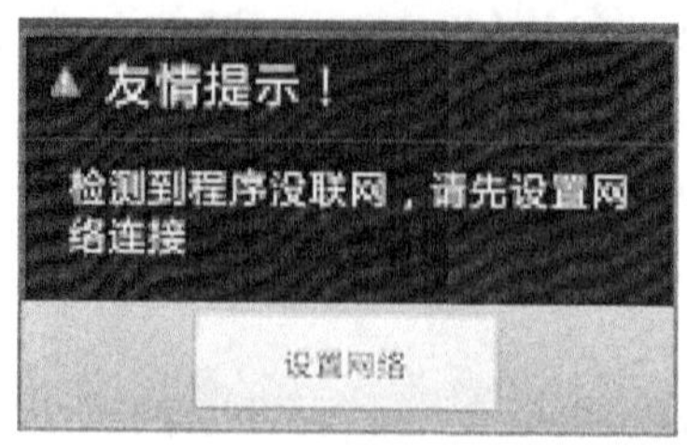

图 10.2.3　检测到没有联网的提示界面

```
//检查移动网络(GPRS)连接
cm =(ConnectivityManager)getSystemService(Context.CONNECTIVITY_SERVICE);
mobileState=cm.getNetworkInfo(ConnectivityManager.TYPE_MOBILE).getState();
if (mobileState==State.CONNECTED||mobileState==State.CONNECTING)
     Toast.makeText(this,"可以使用移动网络",Toast.LENGTH_LONG).show();
else
     Toast.makeText(this,"移动网络未开启!",Toast.LENGTH_LONG).show();

//检查WiFi连接
wifiState=cm.getNetworkInfo(ConnectivityManager.TYPE_WIFI).getState();
if (wifiState==State.CONNECTED||wifiState==State.CONNECTING) {
    Toast.makeText("可以使用WiFi网络!",);
    wifiManager = (WifiManager)getSystemService(Context.WIFI_SERVICE);
    if (wifiManager.isWifiEnabled())
        Toast.makeText(this," 可以得到定位信息！ ",Toast.LENGTH_LONG).show();
}
else
    Toast.makeText(this," 没有WiFi网络！ ",Toast.LENGTH_LONG).show();
```

10.2.3　Android WiFi 定位及实例

使用 WiFi 定位，首先需要在清单文件中注册如下获取位置数据的权限：

```
<uses-permission android:name="android.permission.ACCESS_COARSE_LOCATION" />
```

使用 Android 提供的标准 WiFi 定位的主要代码如下：

```
//得到一个位置管理器对象
LocationManager lm=this.getSystemService(Context.LOCATION_SERVICE);
//得到WiFi管理器对象wfm
```

```
WifiManager wfm=this.getSystemService(Context.WIFI_SERVICE);
String provider="LocationManager.NETWORK_PROVIDER";
Location loc=lm.getLastKnownLocation(provider);
If(loc==null){
        Lm.requestLocationUpdates(provider,0,0,new LocationListener(){
            @Overide
            public void onLocationChanged(Location loc){
                //输出loc.getLongitude();//经度
                //输出loc.getLatitude();//纬度
            }
        });
}
```

【例 10.2.2】使用 WiFi 定位经纬度。

【程序运行】程序运行时，先检测手机是否开启 WiFi，若没有开启 WiFi，则自动打开。待手机的 WiFi 连接正常后(有闪动的图标，需要等待一会儿)，单击“获取位置信息”按钮，则显示手机当前位置的经纬度，如图 10.2.4 所示。

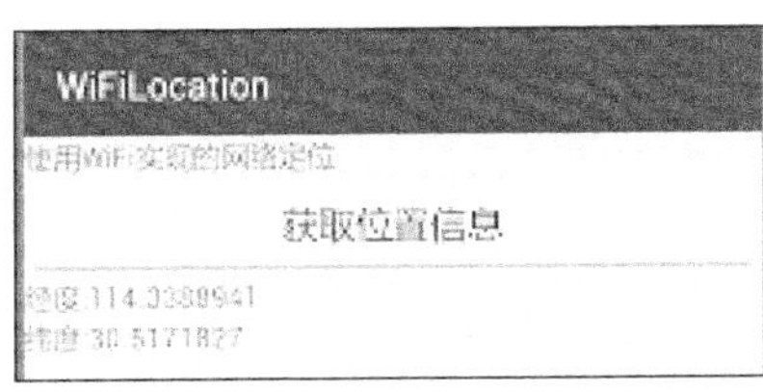

图 10.2.4　WiFi 定位结果

【设计步骤】

(1) 新建名为 WiFiLocation 的 Android 应用工程，勾选创建 Activity，工程文件结构(主要部分)如图 10.2.5 所示。

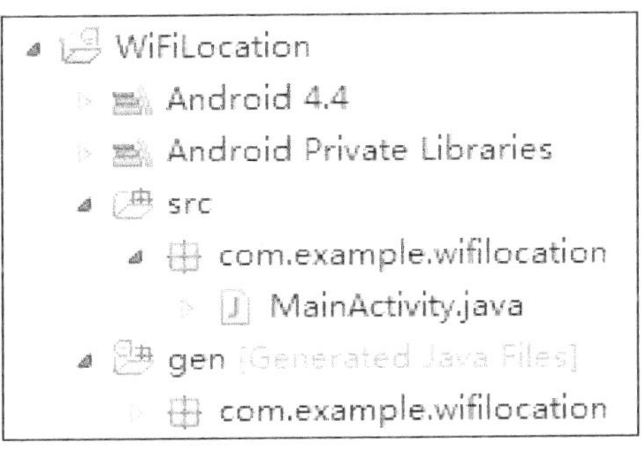

图 10.2.5　WiFiLocation 工程文件结构(主要部分)

(2) 在清单文件中，添加使用 WiFi 定位所需要的权限：

```
<uses-permission android:name=
                        "android.permission.ACCESS_COARSE_LOCATION"/>
```

```
<uses-permission android:name="android.permission.ACCESS_WIFI_STATE"/>
<uses-permission android:name="android.permission.CHANGE_WIFI_STATE"/>
```

(3) 主布局文件与工程 GPSLocation 相同。

(4) 编写源程序，文件 MainActivity.java 的代码如下：

```
package com.example.wifilocation;
import android.app.Activity;
import android.content.Context;
import android.location.Location;   //
import android.location.LocationListener;
import android.location.LocationManager;   //
import android.net.wifi.WifiManager;   //
import android.os.Bundle;
import android.view.View;
import android.view.View.OnClickListener;
import android.widget.Button;
import android.widget.TextView;
import com.example.wifilocation.R;
public class MainActivity extends Activity {
    private Button btn_listen;
    private TextView tv_01,tv_02;
    LocationManager lm;
    WifiManager   wfm ;
    @Override
    protected void onCreate(Bundle savedInstanceState) {
        super.onCreate(savedInstanceState);
        setContentView(R.layout.activity_main);
         lm =(LocationManager) this.getSystemService(Context.LOCATION_SERVICE);//
         btn_listen=(Button) findViewById(R.id.btn_listen);
         tv_01=(TextView)   findViewById(R.id.tv_01);
         tv_02=(TextView)   findViewById(R.id.tv_02);
         wfm = (WifiManager) getSystemService(Context.WIFI_SERVICE);
         if(!wfm.isWifiEnabled())
             wfm.setWifiEnabled(true);
         btn_listen.setOnClickListener(new OnClickListener() {
                 @Override
                 public void onClick(View v) {
                     String provider = LocationManager.NETWORK_PROVIDER;
                     Location loc = lm.getLastKnownLocation(provider);   //
```

```
                if(loc == null){    //可更新
                    lm.requestLocationUpdates(provider, 0, 0, new LocationListener(){
                        @Override
                        public void onLocationChanged(Location loc) {
                            // TODO Auto-generated method stub
                            tv_01.setText("经度： "+loc.getLongitude());
                            tv_02.setText("纬度： "+loc.getLatitude());
                        }
                        @Override
                        public void onProviderDisabled(String arg0) {
                            // TODO Auto-generated method stub
                        }
                        @Override
                        public void onProviderEnabled(String arg0) {
                            // TODO Auto-generated method stub
                        }
                        @Override
                        public void onStatusChanged(String arg0, int arg1,Bundle arg2)
                        {
                            // TODO Auto-generated method stub
                        }
                    });
                }
                tv_01.setText("经度:"+loc.getLongitude());
                tv_02.setText("纬度:"+loc.getLatitude());
            }
        });
    }
}
```

(5) 部署工程到手机，做运行测试。

10.3　百度地图应用开发

为了在 Android 地图应用程序中使用百度定位 SDK 和百度地图 API，需要将相关文件拷贝到工程的文件夹 libs 内，如图 10.3.1 所示。

图 10.3.1　引入百度定位及百度地图所需要的文件

注意：

(1) 导入别人编写的工程时，如果在工程名上出现红色的感叹号，可能是由于别人引用的.jar 包没有拷贝到 libs 文件夹内。此时，解决的办法是：选择工程属性→Java Build Path→Libraries→Add JARs→指定 locSDK4.2.jar。

(2) 如果不将定位结果以地图形式显示，则不需要 baidumapapi_v3_0_0.jar 文件。

10.3.1　百度位置服务开发基础

Baidu Map 不仅提供了定位服务，还提供了地图服务。百度位置包 location 中的主要类与接口如图 10.3.2 所示。

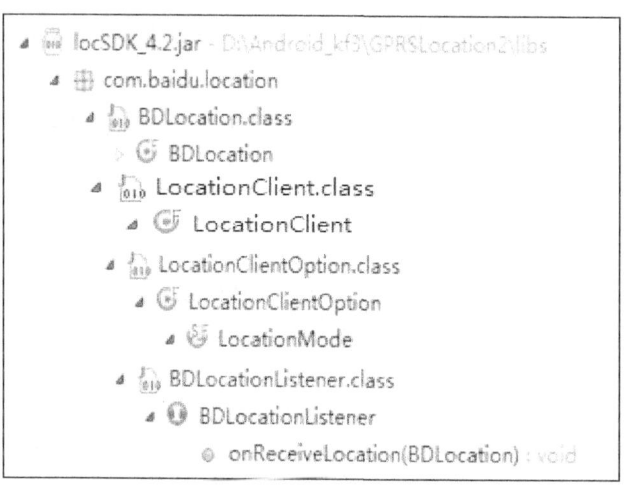

图 10.3.2　百度位置包中的主要类与接口

注意：百度位置 SDK 与 Google Map APIs 基本相似。

1. 百度位置类 BDLocation

百度位置包 location 中的 BDLocation 类是最基础的类，它封装了获取经纬度数据、地名信息等方法，如图 10.3.3 所示。

2. 手机位置客户端类 LocationClient

百度位置包 location 中的 LocationClient 类也是非常重要的一个类，其构造以上下文对象和 LocationClientOption 类型的对象(表示用户端的操作)作为参数，其主要方法如图 10.3.4 所示。

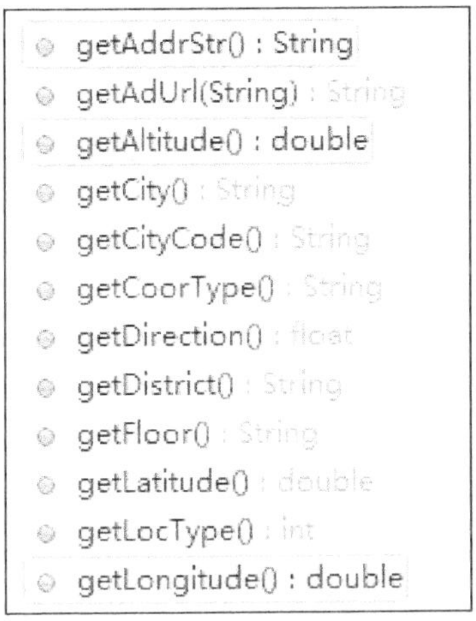

```
getAddrStr() : String
getAdUrl(String) : String
getAltitude() : double
getCity() : String
getCityCode() : String
getCoorType() : String
getDirection() : float
getDistrict() : String
getFloor() : String
getLatitude() : double
getLocType() : int
getLongitude() : double
```

图 10.3.3　百度位置包中 BDLocation 类的主要方法

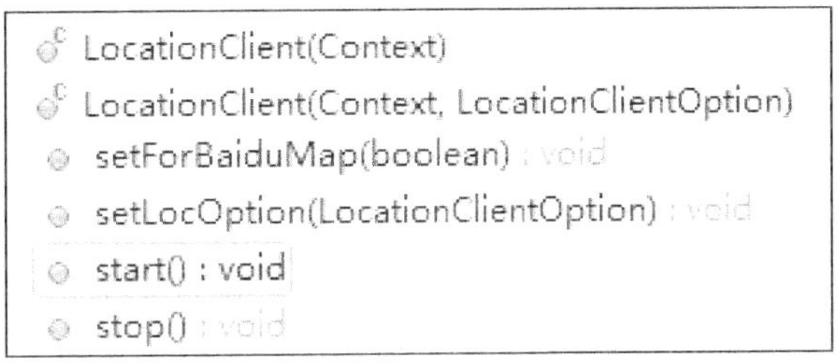

```
LocationClient(Context)
LocationClient(Context, LocationClientOption)
setForBaiduMap(boolean) : void
setLocOption(LocationClientOption) : void
start() : void
stop() : void
```

图 10.3.4　百度位置包中 LocationClient 类的主要方法

3. 手机客户端操作信息类 LocationClientOption

百度位置包 location 中的 LocationClientOption 类中定义了内部类 LocationMode(即 LocationClientOption$LocationMode)，它定义了三种定位模式，如图 10.3.5 所示。

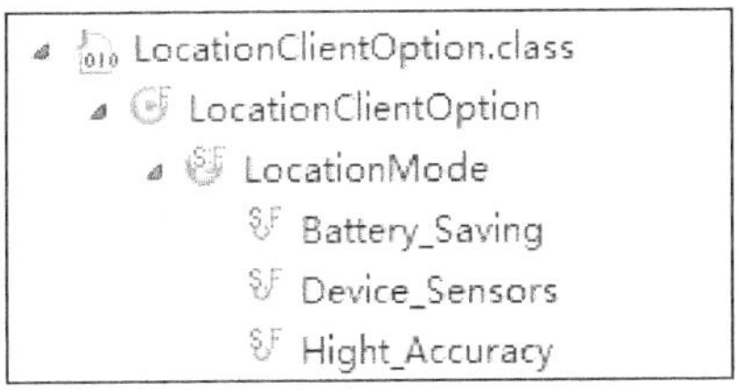

```
LocationClientOption.class
  LocationClientOption
    LocationMode
      Battery_Saving
      Device_Sensors
      Hight_Accuracy
```

图 10.3.5　百度提供的三种定位模式

- 高精度定位模式 Hight_Accuracy：这种定位模式下，会同时使用网络定位和 GPS 定位，优先返回最高精度的定位结果。
- 低功耗定位模式 Battery_Saving：这种定位模式下，不会使用 GPS，只会使用网络定位(WiFi 定位和基站定位)。
- 仅用设备定位模式 Device_Sensors：这种定位模式下，不需要连接网络，只使用 GPS 进行定位，这种模式下不支持室内环境的定位。

4. 位置监听器接口 BDLocationListener

位置监听器接口的作用与标准 Android 中的位置监听器接口类似，其定义如图 10.3.6 所示。

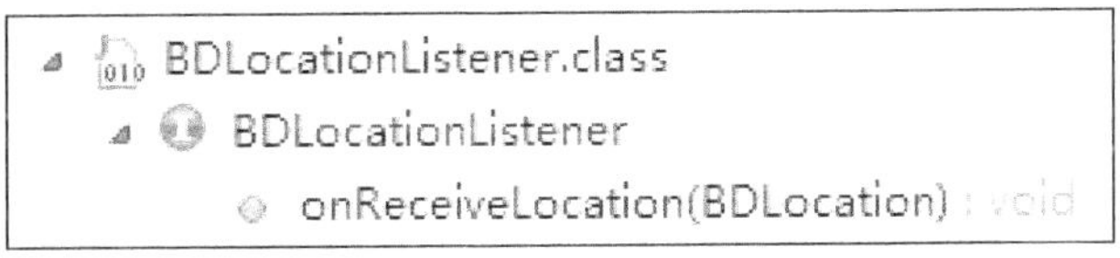

```
BDLocationListener.class
  BDLocationListener
    onReceiveLocation(BDLocation) : void
```

图 10.3.6　百度位置监听器接口

百度地图包中的主要类，如图 10.3.7 所示。

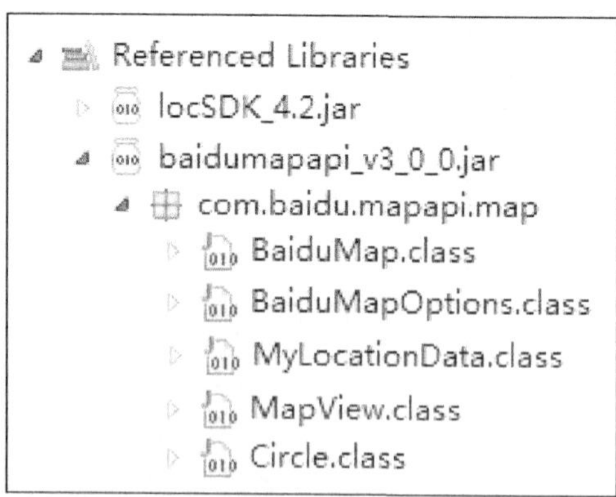

图 10.3.7　百度地图包中的主要类

注意：

(1) 百度地图包的不同版本，提供的类名有差别。

(2) 选择使用百度定位 SDK v3.3 及之前版本的开发者，不需要使用百度 Key。

(3) 若需要在同一个工程中同时使用百度定位 SDK 和百度地图 API，可以共用同一个 Key。

10.3.2　申请定位与地图应用的 Key

作为百度位置或地图应用的开发人员，首先需要申请一个百度开发者帐号。访问百度网站 http://www.baidu.com，单击“登录”按钮，可注册(见界面右下方)或登录，界面如图 10.3.8 所示。

图 10.3.8　百度开发者注册与登录界面

当选择使用 v4.0 以上版本的定位 SDK 时，需要先申请安全码，然后再配置 Android 应用的 Key，并在清单文件中的<application>标签内使用<meta>标签填写这个 Key。

登录百度帐号成功后，访问 http://lbsyun.baidu.com/apiconsole/key，会显示自己已经创建的百度应用，还可以申请创建一个新的百度定位(或地图)应用。

申请创建一个新的百度定位(或地图)应用时，在输入“应用名称”、选择“应用类型”后，还需要在“安全码”文本框内输入“数字签名;应用包名”，如图 10.3.9 所示。

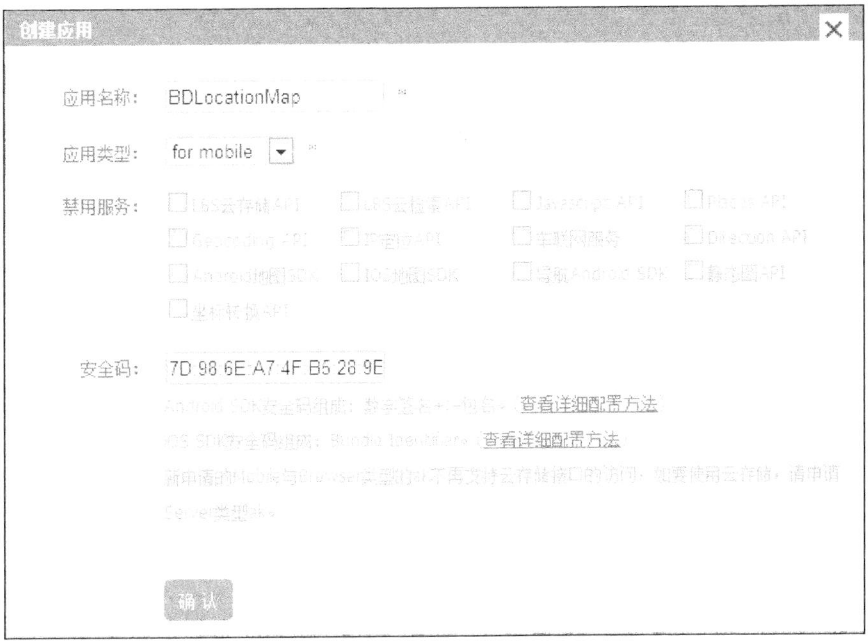

创建应用

应用名称：BDLocationMap

应用类型：for mobile

禁用服务：

安全码：7D 98 6E A7 4F B5 28 9E

查看详细配置方法

查看详细配置方法

确认

图 10.3.9　创建一个百度应用

单击“确认”按钮后，系统自动产生的安全码就是该应用的 Key。其中，数字签名来源于本地的 Android 开发环境，如图 10.3.10 所示。

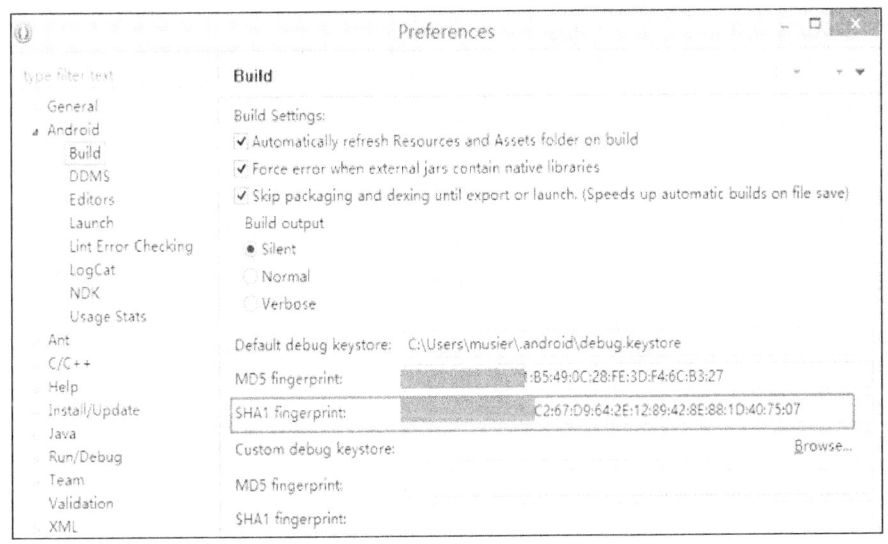

图 10.3.10　生成安全码时所需要的数字签名

注意：

(1) 源文件夹下程序的包名可能存在多个，而应用的包名是唯一的(在清单文件中指定);

(2) 每个 Key 仅且唯一对于一个 Android 应用验证有效。

10.3.3 在清单文件中注册服务、权限及应用 Key

1. 在清单文件中注册服务

由于每个百度定位 App 都拥有自己单独的定位 Service，因此在使用百度定位及地图服务前，应在清单文件的<application>标签中声明 Service 服务组件，代码如下：

```
<service
        android:name="com.baidu.location.f"
        android:enabled="true"
        android:process=":remote" />
```

2. 在清单文件中注册权限

百度地图应用开发需要注册的权限如下：

```
<!-- 这个权限用于网络定位 -->
    <uses-permission android:name="android.permission.ACCESS_COARSE_LOCATION" />
    <!-- 这个权限用于访问GPS定位 -->
    <uses-permission android:name="android.permission.ACCESS_FINE_LOCATION" />
    <!-- 用于访问WiFi网络信息，WiFi信息会用于网络定位 -->
    <uses-permission android:name="android.permission.ACCESS_WIFI_STATE" />
    <!-- 获取运营商信息，用于支持提供运营商信息相关的接口 -->
    <uses-permission android:name="android.permission.ACCESS_NETWORK_STATE" >
    </uses-permission>
    <!-- 这个权限用于获取WiFi，WiFi信息会用来进行网络定位 -->
    <uses-permission android:name="android.permission.CHANGE_WIFI_STATE" />
    <!-- 用于读取手机当前的状态 -->
    <uses-permission android:name="android.permission.READ_PHONE_STATE" />
    <!-- 写入扩展存储，向扩展卡写入数据，用于写入离线定位数据 -->
    <uses-permission android:name="android.permission.WRITE_EXTERNAL_STORAGE"/ >
    <!-- 访问网络，网络定位需要上网 -->
    <uses-permission android:name="android.permission.INTERNET" />
    <uses-permission android:name=
                    "android.permission.MOUNT_UNMOUNT_FILESYSTEMS" />
    <!-- 允许应用读取低级别的系统日志文件 -->
    <uses-permission android:name="android.permission.READ_LOGS" />
```

3. 在清单文件中注册应用 Key

每个百度定位与地图应用的 App，都需要在清单文件的<meta>标签内登记，参见例 10.3.1。

10.3.4　使用百度位置包实现综合定位

虽然标准的 Android API 提供了使用 GPRS 实现定位的功能，但它要访问 Google 网站获取相关信息，因此，我们使用百度位置包来实现 GPRS 定位功能。

在复制位置包到工程的 libs 文件夹后，可以查看相关的类与接口，如图 10.3.11 所示。

前面介绍的以经纬度形式提供的定位结果是不直观的，通常还需要借助地图数据包以地名来直观地表示位置信息。在百度地图应用开发中，通过使用百度 location 包里的 BDLocation 类的相关方法(如 getAddress())来获取地名形式的位置信息。

【例 10.3.1】使用百度位置包实现综合定位并显示地名。

【程序运行】程序运行时，先检测手机是否有网络连接，若没有，则运行手机设置程序。从设置程序返回后，等待一会儿(建立网络连接)，即可显示手机当前位置的经纬度，如图 10.3.12 所示。

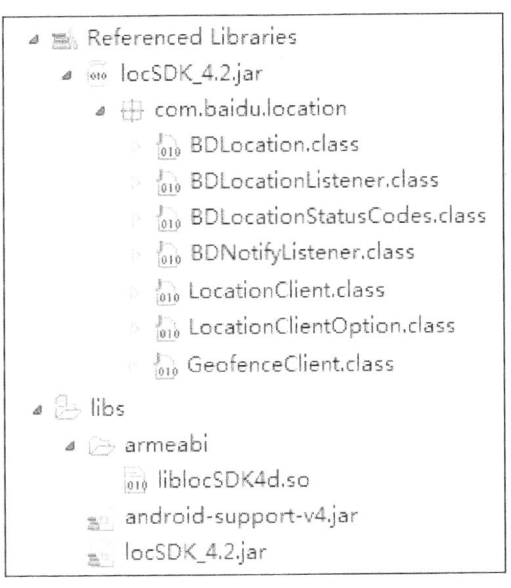

图 10.3.11　百度位置包及其相关类与接口

BaiDuLocation

经度(Longitude): 114.348299
纬度(Latitude): 30.521026

地址: 湖北省武汉市洪山区工大路

图 10.3.12　百度综合定位界面

【设计步骤】

(1) 新建名为 BDLocation 的工程，后面的操作都使用默认值(如包名、创建 Activity、布局文件名等)。

(2) 下载百度位置应用开发包并复制到工程的 libs 文件夹内，工程文件结构(主要部分)如图 10.3.13 所示。

(3) 主布局文件与工程 GPSLocation 相同。

(4) 在清单文件中，添加百度综合定位所需的如下 7 个权限：

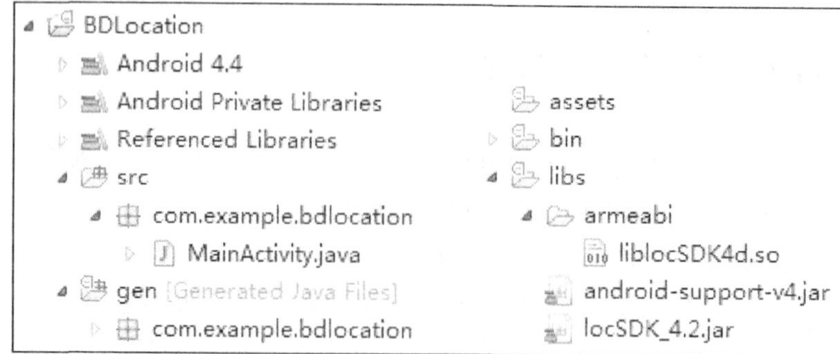

图 10.3.13　BDLocation 工程文件结构(主要部分)

```
<uses-permission android:name="android.permission.ACCESS_FINE_LOCATION"/>
<uses-permission android:name="android.permission.ACCESS_NETWORK_STATE"/>
<uses-permission android:name="android.permission.ACCESS_COARSE_LOCATION"/>
<uses-permission android:name="android.permission.ACCESS_WIFI_STATE"/>
<uses-permission android:name="android.permission.CHANGE_WIFI_STATE"/>
<uses-permission android:name="android.permission.READ_PHONE_STATE"/>
<uses-permission android:name="android.permission.INTERNET"/>。
```

(5) 使用自己的百度帐号申请本定位应用的 API Key。

(6) 在清单文件中，使用<service>标签后，再使用<meta>标签添加刚才申请的 Baidu Map API Key。

```
<service
   android:name="com.baidu.location.f"
   android:enabled="true"
   android:process=":remote" />
<meta-data
   android:name="com.baidu.lbsapi.API_KEY"
   android:value="Id8CEeqtxEgKCg5LFtquU1lH" />
```

(7) 编写源程序，文件 MainActivity.java 的代码如下：

```
package com.example.bdlocation;
/*
 * 百度综合定位工程：BDLocation
 * 高精度定位模式：会同时使用网络定位和GPS定位，优先返回最高精度的定位结果
 * 显示地名(址)信息，需要开启WiFi网络或GPRS网络
 */
import com.baidu.location.BDLocation;   //
import com.baidu.location.BDLocationListener;
import com.baidu.location.LocationClient;   //
import com.baidu.location.LocationClientOption;
import com.baidu.location.LocationClientOption.LocationMode;
```

```
import android.os.Bundle;
import android.app.Activity;
import android.widget.TextView;
import android.widget.Toast;
public class MainActivity extends Activity {
    private TextView tv_LatLonInfo;
    private TextView tv_AddressInfo;
    private LocationClient mLocationClient = null;
    private BDLocationListener myListener = new MyLocationListener();
    @Override
    protected void onCreate(Bundle savedInstanceState) {
        super.onCreate(savedInstanceState);
        setContentView(R.layout.activity_main);
        //检查是否有网络连接(GPRS或WiFi)
        ConnectivityManager cm;
          cm =(ConnectivityManager)getSystemService(Context.CONNECTIVITY_SERVICE);
          NetworkInfo networkInfo = cm.getActiveNetworkInfo();
          if (networkInfo==null) {
            Toast.makeText(this,"检测到没有网络连接，请先打开网络连接...",
                                                    Toast.LENGTH_LONG).show();
            Intent myintent=new Intent(Settings.ACTION_SETTINGS);
        startActivity(myintent);   //运行手机的设置程序
        }
        tv_LatLonInfo = (TextView) findViewById(R.id.LatLonInfo);
        tv_AddressInfo = (TextView) findViewById(R.id.AddressInfo);
        // 声明LocationClient类
        mLocationClient = new LocationClient(getApplicationContext());
        // 注册监听函数
        mLocationClient.registerLocationListener(myListener);
        //设置定位参数：LocationClientOption，
        //包括：定位模式、返回坐标类型、是否打开GPS等
        LocationClientOption option = new LocationClientOption();
        //设置高精度定位模式
        option.setLocationMode(LocationMode.Hight_Accuracy);
        //返回的定位结果是百度经纬度,默认值gcj02
        option.setCoorType("bd09ll");
        //设置发起定位请求的间隔时间为5000ms
        option.setScanSpan(5000);
        //返回的定位结果包含地名(址)信息
```

```
        option.setIsNeedAddress(true);
        //返回的定位结果包含手机机头的方向
        option.setNeedDeviceDirect(true);
        mLocationClient.setLocOption(option);
        //发起定位
        mLocationClient.start();
    }
    public class MyLocationListener implements BDLocationListener {
        @Override
        public void onReceiveLocation(BDLocation location) {
            Toast.makeText(MainActivity.this,"定位方式: "+location.getLocType(),
                                            Toast.LENGTH_LONG).show();
            String result="经度(Longitude): "+location.getLongitude();
            result+="\n纬度(Latitude): "+location.getLatitude();
            tv_LatLonInfo.setText(result);
            tv_AddressInfo.setText("地址: "+location.getAddrStr());
        }
    }
}
```

(8) 部署工程到手机后做运行测试。

注意：经测试发现，只做位置应用(不涉及地图显示)时，使用任何别人申请到的 API Key，都不影响程序的部署和运行。

10.3.5 使用 MapView 显示当前位置

要想以地图形式显示位置信息，则需要使用百度地图包提供的地图显示控件，其定义如图 10.3.14 所示。

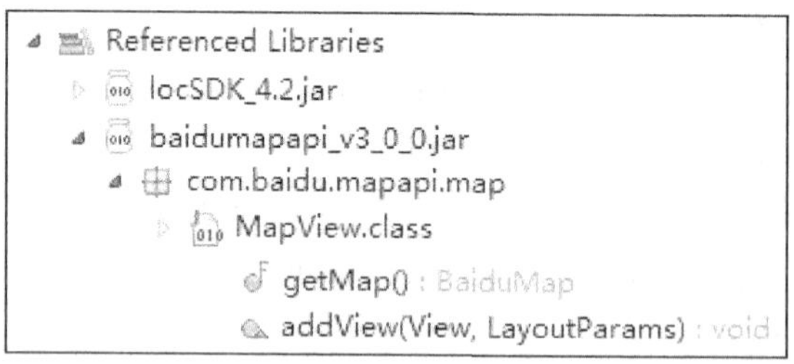

图 10.3.14　百度地图包提供的地图显示控件

【例 10.3.2】显示当前位置的地图。

【程序运行】程序运行时，先检测手机是否开启网络连接，若没有开启，则运行手机设置程序。返回后，逐渐出现定位信息和以当前位置为中心的地图，如图 10.3.15 所示。

图 10.3.15　显示当前位置的地图程序运行效果

【设计步骤】

(1) 新建名为 BDLocationMap 的 Android 应用工程，勾选创建 Activity，工程文件结构(主要部分)如图 10.3.16 所示。

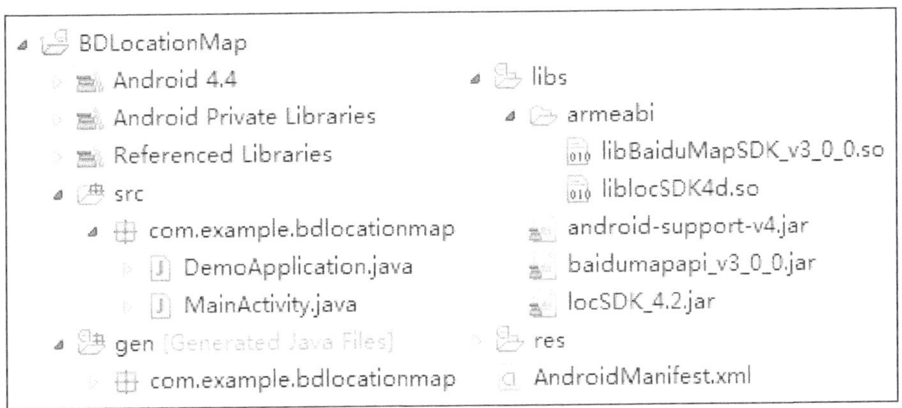

图 10.3.16　BDLocationMap 工程文件结构(主要部分)

(2) 采用工程默认的包名 com.example.bdlocationmap，并在 Eclipse 环境中创建 Map API Key。

(3) 新建源程序文件 DemoApplication.java，代码如下：

```
package com.example.bdlocationmap;
/*
* Baidu MapAK:SUrBhwGjP9Ka6IX7alBUMEH8
```

```
* 定位包版本：4.2，地图包版本：3.0
*/
import android.app.Application;    //
import com.baidu.mapapi.SDKInitializer;    //

public class DemoApplication extends Application {
    @Override
    public void onCreate() {
        super.onCreate();
        // 在使用 SDK 各组件之前初始化 context 信息，传入 ApplicationContext
        SDKInitializer.initialize(this);    //
    }
}
```

(4) 复制百度地图应用所需要的.jar 包至工程的文件夹 libs 内，此时，工程文件结构如图 10.3.16 所示。

(5) 界面设计采用帧布局，实现在 MapView 控件对象之上显示几个 TextView 对象。该布局文件的代码如下：

```
<FrameLayout
      xmlns:android="http://schemas.android.com/apk/res/android"
      xmlns:tools="http://schemas.android.com/tools"
      android:layout_width="match_parent"
      android:layout_height="match_parent" >
<com.baidu.mapapi.map.MapView
     android:id="@+id/bmapView"
     android:layout_width="fill_parent"
     android:layout_height="fill_parent"
     android:clickable="true" />
<LinearLayout
     android:layout_width="fill_parent"
     android:layout_height="wrap_content"
     android:background="#e0000000"
     android:orientation="vertical" >
     <LinearLayout
          android:layout_width="wrap_content"
          android:layout_height="wrap_content"
          android:layout_marginLeft="12dp"
          android:layout_marginTop="20dp"
          android:orientation="horizontal" >
          <TextView
```

```
            android:layout_width="wrap_content"
            android:layout_height="wrap_content"
            android:text="纬度: "
            android:textColor="#ffffff"
            android:textSize="15dp" />
        <TextView
            android:id="@+id/tv_Lat"
            android:layout_width="wrap_content"
            android:layout_height="wrap_content"
            android:text="38"
            android:textColor="#ffffff"
            android:textSize="15dp" />
    </LinearLayout>
    <LinearLayout
        android:layout_width="wrap_content"
        android:layout_height="wrap_content"
        android:layout_marginLeft="12dp"
        android:layout_marginTop="10dp"
        android:orientation="horizontal" >
        <TextView
            android:layout_width="wrap_content"
            android:layout_height="wrap_content"
            android:text="经度: "
            android:textColor="#ffffff"
            android:textSize="15dp" />
        <TextView
            android:id="@+id/tv_Lon"
            android:layout_width="wrap_content"
            android:layout_height="wrap_content"
            android:text="118"
            android:textColor="#ffffff"
            android:textSize="15dp" />
    </LinearLayout>
    <LinearLayout
        android:layout_width="wrap_content"
        android:layout_height="wrap_content"
        android:layout_marginBottom="10dp"
        android:layout_marginLeft="12dp"
        android:layout_marginTop="10dp"
```

```
        android:orientation="horizontal" >
        <TextView
            android:layout_width="wrap_content"
            android:layout_height="wrap_content"
            android:text="地址："
            android:textColor="#ffffff"
            android:textSize="15dp" />
        <TextView
            android:id="@+id/tv_Add"
            android:layout_width="wrap_content"
            android:layout_height="wrap_content"
            android:text="太湖东路常高新大厦"
            android:textColor="#ffffff"
            android:textSize="15dp" />
    </LinearLayout>
</LinearLayout>
```

(6) 编辑工程的清单文件，添加对 Service 的注册和权限注册后，再通过<meta>添加刚才申请的 Baidu Map API Key。

```
    <service
        android:name="com.baidu.location.f"
        android:enabled="true"
        android:process=":remote" />
        <meta-data
          android:name="com.baidu.lbsapi.API_KEY"
            android:value="Id8CEeqtxEgKCg5LFtquU1lH" />
```

(7) 编写工程的源程序，其代码如下：

```
package com.example.mapdemo;
/*
 * MapAK:nonBBaQh9VovWNhjnO39Fmyj
 * 演示地图功能：定位获取经纬度和地图显示经纬度地点
*/
import com.baidu.location.BDLocation;
import com.baidu.location.BDLocationListener;
import com.baidu.location.LocationClient;
import com.baidu.location.LocationClientOption;
import com.baidu.location.LocationClientOption.LocationMode;
import com.baidu.mapapi.SDKInitializer;
```

```
import com.baidu.mapapi.map.BaiduMap;
import com.baidu.mapapi.map.MapStatusUpdate;
import com.baidu.mapapi.map.MapStatusUpdateFactory;
import com.baidu.mapapi.map.MapView;
import com.baidu.mapapi.map.MyLocationConfigeration;
import com.baidu.mapapi.map.MyLocationData;
import com.baidu.mapapi.model.LatLng;
import android.os.Bundle;
import android.app.Activity;
import android.widget.TextView;

public class MainActivity extends Activity {
    private TextView tv_Lat;// 纬度
    private TextView tv_Lon;// 经度
    private TextView tv_Add;// 地址
    private MapView mMapView = null; // 地图
    private BaiduMap mBaiduMap;
    private com.baidu.mapapi.map.MyLocationConfigeration.LocationMode
                mCurrentMode = com.baidu.mapapi.map.
                                MyLocationConfigeration.LocationMode.NORMAL;
    private LocationClientOption option;
    private LocationClient mLocationClient = null;
    private BDLocationListener myListener = new MyLocationListener();
    @Override
    protected void onCreate(Bundle savedInstanceState) {
        super.onCreate(savedInstanceState);
        // 在使用SDK各组件之前初始化context信息，传入ApplicationContext
        // 注意该方法要在setContentView方法之前实现
        SDKInitializer.initialize(getApplicationContext());   //
        setContentView(R.layout.activity_main);
        initView();
    }
    private void initView() {
        tv_Lat = (TextView) findViewById(R.id.tv_Lat);
        tv_Add = (TextView) findViewById(R.id.tv_Add);
        tv_Lon = (TextView) findViewById(R.id.tv_Lon);
        // 地图控件
        mMapView = (MapView) findViewById(R.id.bmapView);
        mBaiduMap = mMapView.getMap();
        // 开启定位图层
```

```
        mBaiduMap.setMyLocationEnabled(true);
        // 声明LocationClient类
        mLocationClient = new LocationClient(getApplicationContext());
        // 注册监听函数
        mLocationClient.registerLocationListener(myListener);
        option = new LocationClientOption();
        setLocModul(option);// 设置定位参数
    }
    private void setLocModul(LocationClientOption option) {
        // 设置定位模式
        option.setLocationMode(LocationMode.Hight_Accuracy);
        // 返回的定位结果是百度经纬度,默认值gcj02
        option.setCoorType("bd09ll");
        // 设置发起定位请求的间隔时间为5000ms
        option.setScanSpan(5000);
        // 返回的定位结果包含地址信息
        option.setIsNeedAddress(true);
        // 返回的定位结果包含手机机头的方向
        option.setNeedDeviceDirect(true);
        mLocationClient.setLocOption(option);
        // 发起定位
        mLocationClient.start();
    }
    public class MyLocationListener implements BDLocationListener {
        @Override
        public void onReceiveLocation(BDLocation location) {
            if (location == null)
                return;
            tv_Add.setText(location.getAddrStr());    //获取定位结果
            tv_Lat.setText("" + location.getLatitude());
            tv_Lon.setText("" + location.getLongitude());
            MyLocationData locData = new MyLocationData.Builder()
                    .accuracy(location.getRadius())
                    // 此处设置开发者获取到的方向信息，顺时针0～360°
                    .direction(100).latitude(location.getLatitude())
                    .longitude(location.getLongitude()).build();
            mBaiduMap.setMyLocationData(locData);
            // 设置定位图层的配置(定位模式、是否允许显示方向信息、用户自定义
              定位图标)
```

```
                mBaiduMap.setMyLocationConfigeration
                            (new MyLocationConfigeration(mCurrentMode, true, null));
                LatLng ll = new LatLng(location.getLatitude(),
                        location.getLongitude());
                MapStatusUpdate u = MapStatusUpdateFactory.newLatLng(ll);
                mBaiduMap.animateMapStatus(u);
            }
        }
        @Override
        protected void onDestroy() {   // 退出时销毁定位
            mLocationClient.stop();
            // 关闭定位图层
            mBaiduMap.setMyLocationEnabled(false);
            mMapView.onDestroy();
            mMapView = null;
            super.onDestroy();
        }
        @Override
        protected void onResume() {
            super.onResume();
            //地图生命周期管理
            mMapView.onResume();
        }
        @Override
        protected void onPause() {
            super.onPause();
            //地图生命周期管理
            mMapView.onPause();
        }
}
```

(8) 部署工程到手机上做运行测试。

习 题 10

一、判断题 5

1. Google Map API 是 Android SDK 的标准库组件，不需要单独下载。
2. 做百度地图开发，必须将百度提供的.jar 包复制到工程的 libs 文件夹内。
3. 做百度地图开发前必须下载 Google APIs。
4. WiFi 定位并不需要权限 android.permission.INTERNET。
5. 建立 GPRS 数据连接的时间比 WiFi 连接所需的时间长。

二、选择题

1. 为了进行 GPS 定位，手机至少需要能接收到____颗卫星信号。

   A. 2　　B. 3　　C. 4　　D. 5

2. 实现 GPS 实时定位，必须实现的方法是____。

   A. onLocationChanged()　　B. onProviderDisabled()
   C. onProviderEnabled()　　D. onStatusChanged()

3. 下列选项中，不是 Android 合法类的是____。

   A. TelephonyManager　　B. WifiManager
   C. GPSManager　　D. ConectivityManager

4. 使用 GPS 定位，必须开启的权限是____。

   A. GLOBAL_SEARCH
   B. ACCESS_FINE_LOCATION
   C. ACCESS_COARSE_LOCATION
   D. ACCESS_LOCATION_EXTRA_COMMANDS

5. 下列选项中，表示网络定位的是____。

   A. Context.LOCATION_SERVICE
   B. LocationManager.GPS_PROVIDER
   C. LocationManager.NETWORK_PROVIDER
   D. Context.WIFI_SERVICE

三、填空题

1. 在标准 Android 提供的 LBS 中，要实现 GPS 的实时定位，需要使用位置管理器对象具有的____方法。
2. 定位精确度最高的是____。
3. 使用 WiFi 定位或 GPRS 定位，需要在清单文件里注册的权限是相同的，该权限名称是____。
4. 开发百度地图应用，在工程中除了需要引入定位包外，还需要引入____。
5. 百度地图包中显示地图的控件名与 Google Map 相同，都是____。

实验 10　位置服务与地图应用开发

一、实验目的

1. 掌握 Android 提供的 GPS 定位的系统服务及相关类。
2. 掌握 Android 提供的 WiFi 定位的系统服务及相关类。
3. 掌握百度提供定位服务的相关类及其方法。
4. 掌握百度地图应用开发的方法。

二、实验内容及步骤

【预备】访问 http://www.wustwzx.com/android/index.html，单击“实验 10”超链接，下载本次实验内容的源代码并解压(得到文件夹 ch10)，供研读和调试使用。

1. GPS 定位程序设计(参见例 10.2.1)。

(1) 再次解压 ch10 内的压缩文件 GPSLocation.zip，导入工程 GPSLocation。

(2) 打开清单文件，查看使用 GPS 定位的权限。

(3) 打开程序文件 MainActivity.java，查看位置监听器接口要实现的方法。

(4) 查看使用 GPS 实时获取经纬度的代码。

(5) 部署工程并做运行测试。

2. WiFi 定位程序设计(参见例 10.2.2)。

(1) 再次解压 ch10 内的压缩文件 WiFiLocation.zip，导入工程 WiFiLocation。

(2) 打开清单文件，查看使用 WiFi 定位的权限。

(3) 打开程序文件 MainActivity.java，查看检测 WiFi 是否打开及开启 WiFi 的代码。

(4) 查看位置管理器方法 requestLocationUpdates()中位置监听器的用法，并与 GPSLocation 工程中的位置监听器的用法相比较。

(5) 查看使用 WiFi 定位获取经纬度的代码。

(6) 部署工程并做运行测试。

(7) 打开清单文件，屏蔽使用 WiFi 中的第二条权限，验证部署工程时会失败。

3. 百度综合定位程序设计(参见例 10.3.1)。

(1) 再次解压 ch10 内的压缩文件 BDLocation.zip，导入工程 BDLocation。

(2) 打开清单文件，查看使用定位的相关权限及百度位置应用的 App Key。

(3) 打开工程里的文件夹 libs，查看使用的百度位置服务的 SDK。

(4) 打开程序文件 MainActivity.java，查看使用获取经纬度和地名信息的代码。

(5) 打开手机的 WiFi 或 GPRS 连接后，部署工程并做运行测试。

(6) 查看源程序中检测网络连接及没有网络连接时的处理代码(运行系统自带的设置程序)，重新在手机上做运行测试。

(7) 打开清单文件，分别选中一些关键代码后，使用“Ctrl+Shift+/”屏蔽，做部署和运行测试，最后使用“Ctrl+Shift+\”还原。

4. 百度地图应用程序设计(参见例 10.3.2)。

(1) 再次解压 ch10 内的压缩文件 BDLocationMap.zip，导入工程 BDLocationMap。

(2) 打开手机的 WiFi 连接。

(3) 部署工程并做运行测试，经纬度和地名信息显示正常，但没有地图。

(4) 访问 http://developer.baidu.com，单击导航条上的“注册”按钮，使用手机注册一个百度开发者帐号。

(5) 访问 http://lbsyun.baidu.com/apiconsole/key，登录后单击“创建应用”按钮，输入应用名称“BDLocationMap”，选择应用类型为“for moble”。

(6) 在安全码文本框内粘贴从 Eclipse 中复制部署 Android 工程使用的 SHA1 fingerprint。

(7) 在安全码文本框内的最后加上“;com.example.bdlocationmap”，单击“确认”按钮，得到本应用的 Key，放入 Windows 的剪贴板内。

(8) 打开清单文件，用刚才得到的 Key 替换<meta-data>标签内的 Map Key。

(9) 部署工程并做运行测试，地图显示正常。

(10) 打开工程里的文件夹 libs，查看使用的百度位置及地图服务的.jar 包。

三、实验小结及思考

(由学生填写，重点写上机中遇到的问题。)

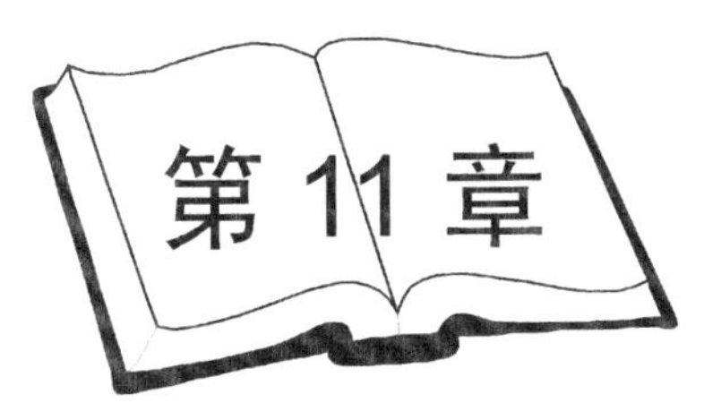

第 11 章 Android 网络编程

在第 9 章，我们学习了 Android 近距离的通信技术，它本质上也属于网络编程的范畴。由于网络分层特性，在 Android 平台中，可以使用多种接口进行网络编程。本章介绍的网络编程不再是近距离的通信，网络编程主要涉及手机客户端与传统的 Web 服务器之间的信息交互。本章学习要点如下：

- 掌握标准 Java 网络编程方式——HttpURLConnection 的使用；
- 掌握使用 Apache 接口 HttpClient 进行网络编程的方法；
- 掌握访问 Web Service 的编程方法；
- 掌握针对 TCP/IP 的 Socket/ServerSocket 编程方法；
- 掌握针对 UDP 的 DatagramSocket /DatagramPackage 编程方法；
- 掌握手机客户端与 Web 服务器通信的编程方法；
- 了解从 Web 服务器向手机推送消息的方法。

11.1 基于 HTTP 协议的标准 Java 网络编程

11.1.1 Android 网络编程概述

Android API 除了包含 Java 提供的标准网络编程方法外，还引入了 Apache 的 http 扩展包，并针对 WiFi、Bluetooth 等设备分别提供了单独的 API。因此，在 Android 平台中，可以使用 Java 标准接口、Apache 接口和 Android 网络接口共三种。

HTTP 是 Hypertext Transfer Protocol 的英文缩写，表示超文本传输协议。HTTP 协议是互联网上应用最多、最为广泛的一种网络协议，是从 WWW 服务器传输超文本到本地浏览器的传送协议。

Android 网络编程可以划分为对应用层的 HTTP 编程和对传输层的 Socket 编程两种，如图 11.1.1 所示。

基于 HTTP 网络编程，既可以使用标准 Java 接口(对应于软件包 java.net)，还可以使用 Apache 接口(对应于软件包 org.apache.http.client 和 org.apache.http.impl)或者 Android 网络接口(对应于软件包 android.net 和 android.net.http)。

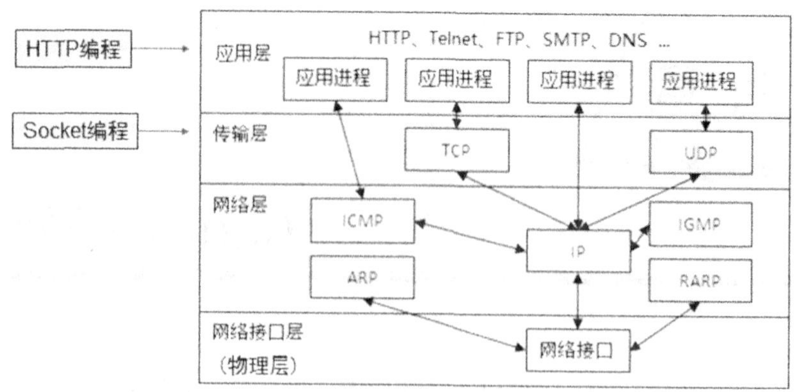

图 11.1.1　HTTP 编程与 Socket 编程

基于 HTTP 请求的 Android 应用，都需要在清单文件中配置如下访问 Internet 的权限：

<uses-permission android:name="android.permission.INTERNET"/>

11.1.2　HTTP 请求与响应

HTTP 请求共有 Get 和 Post 两种请求方式。

Get 请求方式是通过把参数键值对附加在 url 后面来传递的。在服务器端可以直接读取，效率较高，但缺乏安全性，也无法处理复杂的数据，长度受限制。Get 请求方式主要用于传递简单的参数。

在 Post 请求方式中，传输参数会被打包在 http 报头中，可以是二进制的。Post 请求方式便于传送较大的数据，同时因为不暴露数据在浏览器的地址栏中，安全性相对较高，但这样的处理效率会受到影响。

11.1.3　HttpURLConnection 编程

获取请求指定的 URL 后的响应信息的一种方法是，使用终结类 java.net.URL 的 openConnection()方法得到一个 HttpURLConnection 对象，通过该对象建立与服务器的连接，进而获取服务器响应的信息，最后通过 java.io 包提供的相关类进行输出。

抽象类 HttpURLConnection 及相关处理类的定义，如图 11.1.2 所示。

注意：抽象类 HttpURLConnection 是抽象类 URLConnection 的子类，提供了断开 Http 连接的方法 disconnect()，父类不具有该方法。

在 Eclipse 环境中，选中抽象类 HttpURLConnection 后，按快捷键“Ctrl+T”所得到的继承关系，如图 11.1.3 所示。

通过 URL 类的 openConnection()方法创建 HTTP 请求的连接对象时，为了使用断开连接等方法，需要进行类型转换，其代码如下：

```
HttpURLConnection urlConn = (HttpURLConnection) url.openConnection();
```

```
java.net
  URL.class
    URL
      URL(String)
      openConnection() : URLConnection
  URLConnection.class
    URLConnection
      getInputStream() : InputStream
java.io
  InputStream.class
    InputStream
  InputStreamReader.class
    InputStreamReader
      InputStreamReader(InputStream)
  Reader.class
    Reader
  BufferedReader.class
    BufferedReader
      BufferedReader(Reader)
```

图 11.1.2　抽象类 HttpURLConnection 及相关处理类的定义

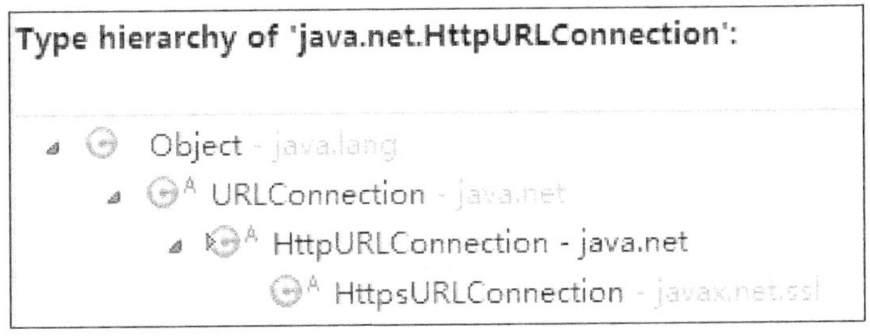

图 11.1.3　抽象类 HttpURLConnection 的继承关系

注意：获取请求指定的 URL 后响应信息的另一种方法是使用 org.apache.http.client.HttpClient 接口(参见第 11.2 节)。

【例 11.1.1】通过创建和使用 URL 对象，访问网络资源(抓取网页)。

【程序运行】程序运行时，单击窗口中的“点击获取数据”按钮后，会在一个 TextView 控件里显示访问本书作者教学网站(http://www.wustwzx.com)主页在浏览器端的 HTML 代码(含客户端脚本)，如图 11.1.4 所示。

【设计步骤】

(1) 新建名为 URLDemo 的 Android 应用工程，勾选创建 Activity，使用默认的主 Activity

和布局文件名，文件工程结构(部分)如图 11.1.5 所示。

图 11.1.4 例 11.1.1 程序运行结果

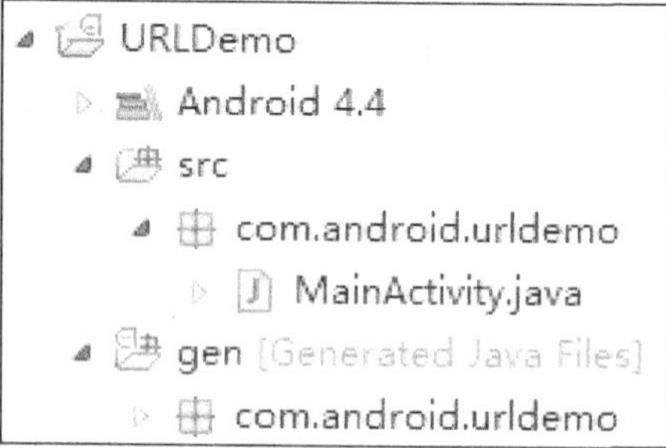

图 11.1.5 URLDemo 工程文件结构(部分)

(2) 编写工程的主 Activity 文件 MainActivity.java，其代码如下：

```
package com.android.urldemo;
/*
 * 手机客户端获取Web服务器数据实例
 * 主要使用Java提供的类HttpURLConnection和URL
 * 在手机上运行程序前，需要先打开移动数据连接或使用WiFi网络
 * 文本框的滑动方法使用android.text.method.ScrollingMovementMethod类型的参数
 * 在布局文件中定义TextView为多行、垂直滑块(任选)
*/
import android.app.Activity;
import android.os.Bundle;
import java.io.BufferedReader;
import java.io.IOException;
import java.io.InputStreamReader; //
import java.net.HttpURLConnection;   //主类
import java.net.MalformedURLException;
import java.net.URL;   //相关类
import android.text.method.ScrollingMovementMethod;   //
import android.view.View;
import android.view.View.OnClickListener;
import android.widget.Button;
import android.widget.TextView;
public class MainActivity extends Activity {
    private TextView textView_HTTP;
    @Override
    public void onCreate(Bundle savedInstanceState) {
```

```
super.onCreate(savedInstanceState);
setContentView(R.layout.main);
textView_HTTP=(TextView) findViewById(R.id.TextView_HTTP);
//设置垂直滑动手势处理
 textView_HTTP.setMovementMethod(ScrollingMovementMethod.getInstance());
//textView_HTTP.setMovementMethod(new ScrollingMovementMethod());
Button button_http = (Button) findViewById(R.id.Button_HTTP);
button_http.setOnClickListener(new OnClickListener() {
    @SuppressWarnings("unused")
    public void onClick(View v){    //事件处理
        String httpUrl = "http://www.wustwzx.com"; //域名字符串
        String resultData = "";  //结果字符串
        URL url = null;  // 定义URL对象
        try {
            url = new URL(httpUrl); // 构造URL对象时需要使用异常处理
        }
        catch (MalformedURLException e) {
            System.out.println(e.getMessage()); //打印出异常信息
        }
        if (url != null) {  //如果URL不为空时
            try {  //有关网络操作时，需要使用异常处理
                HttpURLConnection urlConn = (HttpURLConnection) url
                        .openConnection();   // 打开连接并转型
                InputStreamReader in = new InputStreamReader(urlConn
                        .getInputStream(),"gbk");  // 服务器返回数据
                // 为输出创建BufferedReader
                BufferedReader buffer = new BufferedReader(in);
                String inputLine = null;
                while (((inputLine = buffer.readLine()) != null)){
                    resultData += inputLine + "\n";    //换行
                }
                in.close(); // 关闭InputStreamReader
                urlConn.disconnect(); // 关闭HTTP连接
                if (resultData != null)
                    textView_HTTP.setText(resultData);  //显示
                else
                    textView_HTTP.setText("Sorry,the content is null");
            }
            catch (IOException e) {
```

```
                textView_HTTP.setText(e.getMessage()); }
            }
            else
                textView_HTTP.setText("url is null"); //当url为空时输出
        }
    });
  }
}
```

(3) 部署工程并做运行测试。

11.2 Apache 网络编程与 Web 服务

11.2.1 HttpClient 编程

Apache 实验室针对 Java 网络编程灵活性不足的缺点，对包 java.net 中的相关类做了进一步的封装。Android 引入了 Apache 网络接口中最重要的接口 HttpClient 及其相关类，如图 11.2.1 所示。

```
org.apache.http.client
    HttpClient.class
        HttpClient
org.apache.http.impl.client
    AbstractHttpClient.class
        AbstractHttpClient
    DefaultHttpClient.class
        DefaultHttpClient
```

图 11.2.1　HttpClient 接口及其相关类

其中，AbstractHttpClient 实现了 HttpClient 接口，且是 DefaultHttpClient 类的父类。

注意：这种“接口-抽象类-类”的设计方式，与数据适配器(参见第 4 章)中的“Adapter 接口-AbstractAdapter 抽象类-ArrayAdapter 等类”是一致的。

HttpClient 接口的定义，如图 11.2.2 所示。

```
org.apache.http.client
    HttpClient.class
        HttpClient
            execute(HttpUriRequest) : HttpResponse
            execute(HttpUriRequest, ResponseHandler<? extends T>) <T> : T
            execute(HttpUriRequest, ResponseHandler<? extends T>, HttpContext) <T> : T
            execute(HttpUriRequest, HttpContext) : HttpResponse
            execute(HttpHost, HttpRequest) : HttpResponse
            execute(HttpHost, HttpRequest, ResponseHandler<? extends T>) <T> : T
            execute(HttpHost, HttpRequest, ResponseHandler<? extends T>, HttpContext) <T> : T
            execute(HttpHost, HttpRequest, HttpContext) : HttpResponse
            getConnectionManager() : ClientConnectionManager
            getParams() : HttpParams
```

图 11.2.2　HttpClient 接口的定义

11.2.2　使用 Apache 网络接口调用 Web 服务

Web 服务是一个平台独立、松耦合、基于可编程的 Web 应用程序，可使用开放的 XML 标准来描述、发布、发现、协调和配置这些应用程序，用于开发分布式的互操作的应用程序。简单地说，Web 服务是远程的某个服务器对外公开的某种功能或方法，通过调用该服务以获得用户需要的信息。

注意：由于对于网络状况的不可预见性，很有可能在网络访问的时候造成阻塞 UI 主线程(出现假死)的现象。解决该问题的办法有：① 独立线程； ② 异步线程 AsyncTask； ③ StrictMode 修改默认的策略。

Web 服务及其使用方式，如图 11.2.3 所示。

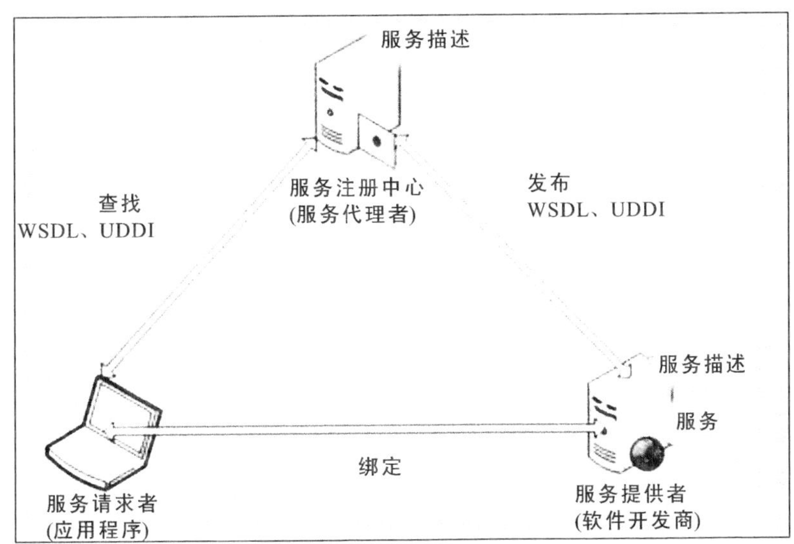

图 11.2.3　Web 服务及其使用方式

【例 11.2.1】使用 HttpClient 调用 Web 服务，查询手机归属地。

【程序运行】程序运行后，输入手机号码(段)前面的 7 位号段，单击“查询”按钮后的程序运行界面，如图 11.2.4 所示。

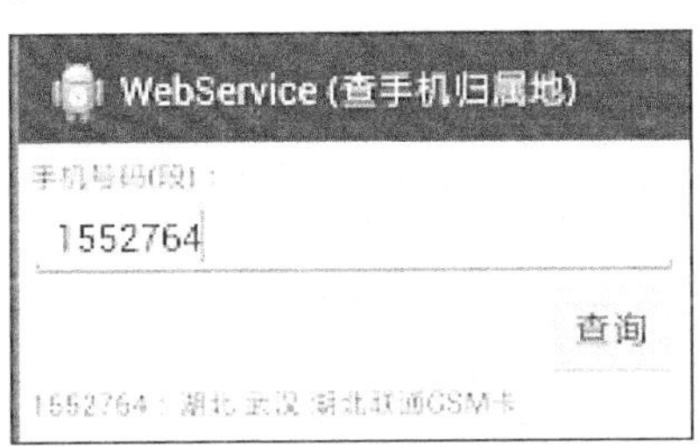

图 11.2.4　例 11.2.1 的程序运行界面

【设计步骤】

(1) 新建名为 WebService 的 Android 应用工程，勾选创建 Activity，使用默认的主 Activity 和布局文件名，工程文件结构(部分)如图 11.2.5 所示。

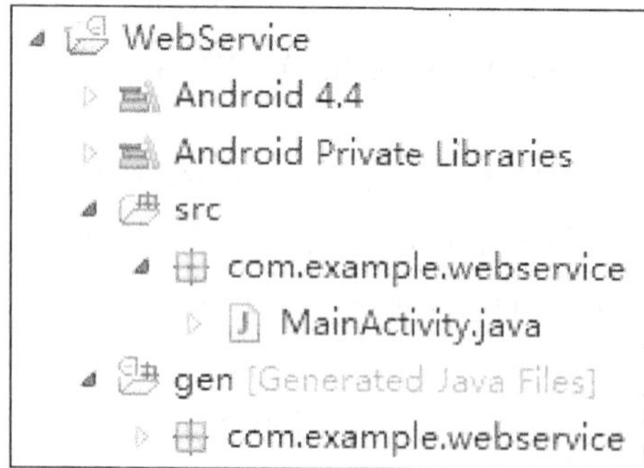

图 11.2.5 WebService 工程文件结构(部分)

(2) 编写工程的主 Activity 文件 MainActivity.java，其代码如下：

```
package com.example.webservice;
/* HttpClient编程：手机调用Web服务，查询国内手机号码归属地
 * 通过HttpClient对象向Web服务器以Post方式发送请求，得到HttpResponse对象
 * Web服务的网址是：http://webservice.webxml.com.cn/WebServices/MobileCodeWS.asmx
*/
import org.apache.http.impl.client.DefaultHttpClient; //主类
import org.apache.http.client.HttpClient;   //主接口1
import org.apache.http.HttpResponse;   //主接口2
import org.apache.http.HttpStatus;
import org.apache.http.NameValuePair;
import org.apache.http.client.entity.UrlEncodedFormEntity;
import org.apache.http.client.methods.HttpPost;   //
import org.apache.http.message.BasicNameValuePair;
import org.apache.http.protocol.HTTP;
import org.apache.http.util.EntityUtils;
import android.os.Bundle;
import android.os.StrictMode;   //
import android.app.Activity;
import android.view.View;
import android.widget.Button;
import android.widget.EditText;
import android.widget.TextView;
import android.widget.Toast;
import java.util.ArrayList;
import java.util.List;
import com.example.webservice.R;
public class MainActivity extends Activity {
    EditText phoneSecEditText;
```

```
TextView resultView ;
@Override
protected void onCreate(Bundle savedInstanceState) {
    super.onCreate(savedInstanceState);
    setContentView(R.layout.activity_main);
    //因为访问网络可能引起主线程阻塞，所以在主线程中强制使用子线程策略
    StrictMode.ThreadPolicy policy = new StrictMode.ThreadPolicy.
                                            Builder().permitAll().build();
    StrictMode.setThreadPolicy(policy);
    phoneSecEditText = (EditText) findViewById(R.id.phone_sec);
    resultView = (TextView) findViewById(R.id.result_text);
    Button queryButton = (Button) findViewById(R.id.query_btn);
    queryButton.setOnClickListener(new View.OnClickListener() {
        @Override
        public void onClick(View v) {
            // 手机号码(段)
            String phoneSec = phoneSecEditText.getText().toString().trim();
            // 判断用户输入的手机号码(段)是否合法
            if ("".equals(phoneSec) || phoneSec.length() < 7) {
                phoneSecEditText.setError("您输入的手机号码不能少于7位！");
                phoneSecEditText.requestFocus();
                // 将显示查询结果的TextView清空
                resultView.setText("");
                return;
            }
            getRemoteInfo(phoneSec); //调用查询手机号码(段)信息的方法
        }
    });
}
public void getRemoteInfo(String phoneSec) {  //手机号码(段)归属地查询
    // 定义待请求的URL
    String requestUrl = "http://webservice.webxml.com.cn/WebServices/" +
                                    "MobileCodeWS.asmx/getMobileCodeInfo";
    HttpClient client = new DefaultHttpClient();  // 创建HttpClient实例
    HttpPost post = new HttpPost(requestUrl);     // Post请求方式
    List<NameValuePair> params = new ArrayList<NameValuePair>();
    // 设置需要传递的参数
    params.add(new BasicNameValuePair("mobileCode", phoneSec));
    params.add(new BasicNameValuePair("userId", ""));
```

```
        try {
            // 设置URL编码
            post.setEntity(new UrlEncodedFormEntity(params, HTTP.UTF_8));
            // 发送请求并获取反馈
            HttpResponse response = client.execute(post);   //
            String result="";
            // 判断请求是否成功处理
            if (response.getStatusLine().getStatusCode() == HttpStatus.SC_OK) {
                // 解析返回的内容
               result = EntityUtils.toString(response.getEntity(),"utf-8");
            }
            else{
                 result ="没有查询结果:网络未连接或输入号段错误.";
                 }
            resultView.setText(filterHtml(result));   //用正则表达式过滤一下
        }
         catch (Exception e) {
            e.printStackTrace();
            Toast.makeText(getBaseContext(), "出错了!",
                                                    Toast.LENGTH_SHORT).show();
        }
    }
    private String filterHtml(String source) {   //使用正则表达式过滤HTML标记
        if(null == source)    return "";
        return source.replaceAll("</?[^>]+>","").trim();
        // 所有HTML标签以<打头，以>结束
        // /?表示匹配0个或1个/字符
        // [^>]+表示匹配不是>的任意一个或多个字符
    }
}
```

11.3 手机客户端程序设计

11.3.1 与 Web 服务器交互的手机客户端

第 11.2.2 小节中的调用 Web 服务，实质上是手机应用程序访问一个特殊的 Web 服务器。

要实现手机应用程序与一般的 Web 服务器之间的信息交互，一般采用 Apache 网络编程接口 HttpClient。

手机客户端是面向 Web 服务器的 Android 应用程序，其原理是客户端程序需要对 Web 服务器返回的结果数据做进一步的处理。

网站的 Web 服务器端有多种选择，例如，由 Tomcat 搭建的 JSP 网站服务器或由 AppServ 搭建的 PHP 服务器。

【例 11.3.1】手机客户端程序与 Web 服务器的信息交互——用户注册与登录。

【程序运行】先开启手机的 GPRS 网络及个人热点，使笔记本电脑通过手机 WiFi 上网，然后启动 JSP 网站服务器 Apache Tomcat(包含了具有用户注册与登录功能的项目)，这样，手机就能根据笔记本电脑此时的 IP 地址访问 JSP 网站。最后，运行具有与 Web 服务器信息交互的手机客户端程序，其界面分别如图 11.3.1 和图 11.3.2 所示。

图 11.3.1　手机注册及注册成功后的界面

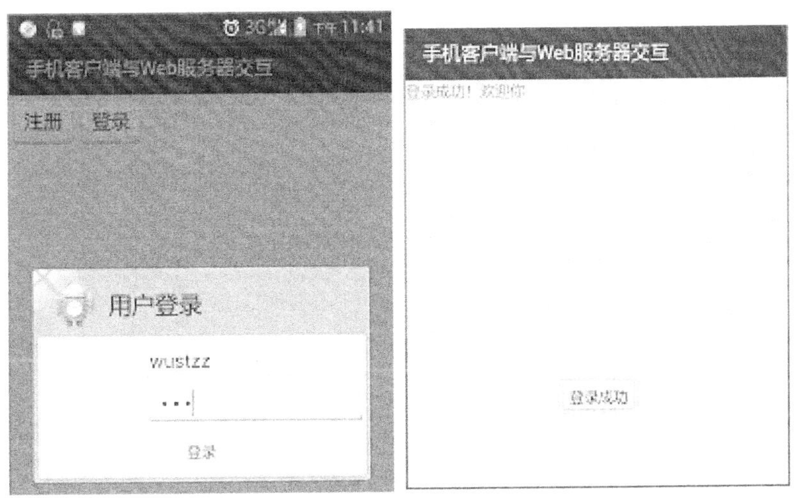

图 11.3.2　手机登录及登录成功后的界面

【设计步骤】

(1) 在 MyEclipse 中新建名为 WuLogin 的 Java Web 工程，采用 MVC 模式开发，完成后

的文件系统如图 11.3.3 所示。

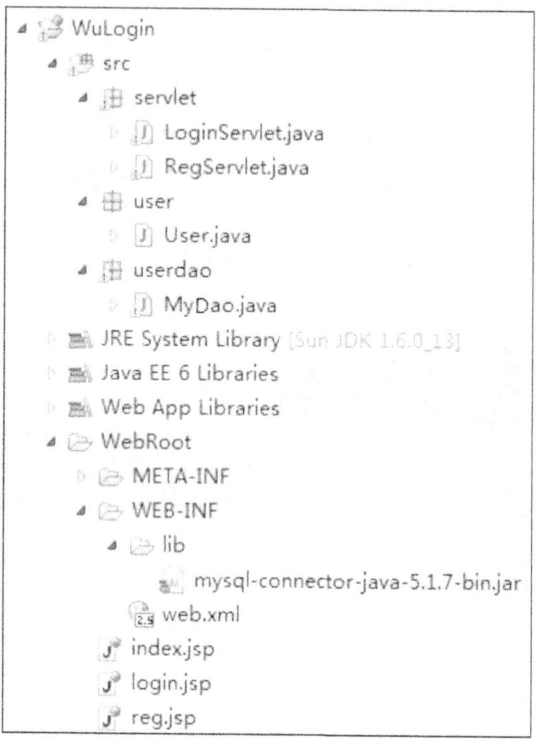

图 11.3.3　一个实现用户注册与登录的 JSP 网站文件系统

其中，工程数据库使用的是 MySQL，其相关信息含于 MySQL 数据库连接文件 MysDao.java 内。

(2) 部署 Java Web 工程 WuLogin 至 Tomcat 服务器，Tomcat 的文件夹 webapps 的文件系统如图 11.3.4 所示。

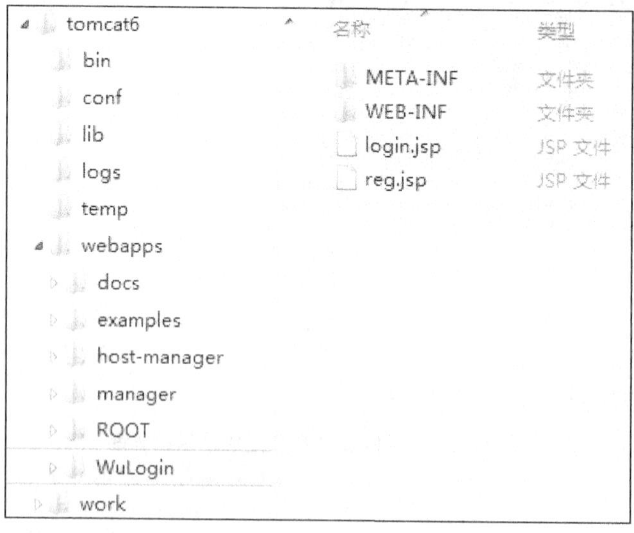

图 11.3.4　webapps 的文件系统

(3) 在浏览器的地址栏中输入 http://localhost:8080/WuLogin/reg.jsp 之后的浏览效果，如图 11.3.5 所示。

(4) 在浏览器的地址栏中输入 http://localhost:8080/WuLogin/login.jsp 之后的浏览效果，如图 11.3.6 所示。

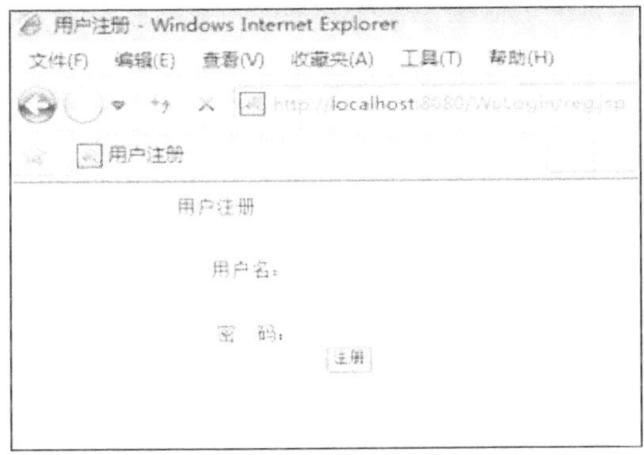

图 11.3.5　访问 JSP 网站注册页面

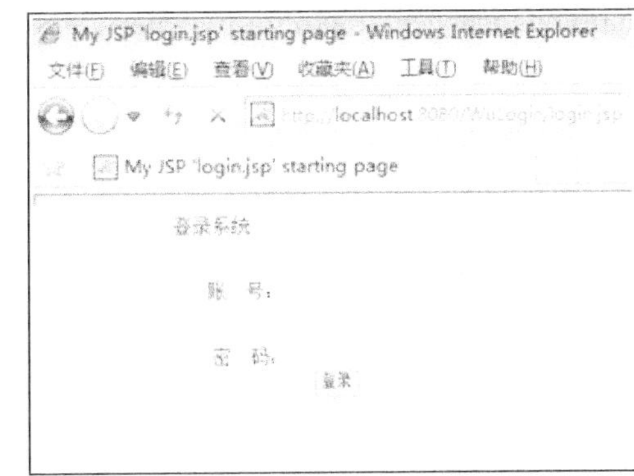

图 11.3.6　访问 JSP 网站登录页面

(5) 在开发 Android 应用的 Eclipse 环境中，新建名为"手机客户端(用户注册与登录)"的 Android 应用工程，其文件系统如图 11.3.7 所示。

图 11.3.7　Android 手机客户端文件系统

(6) 编写 Android 应用工程的主 Activity 文件 MainActivity.java，其代码如下：

```
package com.example.webandandroid;
/*
 * 运行本程序前，要求手机能访问Web服务器
 * 本手机客户端程序的功能是与Web服务器交互信息
 */
import android.app.Activity;
import android.os.Bundle;
```

```
import org.apache.http.HttpResponse;   //
import org.apache.http.client.ClientProtocolException;
import org.apache.http.client.methods.HttpPost;
import org.apache.http.impl.client.DefaultHttpClient;
import org.apache.http.util.EntityUtils;   //
import android.app.AlertDialog;
import android.app.ProgressDialog;
import android.content.Context;
import android.content.DialogInterface;
import android.os.Handler;   //
import android.os.Message;
import android.view.LayoutInflater;   //
import android.view.View;
import android.widget.Button;
import android.widget.EditText;
import android.widget.TextView;
import android.widget.Toast;
import java.io.IOException;
import android.annotation.SuppressLint;

public class MainActivity extends Activity {
    private Button bt_reg ;
    private Button bt_log ;
    private EditText userEdit;
    private EditText passEdit;
    private TextView tv;
    private AlertDialog alertDialogreg;
    private AlertDialog alertDialoglog;
    private ProgressDialog   proDialog_login=null;
    private ProgressDialog   proDialog_reg=null;
    String uri;   //用于存放Web主机地址
    //创建“注册”的Handler对象，重写回调处理函数
    @SuppressLint("HandlerLeak")
    private Handler handler_reg = new Handler(){
        @Override
        public void handleMessage(Message msg) {
            super.handleMessage(msg);   //

            Bundle b = msg.getData();   //得到捆绑数据
            String data = b.getString("key");
```

```
            if(!data.equals("yes")){    //比较字符串内容
                Toast.makeText(MainActivity.this, "注册失败，错误信息："+
                            b.getString("错误信息"),Toast.LENGTH_LONG).show();
                return ;
            }
            Toast.makeText(MainActivity.this, "注册成功，"+
                                    "请登录",Toast.LENGTH_LONG).show();   //
    }};
//创建“登录”的Handler对象，重写回调处理函数
@SuppressLint("HandlerLeak")
private Handler handler_log = new Handler(){   //
        @Override
        public void handleMessage(Message msg) {
            super.handleMessage(msg);
            Bundle b = msg.getData();
            String data = b.getString("key");
            if(!data.equals("yes")){
                Toast.makeText(MainActivity.this, "登录失败"+
                            b.getString("错误信息"),Toast.LENGTH_LONG).show();
                return ;
            }
            Toast.makeText(MainActivity.this, "登录成功",
                                    Toast.LENGTH_SHORT).show();   //
            setContentView(R.layout.fragment_main);   //登录成功后的视图
            tv = (TextView)findViewById(R.id.tv);
            tv.setText("登录成功！ 欢迎你");     //
        }};

@Override
protected void onCreate(Bundle savedInstanceState) {
    super.onCreate(savedInstanceState);
    setContentView(R.layout.activity_main);

    bt_reg = (Button)findViewById(R.id.bt_reg);
    bt_log = (Button)findViewById(R.id.bt_log);
    LayoutInflater layoutInflater = LayoutInflater.from(this);   //实例化
    View view1 = layoutInflater.inflate(R.layout.reg_log, null);   //实例化
    View view2 = layoutInflater.inflate(R.layout.reg_log, null); //公共视图
    //注册操作
    alertDialogreg = new AlertDialog.Builder(this).
```

```
setTitle("用户注册").
setIcon(R.drawable.ic_launcher).
setView(view1).setPositiveButton("注册",
                    new DialogInterface.OnClickListener() {
    @Override
    public void onClick(DialogInterface dialog, int which) {
      userEdit=(EditText)alertDialogreg.findViewById(R.id.user_rid);
            passEdit=(EditText)alertDialogreg.findViewById
                                                (R.id.pass_rid);
      final String user = userEdit.getText().toString().trim();
      final String pwd = passEdit.getText().toString().trim();
      if(user.equals("")||pwd.equals("")){
            Toast.makeText(MainActivity.this,
     "用户名和密码不为空！",Toast.LENGTH_SHORT).show();
            return;
      }
            proDialog_reg=ProgressDialog.show(MainActivity.this,
                        "注册", "正在注册，请稍后......",true);
        new Thread(){
            @Override
            public void run(){
                Message msg = new Message();
                Bundle b = new Bundle();// 存放数据
                 try{
                    String str = reg(MainActivity.this,user,pwd).trim();
                    if(("注册成功").equals(str)){
                       b.putString("key", "yes");
                    }
                    else{
                            b.putString("key","no");
                        }
                 }
                 catch(Exception e){
                       b.putString("key", "no");
                       b.putString("错误信息",
                                   "请检查网络是否正常！");
                       e.printStackTrace();
                 }
                 finally{
```

```
                                        msg.setData(b);
                                        handler_reg.sendMessage(msg);
                                        proDialog_reg.dismiss();
                                    }
                                }
                            }.start();
                        }
                    }).create();

        bt_reg.setOnClickListener(new Button.OnClickListener(){
            @Override
            public void onClick(View v) {
                alertDialogreg.show();
            }
        });

        //登录操作
        alertDialoglog = new AlertDialog.Builder(this).
                setTitle("用户登录").
                setIcon(R.drawable.ic_launcher).
                setView(view2).setPositiveButton("登录",
                                                new DialogInterface.OnClickListener() {
                    @Override
                    public void onClick(DialogInterface dialog, int which) {
                        userEdit=(EditText)alertDialoglog.findViewById(R.id.user_rid);
                        passEdit=(EditText)alertDialoglog.findViewById(R.id.pass_rid);
                        final String user = userEdit.getText().toString().trim();
                        final String pwd = passEdit.getText().toString().trim();
                        if(user.equals("")||pwd.equals("")){
                            Toast.makeText(MainActivity.this, "用户名和密码不为空！",
                                                    Toast.LENGTH_SHORT).show();
                            return;
                        }
                        proDialog_login=ProgressDialog.show(MainActivity.this,
                                                "登录","正在登录，请稍后...",true);
                        new Thread(){
                            @Override
                            public void run(){
                                Message msg = new Message();
```

```
                Bundle b = new Bundle();// 存放数据
        try{
                String str = log(MainActivity.this,user,pwd).trim();   //
            //不能使用==，此处只比较内容
                    if(("登录成功,欢迎你："+user).equals(str))
                    b.putString("key", "yes");
            else
                    b.putString("key","no");
            }
            catch(Exception e){
                b.putString("key", "no");
                b.putString("错误信息","请检查网络是否正常！");
                e.printStackTrace();       }
            finally{
                msg.setData(b);
                handler_log.sendMessage(msg);
                proDialog_login.dismiss();    //取消进度框
            }
        }
    }.start();
  }
}).create();

bt_log.setOnClickListener(new Button.OnClickListener(){
    @Override
    public void onClick(View v) {
        alertDialoglog.show();
    }
});
}

private String reg(Context context,String un, String pwd) {
    //访问Web主机(笔记本电脑)里的注册页面
    uri="http://192.168.43.250:8080/WuLogin/RegServlet?
                        username="+un+"&password="+pwd; //
    return getFromUrl(context,uri);   //请求服务器的返回(响应)值
}

private String login(Context context,String un, String pwd) {
    //访问Web主机(笔记本电脑)里的登录页面
```

```
        uri="http://192.168.43.250:8080/WuLogin/LoginServlet?
                                        username="+un+"&password="+pwd;//
        return getFromUrl(context,uri);
    }
    @SuppressLint("ShowToast")
    private String getFromUrl(Context context, String uri) {
        String strResult="";
        HttpPost httpRequest = new HttpPost(uri);
        try
        {
          HttpResponse httpResponse = new DefaultHttpClient().
                                                  execute(httpRequest);
          if(httpResponse.getStatusLine().getStatusCode() == 200)
          {
            strResult = EntityUtils.toString(httpResponse.getEntity());   //
            return strResult;
          }
          else
            return strResult;
        }
        catch (ClientProtocolException e)
        {
          e.printStackTrace();
          return strResult;
        }
        catch (IOException e)
        {
          e.printStackTrace();
          return strResult;
        }
        catch (Exception e)
        {
          e.printStackTrace();
          Toast.makeText(context, "Error Response: "+
                                        e.getMessage().toString(),0).show();
          return strResult;
        }
    }
}
```

11.3.2 使用激光推送平台 JPush 以 Web 方式向手机推送消息

推送技术的基本思想是将浏览器主动查询信息改为服务器主动发送信息。推送技术是使用一定的技术标准或协议，在互联网上通过定期传送用户需要的信息来减少信息过载的一项新技术。推送技术通过自动传送信息给用户，来减少用于网络上搜索的时间。它根据用户的兴趣来搜索、过滤信息，并将其定期推给用户，帮助用户高效率地发掘有价值的信息。

手机推送服务是指服务器单向将信息实时送达手机的服务，其原理就是通过建立一条手机与服务器的连接链路，当有消息需要发送到手机时，通过此链路发送。服务器发送一批数据，浏览器显示这些数据，同时保证与服务器的连接。当服务器需要再次发送一批数据时，浏览器显示数据并保持连接。以后，服务器仍然可以发送批量数据，浏览器继续显示数据，依此类推。

注意：

(1) 手机客户端软件需要安装且运行一次，否则无法接收推送的消息；

(2) 手机需要开启 GPRS 网络或使用 WiFi，否则无法接收消息；

(3) Web 服务器向手机推送信息，不同于手机短信广播，但可以认为它是“Web 广播”。

鉴于 Web 服务器的开发难度大，不建议小团队开发，小团队可使用稳定的第三方推送方案实现 Android 推送服务，如使用极光推送(https://www.jpush.cn)等。

首次使用极光推送服务，需要先访问网站 http://www.jpush.cn，注册极光推送开发者账号。

注意： 注册时，用户只有通过了邮箱验证才能登录。

JPush 用户登录后，单击下方的“控制台”按钮，出现的界面如图 11.3.8 所示。

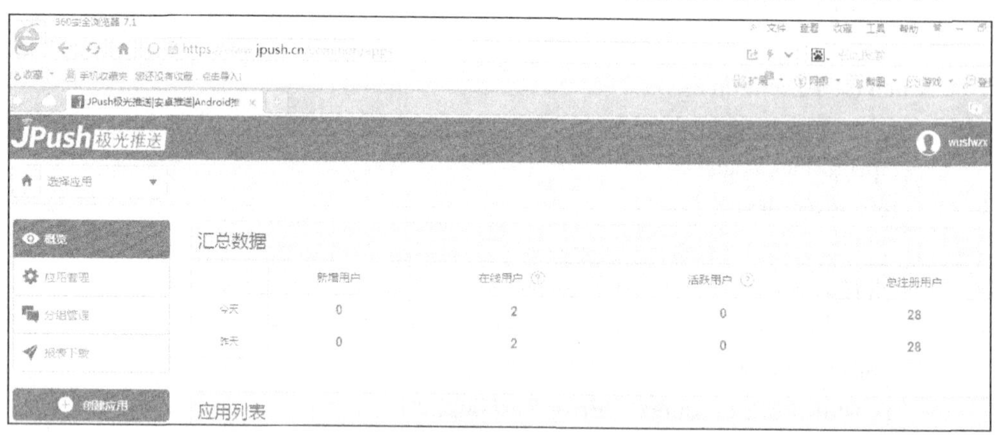

图 11.3.8 登录 JPush 网站进入控制台

单击左下方的“创建应用”按钮，出现创建 JPush 应用的界面，需要开发者填写应用名称及应用包名，如图 11.3.9 所示。

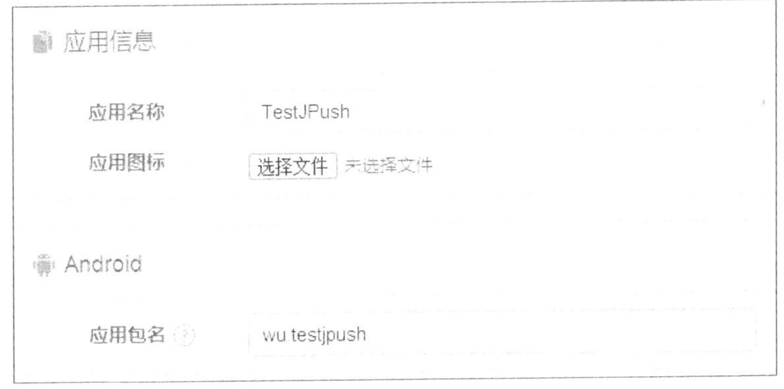

图 11.3.9　创建 JPush 应用的界面

单击下方的“创建我的应用”按钮后，网站系统自动生成 AppKey 和该 Android 应用工程的下载链接，界面如图 11.3.10 所示。

单击“下载 Android Example”按钮，得到该 Android 应用工程的压缩文件。解压文件并导入该工程后的文件系统(主要部分)，如图 11.3.11 所示。

图 11.3.10　JPush 应用的 AppKey 和工程源码

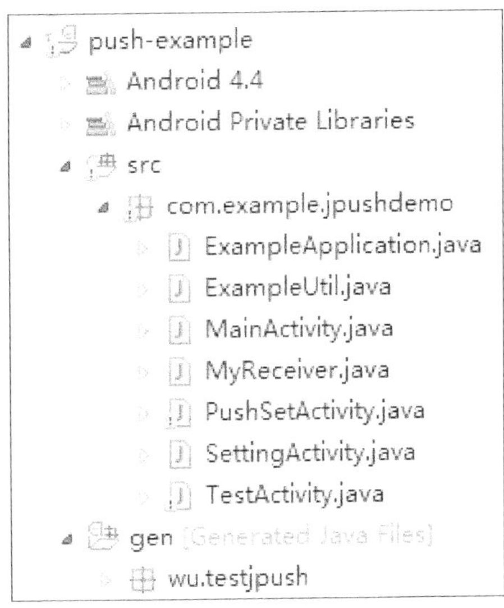

图 11.3.11　JPush 应用工程的文件系统(主要部分)

部署工程到手机后，只要手机有网络(使用 GPRS 或 WiFi)，就可以接收从 JPush 网站推送的消息了。

注意：

(1) 使用 JPush 创建的应用，实质上是一个手机客户端程序。

(2) 推送消息不同于短信，它是以通知的形式出现在手机的通知栏里的。

在 JPush 网站里，选择自己创建的某个应用，单击“发送通知”按钮后，就可编辑推送内容，最后单击“立即发送”按钮完成信息的推送。

在 JPush 网站里，选择某个应用，单击“推送”按钮后，即可查询已经推送的历史记录，包括发送时间、内容、类型和客户端接收人数等，其界面如图 11.3.12 所示。

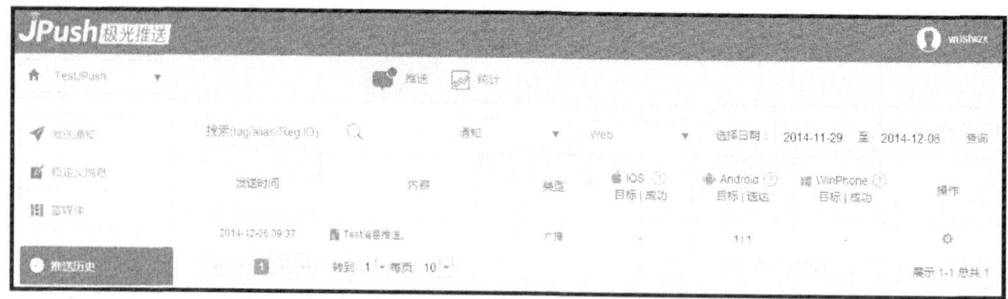

图 11.3.12　查看 JPush 应用的推送历史

11.3.3　使用百度 LBS 云服务器

在第 11.3.1 小节中，手机客户端使用的数据通过 Web 服务器存放在数据库服务器里。为了方便关于位置服务方面的应用，百度提供了 LBS 云服务器，用以存储用户地理位置的信息数据。

百度开发者访问 http://developer.baidu.com/map/index.php?title=lbscloud，即可进入百度地图 LBS 云平台，如图 11.3.13 所示。

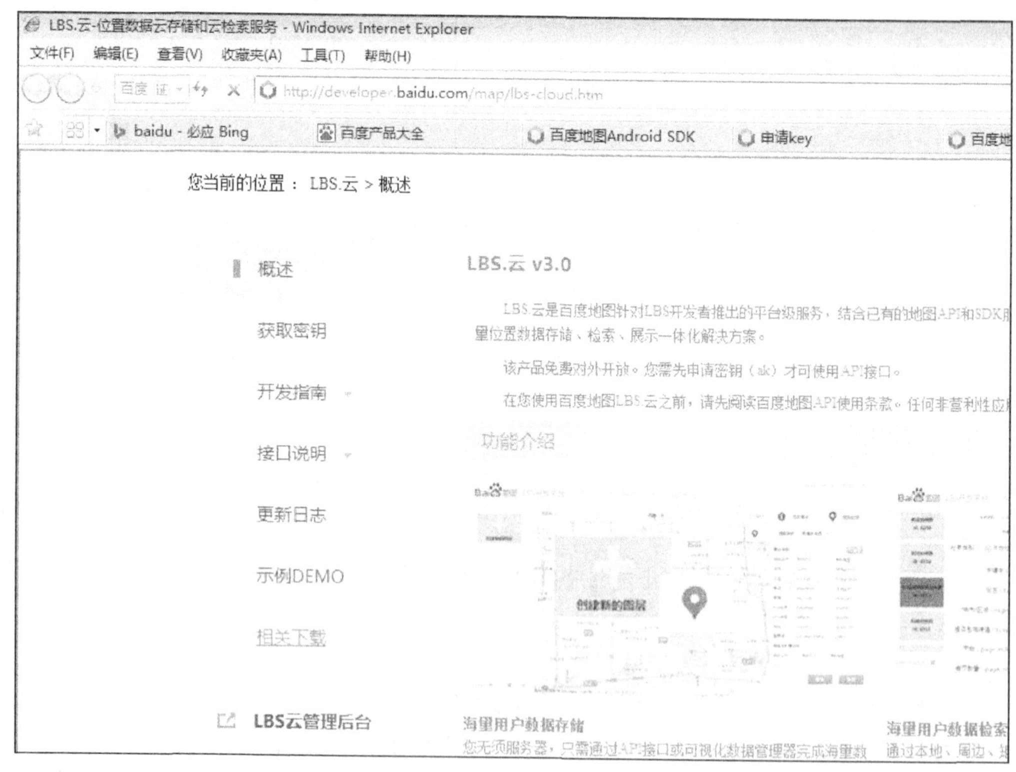

图 11.3.13　百度 LBS 云平台

进入“LBS 云管理后台”后，可进行位置数据的空间数据库管理和维护，通过它可以实现服务器零成本，让开发者更加注重前端的实现。

开发者使用 LBS·云存储 API 可以对用户自有数据的字段进行设计和存储管理；在云端，百度将开放地图引擎的运算能力，对存储的数据进行实时索引，使用 LBS·云检索 API 返回各类基于位置数据的检索请求。同时，使用百度地图 API 可以实现丰富的地图展现功能。

在存储数据前，百度为开发者精心设计了位置数据存储容器——databox，类似于数据库当中的一个表，即用于存储位置数据的字段定义。databox 包括两个元素：基本字段和扩展字段。每个开发者目前最多支持 30 个 databox。

注意：百度 LBS 云服务器只是提供数据存储以及 JSON(Javascript object notation)格式的数据返回。JSON 是一种独立于语言、轻量级的数据交换格式，采用键值对形式存储数据。

11.4　基于 TCP/IP 协议的标准 Java Socket 网络编程

11.4.1　TCP/IP 协议基础

TCP/IP 协议是 Internet 最基本的协议，也是 Internet 的基础。TCP/IP 协议定义了电子设备如何连入 Internet 及数据如何在它们之间进行传输的标准。

TCP 是 transmission control protocol 的英文缩写，表示传输控制协议。IP 是 Internet protocol 的英文缩写，表示因特网互联协议。

注意：TCP/IP 协议不只是 TCP 和 IP 这两个协议，而是指整个 TCP/IP 协议簇。

11.4.2　基于 TCP 或 UDP 的 Socket 网络通信

Socket 的英文含义是“孔”或“插座”，在计算机中翻译为“套接字”，用于描述 IP 地址和端口。

主机一般运行了多个服务器软件，同时还提供了多种服务。每种服务都使用一个 Socket，并绑定到某个端口(即不同的端口对应不同的服务)上。

应用程序通过套接字向网络发出请求或者应答网络请求，两个应用程序之间使用 Socket 并按照一定的协议通信，其模型如图 11.4.1 所示。

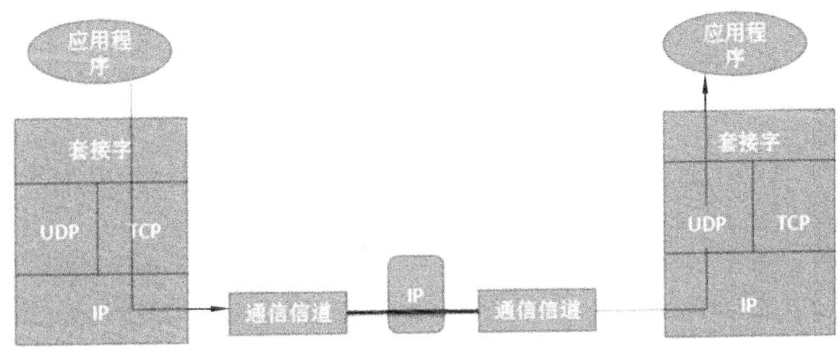

图 11.4.1　Socket 通信模型

显然，Socket 属于 C/S 结构，因此，发出请求的一端使用 Socket 类描述，应答网络请求的那一端使用 ServerSocket 类描述，两个类的定义如图 11.4.2 所示。

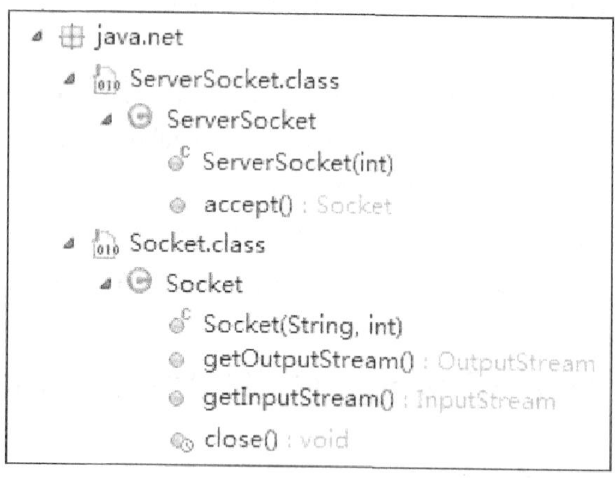

图 11.4.2　Socket 通信的两个主要类

注意：当服务器程序有变动时会影响客户端程序，反之亦然。

通常情况下，服务器端的 ServerSocket 对象建立后需要不断监听指定端口，当有客户端请求时调用 accept()方法接受请求并做相应处理。accept()方法被调用后，将阻塞进程，等待客户的请求，直到有客户端请求启动并连接到该端口，然后返回一个对应于客户的 Socket。

根据通信时是否建立信任连接，Socket 通信可分为有连接的 TCP 通信和无连接的 UDP 通信两种。

在 Socket 通信中，传递的信息分别用抽象类 OutputStream 和 InputStream 表示，它们是相对于客户端或服务器端而言的，如图 11.4.3 所示。

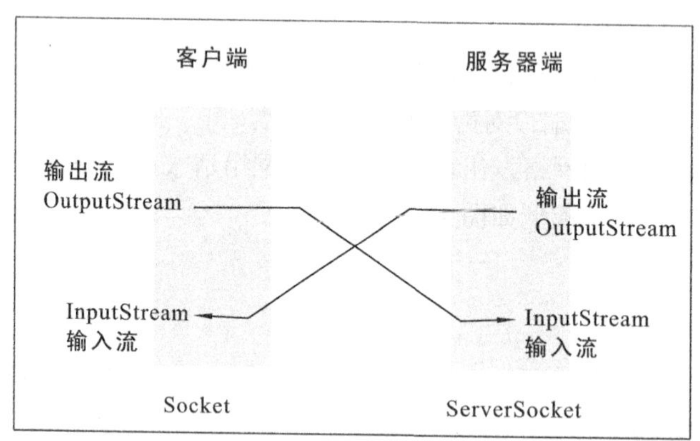

图 11.4.3　Socket 通信中的输入流和输出流

在标准的 Java 包 java.io 内，定义了类 DataOutputStream 和 DataInputStream，它们分别表示输出和输入的信息，其定义如图 11.4.4 所示。

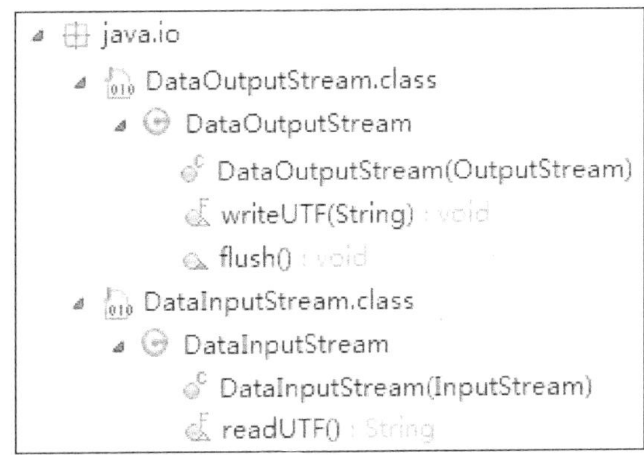

图 11.4.4 Socket 通信输出和输入的两个类

【例 11.4.1】基于 TCP 的简单 Socket 编程。

【程序运行】运行服务器程序，控制台信息输出如图 11.4.5 所示。

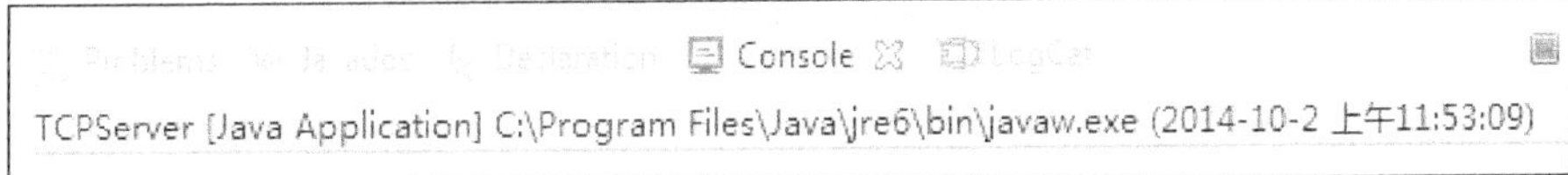

图 11.4.5 运行 TCP Server 时的信息输出

运行服务器程序，控制台信息输出如图 11.4.6 所示。

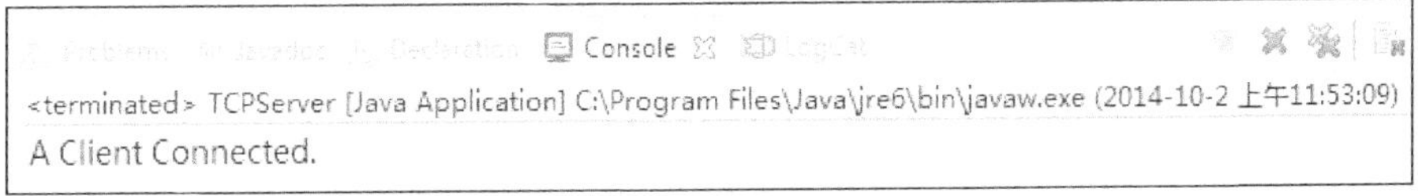

图 11.4.6 运行 TCP Client 时的信息输出

【设计步骤】

(1) 新建名为 SocketTCP1 的 Java 工程，分别创建 TCPServer.java 和 TCPClient.java 文件，工程文件结构如图 11.4.7 所示。

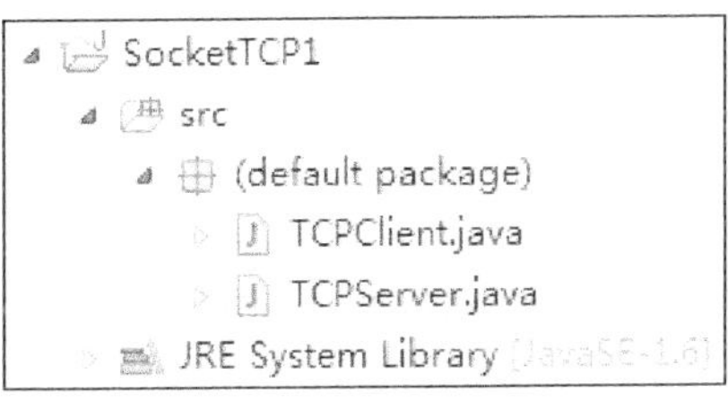

图 11.4.7 SocketTCP1 工程文件结构

(2) Socket 通信的服务器程序 TCPServer.java 的代码如下：

```
/*
 * 先运行TCPServer程序，则在指定的端口处于监听状态，后运行TCPClient
 * 由于没有循环监听，所以重复运行TCPClient会出现运行异常
* 解决重复运行出现的异常，见工程SocketTCP2
```

```
 */
import java.io.IOException;
import java.net.ServerSocket;   //
@SuppressWarnings("all")
public class TCPServer {
    public static void main(String[] args) throws IOException {
        // TODO Auto-generated method stub
        ServerSocket ss= new ServerSocket(8888);   //设置监听端口
        ss.accept();   //监听,阻塞(等待)
        System.out.println("A Client Connected.");   //建立连接后输出
    }
}
```

(3) Socket 通信的客户端程序 TCPClient.java 的代码如下：

```
import java.io.IOException;
import java.net.Socket;      //主类
import java.net.UnknownHostException;

public class TCPClient {
    public static void main(String[] args) {
        // TODO Auto-generated method stub
        try {
            @SuppressWarnings("unused")
            Socket s=new Socket("127.0.0.1",8888);
            //第一个参数为服务器IP；第二个参数为服务器端口
        } catch (UnknownHostException e) {
            // TODO Auto-generated catch block
            e.printStackTrace();
        } catch (IOException e) {
            // TODO Auto-generated catch block
            e.printStackTrace();
        }
    }
}
```

【例 11.4.2】基于 TCP 连接且带数据传送的 Socket 编程。

【程序运行】首先运行服务器程序，然后运行客户端程序两次，此时控制台信息输出如图 11.4.8 所示。

Problems　Javadoc　Declaration　Console　LogCat

TCPServer (1) [Java Application] C:\Program Files\Java\jre6\bin\javaw.exe (2014-10-2 下午1:09:23)

Hello Server!

Hello Server!

图 11.4.8　运行 TCPServer 时的信息输出

【设计步骤】

(1) 新建名为 SocketTCP2 的 Java 工程，分别创建 TCPServer.java 和 TCPClient.java 文件，工程文件结构如图 11.4.9 所示。

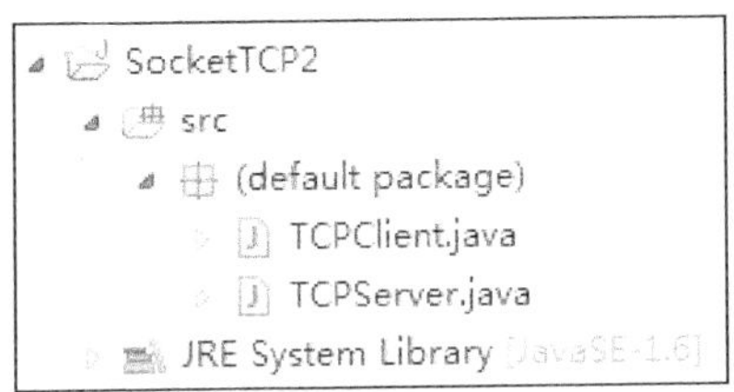

图 11.4.9　SocketTCP2 工程文件结构

(2) Socket 通信的服务器程序 TCPServer.java 的代码如下：

```
/*
 * 通过Socket实现服务器与客户端通信实例
 * 先运行TCPServer，然后再运行TCPClient
 * 当第二个程序运行后，则由第一个程序输出来自客户端的信息
*/
import java.io.DataInputStream;
import java.net.ServerSocket;   //
import java.net.Socket;   //

public class TCPServer {
    public static void main(String[] args) throws Exception {
        ServerSocket ss= new ServerSocket(7777);   //设置监听端口
        while(true){   //始终监听
            //监听客户端请求,可能产生阻塞(等待)的原因是没有客户端请求
            Socket s = ss.accept();
            DataInputStream dis = new DataInputStream(s.getInputStream());//
            System.out.println(dis.readUTF()); //客户端迟迟不发消息会引起阻塞!
            //输出来自客户端传送的信息
            dis.close();
            s.close();
        }
    }
}
```

(3) Socket 通信的客户端程序 TCPClient.java 的代码如下：

```
/*
 * 本程序是客户端运行的程序，用于向服务器发送信息
 * 实际应用时要修改服务器IP及端口
*/
import java.io.DataOutputStream;
import java.net.Socket;
public class TCPClient {
    public static void main(String[] args)throws Exception {
            Socket s=new Socket("127.0.0.1",7777);
            //第一个参数为服务器IP；第二个参数为服务器监听端口
            DataOutputStream dos = new DataOutputStream(s.getOutputStream());
            dos.writeUTF("Hello Server!");   //向服务器传送信息
            dos.flush();
            dos.close();   //超类方法
            s.close();
    }
}
```

【例 11.4.3】基于 UDP 连接且带数据传送的 Socket 编程。

【程序运行】首先运行服务器程序，然后运行客户端程序，此时控制台信息输出如图 11.4.10 所示。

```
Problems  Javadoc  Declaration  Console  LogCat
UDPServer [Java Application] C:\Program Files\Java\jre6\bin\javaw.exe (2014-10-5 上午10:17:49)
Hello
```

图 11.4.10　运行 UDPServer 时的信息输出

【设计步骤】

(1) 新建名为 SocketUDP 的 Java 工程，分别创建 UDPServer.java 和 UDPClient.java 文件，工程文件结构如图 11.4.11 所示。

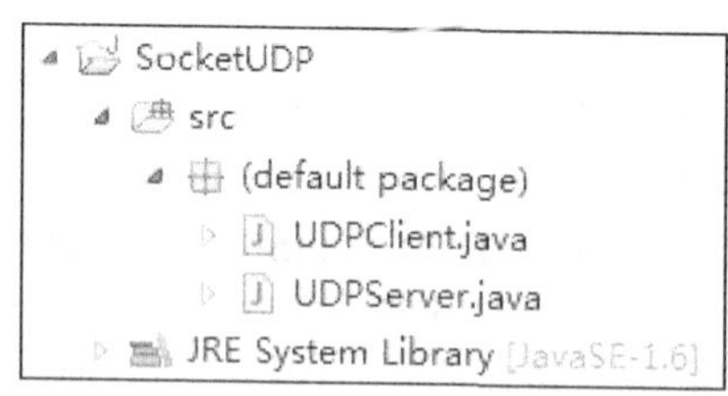

图 11.4.11　SocketUDP 工程文件结构

(2) 创建 Socket 通信的服务器程序 UDPServer.java 的代码如下：

```
/*
 * 本程序是面向无连接的Socket通信，使用UDP传输协议
 * 本程序为UDP Socket通信的服务器端程序
```

```
 */
import java.net.DatagramPacket;
import java.net.DatagramSocket;
public class UDPServer {
    public static void main(String[] args) throws Exception{
        byte buf[]=new byte[1024];
        //设置数据报文包
        DatagramPacket dp=new DatagramPacket(buf,buf.length);
        //创建报文接收端口
        DatagramSocket ds=new DatagramSocket(5678);
        while(true){
            ds.receive(dp);   //阻塞式的接收
            System.out.println(new String(buf,0,dp.getLength()));
        }
    }
}
```

(3) 创建 Socket 通信的客户端程序 UDPClient.java 的代码如下：

```
/*
 * UDP Socket的客户端
 * 在运行本程序前需要先运行服务器程序
 * 使用UDP协议传输信息
 */
import java.net.DatagramPacket;
import java.net.DatagramSocket;
import java.net.InetSocketAddress;
public class UDPClient {
    public static void main(String[] args) throws Exception{
        byte[] buf=(new String("Hello")).getBytes();
        //定义数据报文包DatagramPacket对象
        DatagramPacket dp=new DatagramPacket(buf,buf.length,
                                 new InetSocketAddress("127.0.0.1",5678));
        //每个报文包中包含了服务器地址及端口
        //InetSocketAddress继承抽象类SocketAddress
        //在客户端创建DatagramSocket对象
        DatagramSocket ds=new DatagramSocket(9999);
        ds.send(dp);   //发送到指定的服务器端口
        ds.close();
    }
}
```

习 题 11

一、判断题

1. 使用 HttpURLConnection 和 Socket 的网络编程，都是标准的 Java 网络编程。
2. 抽象类 HttpURLConnection 和接口 HttpClient 位于相同的软件包中。
3. HttpURLConnection 编程和 HttpClient 编程都是基于 HTTP 请求的网络编程。
4. 调用 Web 服务的 Android 应用程序应使用 Socket 编程方式。
5. 创建 JPush 应用，实质上是创建访问 Web 服务器的手机客户端程序。

二、选择题

1. 使用 HttpURLConnection 实现移动互联时，设置读取超时属性的方法是____。

   A. setTimeout()　　B. setReadTimeout()

   C. setConnectTimeout()　　D. setRequestMethod()

2. HttpClient 接口是由____包提供的。

   A. 标准 Java 包　B. 扩展 Java 包　C. Apache 包　D. Android 包

3. 使用 HttpClient 的 Get 方式请求时，可以使用____来创建 Http 请求对象。

   A. HttpResponse　B. HttpClient　C. HttpGet　D. HttpPost

4. Android 中网络互联需要获取状态码，根据状态码来判断请求是否已经完成，下列状态码表示请求完成的是____。

   A. 100　B. 202　C. 404　D. 200

5. 下列关于基于 TCP 连接的 Socket 通信的说法中，正确的是____。

   A. 服务器端使用 ServerSocket，且只需要设置端口号

   B. 服务器端使用 ServerSocket，且需要设置端口号和 IP 地址

   C. 客户端使用 Socket，且只需要设置端口号

   D. 客户端使用 ServerSocket，且需要设置端口号

三、填空题

1. 使用 URL 对象的 openConnection()方法得到的是____对象。
2. 通过使用 HttpClient 对象的____方法，可以得到一个 HttpResponse 对象。
3. 如果要发起网络连接，不仅要知道远程主机的 IP 或域名，还要约定通信的____。
4. 通过 ServerSokcet 提供的 accept()方法可以得到____类型的对象。
5. Socket 通信的连接方式有 TCP 和____两种。

实验 11　Android 网络编程

一、实验目的

1. 掌握使用 HttpURLConnection 访问 Web 服务器的方法。
2. 掌握使用 HttpClient 访问 Web 服务的方法。
3. 掌握手机客户端与 Web 服务器通信的编程方法。
4. 了解使用 JPush 实现 Web 服务器向手机推送信息的实现原理和方法。
5. 了解基于 TCP 或 UDP 的 Socket 网络通信。

二、实验内容及步骤

【预备】访问 http://www.wustwzx.com/android/index.html，下载本次实验内容的源代码(压缩文件)并解压，得到文件夹 ch11，供下面的实验使用。

1. 使用 Java 提供的 net 包，通过创建和使用 URL 对象，进而得到 HttpURLConnection 对象，访问网络资源(抓取网页)(参见例 11.1.1)。

(1) 再次解压 ch11 文件夹里的压缩文件 URLDemo.zip，得到 Android 工程文件夹 URLDemo。

(2) 在 Eclipse 中导入 Android 工程 URLDemo。

(3) 查看源文件 MainActivity.java 中按钮单击事件的处理代码，总结获取从 Web 服务器返回页面的 HTML 代码的实现步骤。

(4) 打开手机的网络连接。

(5) 部署工程到手机，单击按钮后得到访问 Web 服务器后的客户端 HTML 代码。

(6) 将返回服务器数据的方法中的第二参数“gbk”改为“utf-8”，再次部署项目并访问，可观察到出现了中文乱码。

(7) 验证两种设置垂直滑动手势处理方法的等效性。

2. 使用 Apache 提供的网络编程接口，创建 HttpClient 对象，以 Post 方式调用 Web 服务，得到 HttpResponse 对象，实现手机归属地查询(参见例 11.2.1)。

(1) 再次解压 ch11 文件夹里的压缩文件 WebService.zip，得到 Android 工程文件夹 WebService。

(2) 在 Eclipse 中导入 Android 工程 WebService。

(3) 打开源文件 MainActivity.java，查看按钮单击事件的处理代码。

(4) 打开手机的网络连接(使用 WiFi 或 GPRS)。

(5) 部署工程到手机后做运行测试。

(6) 去掉源程序文件中过滤标签的正则处理方法，再次部署工程，观察查询信息(XML 文档)。

(7) 验证必须在清单文件中配置访问 Internet 的权限。

(8) 验证在 MainActivity.java 中必须使用子线程。

3. 手机客户端程序与 Web 服务器的信息交互——用户登录与注册(参见例 11.3.1)。

(1) 再次解压 ch11 文件夹里的压缩文件“手机客户端与 Web 服务器交互(注册与登录).zip”，得到一个名为“WuLogin”的 Java Web 工程和一个名为“手机客户端(用户注册与登录)”的 Android 工程。

(2) 在 MyEclipse 中导入名为“WuLogin”的 Java Web 工程(不是 Android 应用工程！)。

(3) 安装 MySQL 数据库服务软件(可从本书作者教学网站中下载)，且设置用户名和密码都是“root”，端口为“3306”。

(4) 启动 MySQL，创建名为“test”的数据库。

(5) 使用 WuLogin 的数据库脚本文件，创建名为“user”的数据表。

(6) 启动 Tomcat 服务器(可从本书作者教学网站中下载免安装版)。

(7) 在 MyEclipse 中，部署工程 WuLogin 到 Tomcat 后，先浏览 JSP 网站的注册页面 reg.jsp，后浏览登录页面 login.jsp。

(8) 导入 Android 工程“手机客户端(用户注册与登录)”。

(9) 打开手机的移动数据连接和个人热点，让笔记本电脑通过 WiFi 方式上网，在命令行方式下使用 ipconfig 命令查看笔记本电脑的 IP 地址。

(10) 验证在手机上通过输入“http://Web 服务器地址:8080/WuLogin/reg.jsp”的方式，可以访问 JSP 网站中的注册页面。

(11) 打开 Android 工程的源程序 MainActivity.java，并修改 Web 服务器 IP 为刚才笔记本电脑的 IP 地址。

(12) 部署 Android 应用工程到手机，通过单击“注册”按钮，实现使用手机客户端程序注册一个新用户并通过 Web 服务器程序写入数据库服务器的数据库表里。

(13) 通过单击“登录”按钮，实现使用手机客户端程序登录 Web 服务器。

(14) 查看源程序 MainActivity.java 中实现手机客户端与 Web 服务器交互的过程及代码。

4. 掌握使用激光推送平台 JPush 实现 Web 服务器向手机推送消息的方法及原理。

(1) 打开 360 浏览器，访问网站 http://www.jpush.cn。

(2) 注册一个开发者账号。

(3) 登录后，单击“控制台”按钮，填写 Android 应用名称及应用包名后，单击“创建我的应用”按钮。

(4) 单击“下载 Android Example”按钮，得到 Android 应用工程源代码的压缩包。

(5) 解压后，将其导入该 Android 工程。

(6) 部署工程到手机后按返回键退出该应用。

(7) 在网站 http://www.jpush.cn 里，选择刚才创建的应用，单击“发送通知”按钮完成消息的推送。

(8) 在手机的通知栏里查看由 JPush 的 Web 服务器推送过来的消息。

(9) 打开相关文件，查看和分析手机客户端工程的实现过程及原理。

5. 了解基于 TCP 或 UDP 的 Socket 通信编程(参见例 11.4.1、例 11.4.2 和例 11.4.3)。

(1) 再次解压 ch11 文件夹里的压缩文件 Socket.zip，得到一个包含 3 个 Java 工程的文件夹 Socket。

(2) 在 Eclipse 中导入 Socket 文件夹里的 Java 工程 SocketTCP1。

(3) 打开源程序文件 TCPServer.java，查看创建 ServerSocket 对象的方法并运行。

(4) 打开源程序文件 TCPClient.java，查看创建 Socket 对象的方法并与创建 ServerSocket 对象的方法进行比较。

(5) 运行客户端程序，查看控制台输出的信息。

(6) 导入 Socket 文件夹里的 Java 工程 SocketTCP2。

(7) 查看源程序文件 TCPServer.java，查看 ServerSocket 获取客户端信息流的过程。

(8) 运行客户端程序，查看控制台输出的信息。

(9) 导入 Socket 文件夹里的 Java 工程 SocketUDP。

(10) 分别查看服务器程序 UDPServer.java 和客户端程序 UDPClient.java 中使用 DatagramPacket 对象和 DatagramSocket 对象对数据报文的发送和接收处理代码。

(11) 分别运行服务器程序和客户端程序，查看控制台的输出结果。

(12) 总结基于 TCP 和 UDP 连接的 Socket 通信的用法差别。

三、实验小结及思考

(由学生填写，重点写上机中遇到的问题。)

附录A 在线测试

Android 应用虽然重在编程实践，但指导实践的是理论，理论也来源于实践。设计完成后，要即时总结。为此，作者设计了一套综合的在线测试题，在提交后能立即显示答题者的成绩和每道题的正误，以方便学生练习。

该测试题含有判断题、单选题和多选题三种题型。其中：判断题共 15 题，每小题 2 分，共 30 分；单选题共 20 题，每小题 2 分，共 40 分；多选题共 10 题，每小题 3 分，共 30 分。

使用在线测试，请访问 http://www.wustwzx.com/ android/index.html。

附录 B　三次实验报告

在完成某个阶段的学习后，要写一次综合性的实验报告。本书共设计了三次实验报告：第一次实验报告对应于第 1～5 章的内容；第二次实验报告对应于第 6～8 章的内容；第三次实验报告对应于第 9～11 章的内容。

实验报告分为实验目的、实验内容及步骤和实验小结及思考三个部分，后两部分要求学生填写。学生可以先将实验报告的文本打印出来，以供在实验前进行分析和思考。

实验报告文本的下载地址：http://www.wustwzx.com/android/index.html。

实验名称：Android 应用开发基础

一、实验目的

1. 掌握 在 Eclipse 中搭建 Android 开发环境的方法。
2. 了解 Android ADT 的作用，掌握 SDK Manager 的使用。
3. 掌握 Android 应用程序的执行过程和 Dalvik 虚拟机的工作原理。
4. 掌握 Android 工程的目录结构及 Android 平台的调试方法。
5. 了解 Android 应用的签名机制。
6. 掌握 Activity 组件的使用和布局文件的设计方法。
7. 掌握 Intent、Bundle 和 SharedPreferces 的用法。
8. 掌握常用 Widget 控件、单击事件监听器和数据适配器的使用。
9. 掌握通知、菜单和文件读写的设计方法。

二、实验内容及步骤

(提示：根据实验目的，组织教材中的相关示例代码，说明其用法。)

三、实验小结及思考

(如内部文件读写与外部文件读写的用法区别等。)

实验名称：Service、BroadcastReceiver 和 ContentProvider 的应用 (含 SQLite 数据库和数据适配器 SimpleAdapter 的使用)

一、实验目的

1. 掌握 Service 的两种调用方式(显示与隐式)的用法区别。

2. 掌握 Service 的两种启动方式(绑定与非绑定)的用法区别。
3. 掌握本地服务调用与远程服务调用的用法区别。
4. 掌握 Android 的广播机制和常用的系统广播。
5. 掌握注册广播接收者的两种用法(静态注册与动态注册)。
6. 掌握 SQLite 数据库数据适配器 SimpleAdapter 的使用。
7. 掌握使用 ContentProvider 建立内容提供者的方法。
8. 掌握通过 ContentResolver 访问 ContentProvider 的方法。
9. 掌握使用手机联系人的编程方法。

二、实验内容及步骤

(提示：根据实验目的，组织教材中的相关示例，说明相关用法。)

三、实验小结及思考

(如总结三个组件 Service、BraodcastReceiver 与 Activity 的不同点等。)

实验名称：Android 位置服务、通信和网络编程

一、实验目的

1. 掌握 Android 中的三种定位方式的特点。
2. 掌握 GPS 定位原理及编程实现。
3. 掌握 WiFi 定位原理及编程实现。
4. 掌握百度综合定位及地图显示的设计方法。
5. 掌握 Bluetooth 通信的特点及实现方法。
6. 掌握使用 HTTP 通信访问指定 URL 网络资源的方法。
7. 掌握在 Android 应用中使用 Web Service 的方法。
8. 掌握手机客户端与 Web 服务器通信的设计方法。
9. 了解 Socket 通信的原理及实现方法。

二、实验内容及步骤

(提示：根据实验目的，组织教材中的相关示例，说明相关用法。)

三、实验小结及思考

(如 Java 网络编程与 Android 网络编程的区别与联系等。)

附录 C　模拟试卷及参考答案

本课程在不同的学校有不同的考核方式，一般有两种。其一是使用传统的出试卷的方式。另一种是提交设计的方式。作者认为，以试卷方式考核，有利于学生掌握基本理论和设计技巧。

模拟试卷下载，可访问 http://www.wustwzx.com/android。

一、单选题(每题 2 分，共 40 分)

1. 下面不属于智能手机操作系统范畴的是(　　)。

A. Android　　B. iOS　　C. Windows Phone　　D. Chrome

2. Activity 启动后第一个被调用的函数是(　　)。

A. onCreate　　B. onStart　　C. onResume　　D. onRestart

3. Android 模拟器如要访问本机的 Web 服务器，则使用的地址是(　　)。

A. 0.0.0.0　　B. 127.0.0.1　　C. 255.255.255.255　　D. 10.0.2.2

4. 在为选项菜单添加菜单功能代码时，需要重写的方法是(　　)。

A. onCreateOptionsMenu()　　B. onCreateContextMenu()

C. onOptionsItemSelected()　　D. onContextItemSelected()

5. Android 项目工程下面的 drawable 目录的作用是(　　)。

A. 放置应用程序的图片资源

B. 放置应用程序的音、视频资源

C. 放置字符串等常量数据

D. 放置布局资源文件

6. 如果将一个 TextView 的 android:layout_width 属性值设置为 fill_parent，那么该组件将是以下哪种显示效果？(　　)。

A. 该文本域的宽度将填充父容器宽度

B. 该文本域的宽度仅占据该组件的实际宽度

C. 该文本域的高度将填充父容器高度

D. 该文本域的高度仅占据该组件的实际高度

7. 创建数字密码框时，应设置 android:inputType 属性值为(　　)。

A. textPassword　　B. numberPassword

C. number　　D. textMultiLine

8. AndroidManifest.xml 中对某个 Activity 进行如下定义：

```
<intent-filter>
    <action android:name="android.intent.action.MAIN" />
    <category android:name="android.intent.category.LAUNCHER" />
</intent-filter>
```

这样的描述代表什么含义？(　　)。

A. 无明确含义，每个 Activity 都需要这样定义

B. 表明该 Activity 是程序的入口，并且显示在程序列表里

C. 表明该 Activity 将在程序列表里建立图标并启动

D. 表明该 Activity 的优先级高于其他的 Activity

9. 在使用 RadioButton 时，要想实现互斥的选择需要(　　)。

A. ButtonGroup　　B. RadioButtons

C. CheckBox　　D. RadioGroup

10. 在多个应用程序中读取共享数据时，需要用到（　　）对象的 query 方法。

A. ContentResolver　　B. ContentProvider

C. Cursor　　D. SQLiteHelper

11. 仅创建 SharedPreferences 的程序有权限对其进行读取或写入的模式是(　　)。

A. MODE_PRIVATE　　B. MODE_WORLD_READABLE

C. MODE_APPEND　　D. MODE_WORLD_WRITEABLE

12. 有关 Intent 的作用描述正确的是(　　)。

A. 处理一个应用程序整体性的工作

B. 是一段长的生命周期，没有用户界面的程序

C. 实现应用程序间的数据共享

D. 连接四大组件的纽带，可以包含动作和动作数据。

13. Android 是以(　　)方式组织 Activity 的。

A. 栈　　B. 队列　　C. 树形　　D. 链式

14. 在使用 SQLiteOpenHelper 这个类时，用来实现版本升级之用的方法是(　　)。

A. onCreate()　　B. 构造函数　　C. onUpdate()　　D. onUpgrade()

15. SharedPreferences 数据存放的目录是(　　)。

A. /data/data/应用程序包名/shared_prefs/

B. /data/data/应用程序包名/databases/

C. \system\usr\share

D. \ system\etc\security

16. 下面的 Intent 动作能够实现直接拨打电话的是(　　)。

A. Intent.ACTION_VIEW　　B. Intent.ACTION_DIAL

C. Intent.ACTION_CALL　　D. Intent.ACTION_CALL_BUTTON

17. 不能表示全部联系人信息的 URI 串是(　　)。

A. content://contacts/people/

B. content://contacts/people/1

C. content://com.android.contacts/contacts/

D. ContactsContract.Contacts.CONTENT_URI

18. 允许程序读取联系人的权限的是(　　)。

A. <uses-permission android:name="android.permission.CALL_PHONE" />

B. <uses-permission android:name="android.permission.READ_CONTACTS" />

C. <uses-permission android:name="android.permission.WRITE_CONTACTS" />

D. <uses-permission android:name="android.permission.INTERNET" />

19. 使用 HttpClient 的 Get 方式请求数据时，可以（　　）来创建 Http 请求对象。

A. HttpResponse　　B. HttpClient

C. HttpGet　　D. HttpPost

20. 对于 Toast 的描述正确的是(　　)。

A. Toast 能自定义布局

B. Toast 的显示不需要调用 show 方法

C. Toast 不会获得焦点

D. Toast 会影响用户的输入

二、填空题(每题 1 分，共 15 分)

1. Eclipse 开发 Android 程序的插件简称________，运行 Android 程序的模拟器简称是________。

2. Android 四大组件分别是：________、_________、________和________。

3. Android 应用程序的有 3 种菜单分别是：________、________和子菜单。

4. 下面代码是隐式启动一个浏览器打开 163.com 网站。请完成：

```
Intent intent = new Intent( ___________ , ______________ );
startActivity(intent);
```

5. 下面代码调用 Web 服务来查询手机号码信息。请完成程序：

```
public void getRemoteInfo(String phoneSec) {
        String requestUrl ="http://webservice.webxml.com.cn/
                                WebServices/MobileCodeWS.asmx/getMobileCodeInfo";
        ________ client = new DefaultHttpClient();
        HttpPost post =____________;
        List<NameValuePair> params = new ArrayList<NameValuePair>();
        params.add(new BasicNameValuePair("mobileCode", phoneSec));
        params.add(new BasicNameValuePair("userId", ""));
        try {
            post._______(new UrlEncodedFormEntity(params, HTTP.UTF_8));
            HttpResponse response =____________;
            String result="";
            if (response.getStatusLine().getStatusCode() == HttpStatus.SC_OK) {
                result = EntityUtils.toString( ___________ , "utf-8");
            }
            else{
            result ="没有查询结果";
            }
            resultView.setText(result);
```

```
            } catch (Exception e) {
                e.printStackTrace();
                Toast.makeText(getBaseContext(), "出错了!",
                                                    Toast.LENGTH_SHORT).show();
            }
    }
```

三、简答题(每题 5 分，共 25 分)

1. 写出横竖屏切换时 Activity 的生命周期回调函数执行顺序。
2. 简要描述 LinearLayout 布局基本特点。
3. 简要描述 res/raw 文件夹特点。
4. 什么是 ContentProvider？
5. 简要说明启动 Activity 的方式。

四、编程题(每题 10 分，共 20 分)

1. 在 LoginActivity 中输入用户名和密码(下图左所示)，单击“提交”按钮，在 InfoShowActivity 中显示录入的信息(下图右所示)。写出主要代码即可。

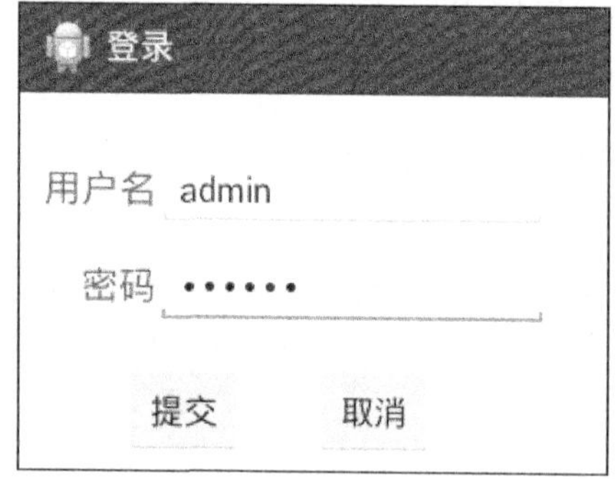

LoginActivity.java

InfoShowActivity.java

2. 编写一个 DBHelper 类，其父类是 SQLiteOpenHelper，该类实现：

(1) 创建一个 “diary” 表，其结构如下：

字段名称	描述
_id	整形，自增，主键
topic	varchar(100)
content	varchar(1000)

(2) 在数据库版本变化时删除原来的 diary 表，并重建 diary 表。

然后在主程序中使用 DBHelper 类来创建一个版本为 1 的“test.db”的数据库，并添加一条记录到 diary 表。

参考答案及评分标准

一、单选题(每题 2 分，共 40 分)

1～5： D A D C A　　6～10：A B B D A

7～15：A D A D A　　16～20：C B B C C

二、填空题(每题 1 分，共 15 分)

1. ADT、 AVD

2. Activity、Service、Content Provider、Broadcast Receiver

3. 选项菜单、 上下文菜单

4. Intent.ACTION_VIEW、 Uri.parse("http://www.163.com")

5. HttpClient、new HttpPost(requestUrl)、setEntity、client.execute(post)、response.getEntity()

三、简答题(每题 5 分，共 25 分)

1. 写出横竖屏切换时 Activity 的生命周期回调函数执行顺序。

答：onPause → onStop → onDestroy → onCreate → onStart → onResume

2. 简要描述 LinearLayout 布局基本特点。

答：在 LinearLayout 布局中，所有的子元素都按照垂直或水平的顺序在界面上排列，如果垂直排列，则每行仅包含一个界面元素，如果水平排列，则每列仅包含一个界面元素。

3. 简要描述 res/raw 文件夹特点。

答：res/raw 文件夹通常存放资源文件，如音频、视频等资源，在打包后会原封不动的保存在 apk 包中，res/raw 中的文件会被映射到 R.java 文件中，访问的时候直接使用资源 ID，res/raw 不可以有目录结构。

4. 什么是 ContentProvider？

答：Android 应用程序运行在不同的进程空间中，因此不同应用程序的数据是不能够直接访问的。ContentProvider 提供了应用程序之间共享数据的方法。应用程序通过 ContentProvider 访问数据而不需要关心数据具体的存储及访问过程，这样既提高了数据的访问效率，同时也保护了数据。

5. 简要说明启动 Activity 的方式。

答：显式启动，程序必须在 Intent 中指明启动的 Activity 所在的类。

隐式启动，无需指明具体启动哪一个 Activity，Android 系统会根据 Intent 的动作和数据来决定启动哪一个 Activity。

四、编程题(共 20 分)

1. (本题 10 分)

答案：

LoginActivity：onCreate()方法　　-----6 分

```
    Button bt=(Button)findViewById(R.id.button1);
    bt.setOnClickListener(new View.OnClickListener() {
```

```
                @Override
                public void onClick(View arg0) {
                    String name=((EditText)findViewById(R.id.editText1))
                                                            .getText().toString();
                    String psd=((EditText)findViewById(R.id.editText2))
                                                            .getText().toString();
                    Intent intent = new Intent(MainActivity.this, InfoShowActivity.class);
                    Bundle bundle=new Bundle();
                    bundle.putString("username", name);
                    bundle.putString("password", psd);
                    intent.putExtras(bundle);
                    startActivity(intent);
                }
            });
```

InfoShowActivity：onCreate()方法　　　-----4 分

```
        Intent intent=this.getIntent();
        Bundle bundle=intent.getExtras();
        String un=bundle.getString("username");
        String psd=bundle.getString("password");
        TextView tv=(TextView)findViewById(R.id.textView1);
        tv.setText("你输入的用户名是："+un+"\n 密码是："+psd);
```

2. (本题 10 分)

publicclass DBHelper　extends SQLiteOpenHelper{　　　　-----7 分

```
    public DBHelper(Context context,Stringname,
                                        CursorFactory factory,int version){
            super(context, name, factory,version);
    }
    public void onCreate(SQLiteDatabase db){
      String sql ="create table diary(_id integer primary key auto
                                increment,topic varchar(100), content varchar(1000))";
          db.execSQL(sql);
    }
    public void onUpgrade(SQLiteDatabase db,int oldVersion,int newVersion){
            String sql = "drop table if exists diary";
            db.execSQL(sql);
            onCreate(db);
    }
}
```

主程序：-----3 分

```
    DBOpenHelper helper = new DBOpenHelper(getApplicationContext(),
                                                              "test.db", null,1);
    SQLiteDatabase db=helper.getWritableDatabase();
db.execSQL("INSERT INTO diary VALUES (NULL, ?, ?)",
                              new Object[ ] { "about Andorid", "This is a test topic." } );
db.close();
```

习 题 答 案

习 题 1

一、判断题(正确用“T”表示，错误用“F”表示)

1～5：FTFFT　6～10：FFFTT

二、选择题

1～5：CDBDA　6～9：BCAA

三、填空题

1. Ctrl+T　2. 自闭　3. 4　4. .odex　5. Linux　6. Util　7. Ctrl+Shift+\
8. hasNext()　9. newInstance()　10. invoke()　11. sysout+回车

习 题 2

一、判断题(正确用“T”表示，错误用“F”表示)

1～5：TTTFF　6～10：TTTFT

二、选择题

1～5：ACABD

三、填空题

1. android.jar　2. 动态调试　3. ADT　4. Show View　5. 5554

习 题 3

一、判断题(正确用“T”表示，错误用“F”表示)

1～5：FTTTF　6～8：FTF

二、选择题

1～5：CCABA　6～10：BCADC

三、填空题

1. Google　2. Linux　3. android:label　4. 主　5. 类　6. .odex
7. 寄存器　8. assertEquals()　9. assets

习 题 4

一、判断题(正确用“T”表示，错误用“F”表示)

1～5：TTFTT　6～9：FTFFT　11～15：FFFTT

二、选择题

1～5：AACDC　　6～10：ACDDB

三、填空题

1. android:orientation　2. android:gravity　3. sp　4. onStop()　5. hint

6. SharedPreferences　7. android.app　8. XML　9. layout_weight

10. onClick()　11. getItem()

习　题　5

一、判断题(正确用“T”表示，错误用“F”表示)

1～6：FFTTFT

二、选择题

1～5：DDCAC

三、填空题

1. android.net　2. Intent.ACTION_CALL　3. 3　4. startActivityForResult()

5. 字符串

习　题　6

一、判断题(正确用“T”表示，错误用“F”表示)

1～5：TTTTF　　6～8：TFT

二、选择题

1～5：BDACC

三、填空题

1. android.app　2. 隐式　3. <intent-filter>　4. IBinder　5. AIDL

6. Intent　7. onBind()

习　题　7

一、判断题(正确用“T”表示，错误用“F”表示)

1～5：TTTFT　6～7：TT

二、选择题

1～5：CDACD　　6～7：BD

三、填空题

1. db　2. SQLiteOpenHelper　3. SQLiteDatabase　4. void　5. onUpgrade()

6. setAdapter()　7. boolean　8. Linux

习　题　8

一、判断题(正确用“T”表示，错误用“F”表示)

1～5：FFFFF

二、选择题

1～6：CDCCAC

三、填空题

1.　<provider>　2. 逻辑　3. content　4. ContactsContract　5. handleMessage()

习　题　9

一、判断题(正确用“T”表示，错误用“F”表示)

1～5：TTFFT　6～7：FT

二、选择题

1～4：ACAD

三、填空题

1. 4　2. startScan()　3. getScanResult()　4. getConnectionInfo()
5. getDefaultAdapter()　6. 流

习　题　10

一、判断题(正确用“T”表示，错误用“F”表示)

1～5：FFFTF

二、选择题

1～5：CACBC

三、填空题

1. requestLocationUpdates()　2. GPS　3. ACCESS_COARSE_LOCATION
4. 地图包　5. MapView

习　题　11

一、判断题(正确用“T”表示，错误用“F”表示)

1～5：TFTFT

二、选择题

1～5：BCCDA

三、填空题

1. URLConnection　2. execute()　3. 端口　4. Socket　5. UDP

参 考 文 献

[1] 吴志祥. 网页设计理论与实践[M]. 北京：科学出版社. 2011.

[2] 吴志祥，李光敏，郑军红. 高级 Web 程序设计——ASP.NET 网站开发[M]. 北京：科学出版社. 2013.

[3] 吴志祥，王新颖，曹大有. 高级 Web 程序设计——JSP 网站开发[M]. 北京：科学出版社. 2013.

[4] 李波，史江萍，王祥凤. Android 4.X 从入门到精通[M]. 北京：清华大学出版社. 2012.

[5] 王向辉，张国印，赖明珠. Android 应用程序开发[M]. 2 版. 北京：清华大学出版社. 2012.

[6] 李鸥，等. 实战 Android 应用开发[M]. 北京：清华大学出版社. 2012.

[7] 陈会安. Android SDK 程序设计与开发范例[M]. 北京：清华大学出版社. 2013.